EXAMINING ECOLOGY

EXAMINING ECOLOGY
Exercises in Environmental Biology and Conservation

PAUL A. REES

School of Environment & Life Sciences,
University of Salford, Manchester, United Kingdom

ACADEMIC PRESS
An imprint of Elsevier

Academic Press is an imprint of Elsevier
125 London Wall, London EC2Y 5AS, United Kingdom
525 B Street, Suite 1800, San Diego, CA 92101-4495, United States
50 Hampshire Street, 5th Floor, Cambridge, MA 02139, United States
The Boulevard, Langford Lane, Kidlington, Oxford OX5 1GB, United Kingdom

Notices
Knowledge and best practice in this field are constantly changing. As new research and experience broaden our understanding, changes in research methods, professional practices, or medical treatment may become necessary.

Practitioners and researchers must always rely on their own experience and knowledge in evaluating and using any information, methods, compounds, or experiments described herein. In using such information or methods they should be mindful of their own safety and the safety of others, including parties for whom they have a professional responsibility.

To the fullest extent of the law, neither the Publisher nor the authors, contributors, or editors, assume any liability for any injury and/or damage to persons or property as a matter of products liability, negligence or otherwise, or from any use or operation of any methods, products, instructions, or ideas contained in the material herein.

British Library Cataloguing-in-Publication Data
A catalogue record for this book is available from the British Library

Library of Congress Cataloging-in-Publication Data
A catalog record for this book is available from the Library of Congress

ISBN: 978-0-12-809354-2

For Information on all Academic Press publications
visit our website at https://www.elsevier.com/books-and-journals

www.elsevier.com • www.bookaid.org

Publisher: Sara Tenney
Acquisition Editor: Kristi Gomez
Editorial Project Manager: Pat Gonzalez
Production Project Manager: Mohanambal Natarajan
Designer: Matthew Limbert

Typeset by MPS Limited, Chennai, India

For
Harry David Clark,
George Arthur Clark and
Elliot Henry Clark

CONTENTS

CONTENTS

Chapter 9 Conservation Biology 237

Chapter 10 Statistics 275

PREFACE

The purpose of this book is to help students to learn about the principles of ecology by completing a series of problems. It is primarily intended for students at the end of their secondary education and those at the beginning of their tertiary education. It assumes little prior knowledge of the subject but requires users to be able to perform simple calculations (largely arithmetic) and draw graphs.

Most ecology textbooks provide a great deal of information in the form of text, diagrams, graphs, and tables of data, with, perhaps, a few exercises for the students to attempt at the end of each chapter. This book takes the opposite approach. Very little background text is provided and students are expected to built up their knowledge of ecology by completing exercises, many of which are based on the work of pioneering ecologists who built the foundations of the subject.

Each chapter begins with a brief account of general principles. This is followed by a list of the intended learning outcomes for the chapter and then a series of exercises. At the end of the book there is a series of multiple choice tests covering the material in the individual chapters. The exercises themselves vary in difficulty. Some consider very basic concepts, such as food chains, while others require some understanding of mathematics, for example, matrix algebra.

Ecology is a practical subject which should be taught, at least in part, by requiring students to engage in fieldwork. There is no substitute for this. However, financial and time constraints often preclude teachers and lecturers from engaging in extensive (and expensive) field trips. Some of the exercises presented here are intended to help students to understand field techniques without the need to spend time getting cold, wet, and tired. With the best will in the world, most young people who are required to study some ecology as part of their biological education do not intend to work as ecologists and may be much happier in a warm, dry laboratory.

Almost all of the exercises in this book are based on real studies. In some cases they require the user to reconstruct graphs produced by others from points generated by inspection of the original published graphs. In effect the user has to draw a graph from raw data in much the same way as the original researchers. Some exercises are based on simulated data that I have produced using various computer programs.

Many of the exercises presented here have been used to teach a very wide range of students in further and higher education colleges, at university, and mature students working in industry. My overall impression was that they preferred doing these exercises to listening to me talking about ecology. I hope you enjoy them and that they inspire you to go outside, get cold and wet (perhaps even very hot), and do some real ecology in the field.

Paul A. Rees

ACKNOWLEDGEMENTS

This book began its life around 1985 when I wrote some exercises for students as part of the assessment required for my Certificate in Education. Over a period of years I added to these exercises until I had several dozen and they became part of the way I taught ecology to college students and, later, undergraduates.

A number of people and organisations have supplied materials used in this book. Sarah Proctor, Community Science Project Manager, *Moors For the Future* kindly supplied the images used for Fig. 4.5. I have used images kindly made available by Openclipart (www.openclipart.org) in a number of figures (Figs. 4.2, 4.17, 4.21, 6.4, 6.7, 7.1, 8.3, 8.14, and 9.24) and I am grateful to various contributors for allowing unrestricted use of their work. A number of photographs of historical interest have been obtained from the US Library of Congress (Figs.1.3, 7.14, 8.7, 9.22 and 9.23) which has indicated that there is no known restriction on their publication. All of the other images are my own. Dr Alan Woodward kindly checked some of the content for errors.

Many of the exercises in this book were inspired by my teachers. At the University of Liverpool, Prof. Anthony Bradshaw taught me about heavy metal tolerant grasses, Prof. Philip Sheppard and his colleague Dr Jim Bishop taught me about the genetics of *Biston betularia*, Prof. Arthur Cain taught me about evolution and taxonomy, and Prof. Michael Begon supervised my research on the assimilation efficiency of African elephants. At the University of Bradford, Prof. Michael Delany supervised my doctoral research on the ecology and behaviour of feral cats.

Many students have unwittingly tested many of the exercises presented in this book and I am grateful to them and to Dr Louise Taylor who coauthored one of the papers that has been used as the basis of several exercises.

At Elsevier I am indebted to Kristi Gomez (Life Sciences Acquisitions Editor) and Pat Gonzalez (Senior Editorial Project Manager) for believing in this project and seeing it through the production process to publication. In India, I am grateful to Mohanambal Natarajan for overseeing the typesetting of the work.

Finally, I must acknowledge the tolerance of my wife Katy who only complained about the piles of papers and books that were scattered around our home during the writing of this book when she genuinely believed her life to be in danger.

CHAPTER 1

BIODIVERSITY AND TAXONOMY

This chapter contains exercises concerned with the diversity of living things, their identification and their classification.

Northern gannet (*Morus bassanus*)

Examining Ecology. DOI: http://dx.doi.org/10.1016/B978-0-12-809354-2.00001-4

INTENDED LEARNING OUTCOMES

On completion of this chapter you should be able to:

- Explain the importance of taxonomy in ecological studies.

- Distinguish between the vernacular and scientific names of organisms.

- Explain why scientists use scientific names for organisms.

- List the major taxonomic groups of animals in the correct hierarchical sequence from phylum to species.

- Construct a simple dichotomous key.

- Identify patterns in species diversity and abundance.

- Explain the variety of ways in which new species may be identified in the wild and from among existing collections of organisms.

- Identify reasons why island species are especially vulnerable to extinction.

- Discuss problems with the accuracy of historical estimates of the sizes of wild animal populations.

- Discuss methods for protecting indigenous species from introduced species.

- List the benefits and shortcomings of citizen science projects designed to investigate ecological problems.

- Use the lognormal distribution to investigate the pattern of species diversity.

- Explain the ecological value of ancient ecosystems

INTRODUCTION

THE NAMING AND CLASSIFICATION OF ORGANISMS

Each living thing has a scientific name that consists of two parts: the generic name and the specific name. For example, *Panthera leo* is the lion. *Panthera* is the name of the genus to which the animal belongs, and *leo* is the name of the species within this genus. The system used for naming organisms is called the binomial system and was devised by the Swedish biologist and physician Carl Linnaeus (Fig. 1.1). Where subspecies are recognised a trinomial name is used. For example *Panther leo persica* is the subspecies of the lion which occurs in India, the

Figure 1.1 Carl Linnaeus.

Asiatic lion. Once the full scientific name has been used it may be abbreviated: *Panthera leo* becomes *P. leo* and *Panthera leo persica* becomes *P. l. persica*. The individual specimen of a species upon which the first scientific description of that species was based is called the holotype.

Animals are classified into taxa in a hierarchical manner. For example, the grey wolf is classified as follows:

Kingdom	Animalia
Phylum	Chordata
Class	Mammalia
Order	Carnivora
Family	Canidae
Genus	*Canis*
Species	*lupus*

Well-known species of organisms have common or vernacular names. These are often locally used names so the name of a particular species varies from place to place, from language to language and may vary with time. Due to their inconsistency, vernacular names are not widely used by scientists unless the species concerned is also identified by its scientific name. Vernacular names do not indicate taxonomic relationships, so unrelated species may have similar vernacular names.

THE IDENTIFICATION OF SPECIES

The ability to identify species is essential if we are to measure biodiversity and monitor changes in biodiversity with time. In the field and laboratory the identification of an unidentified organism is often achieved by using an identification guide. Such guides often contain keys that require the user to examine the characteristics of the specimen in a sequence of steps until identification is achieved. When using a dichotomous key the user is required to answer a series of questions about the unidentified specimen about its size, shape, colour and other aspects of its morphology. Each question in the key may only have one of two possible answers, for example, 'Does the organism have three pairs of walking legs or more than three pairs of walking legs?'

BIODIVERSITY HOTSPOTS VERSUS COLD SPOTS

Trends in biodiversity occur across the planet. Biodiversity tends to be higher in tropical areas and lower in the polar regions. For many taxa there are clear north–south gradients in biodiversity.

Biodiversity hotspots are regions that have high concentrations of endemic species and have suffered a high level of habitat destruction (Fig. 1.2). The term was first coined by Norman

Figure 1.2 Costa Rica is located within the Mesoamerica hot spot. The jaguar (*Panthera onca*) is one of its iconic apex predators.

Myers almost 30 years ago (Myers, 1988). Myers, Mittermeier, Mittermeier, da Fonseca and Kent (2000) argue that limited conservation funds should be directed towards biodiversity hotspots because we cannot save everything and 'as many as 44% of all species of vascular plants and 35% of all species in four vertebrate groups are confined to 25 hotspots comprising only 1.4% of the land surface of the Earth.' Since the publication of this paper a total of 36 hotspots have been identified.

Some conservationists believe that we should not just concentrate our efforts (and spend all our money) on hotspots, but we should prioritise the protection of nonhotspot areas or 'cold spots': areas which do not have high biodiversity but which provide valuable ecosystem services. These areas make up the remaining 98.6% of the land surface of the Earth and they contain rare species, encompass important large wilderness areas, provide habitat for many wide-ranging species and provide ecological services that are of local and global importance. These services include flood protection, carbon sequestration, water purification, pest control, waste detoxification, raw materials, food, genetic resources and medicines.

Citizen Science

Some areas of science have always attracted the interest of keen amateurs. The English amateur astronomer Sir Patrick Moore became an expert on the Moon, published dozens of astronomy books and presented the BBC's long-running television programme *The Sky at Night*. Many other ordinary people, with no formal training in their subject, have turned a hobby into a lifelong interest and become acknowledged experts in entomology, plant identification, bird identification and many other fields.

In order to tap the amateur enthusiasm for natural history some individual scientists and organisations have devised projects that use members of the public, including children, to collect data. In the United Kingdom, the Royal Society for the Protection of Birds (RSPB) has been organising the *Big Garden Birdwatch* each year since 1979. Over half a million people take part each year counting birds over a period of three days. In 2016 they counted over 8 million birds (BBC, 2016).

The following exercises examine the way in which scientists name, identify and classify organisms, patterns in biodiversity, factors affecting extinction and opportunities for the public to engage in biodiversity projects.

References

BBC. (2016). Big Garden Birdwatch. <https://rspb.org.uk/get-involved/activities/birdwatch/results.aspx>. Accessed 22.11.16.

Myers, N. (1988). Threatened biotas: "Hot spots" in tropical forests. *Environmentalist, 8*, 187–208.

Myers, N., Mittermeier, R. A., Mittermeier, C. G., da Fonseca, G. A. B., & Kent, J. (2000). Biodiversity hotspots for conservation priorities. *Nature, 403*, 853–858.

ECOLOGY AND TAXONOMY

Ecology is a relatively new science. The term 'ecology' is derived from the word 'oecologie' coined in 1866 by the German zoologist Ernst Haeckel in his book *Generelle Morphologie der Organismen* (Haeckel, 1866) (Fig. 1.3). Since then many other definitions have been used some of which are listed in Table 1.1.

The British ecologist Charles Elton called ecology 'scientific natural history'. In his book *Animal Ecology*, Elton (1927) claimed that:

It is a fact that natural history has fallen into disrepute among zoologists, at least in England, and since it is a very serious matter that a third of the whole subject of zoology should be neglected by scientists, we may ask for reasons.

Elton went on to explain that zoologists were turning to the study of the morphology and physiology of organisms in the laboratory and away from fieldwork at that time. These disciplines do not require – at least according to Elton – a detailed knowledge of the taxonomy of the organisms being studied.

Almost 70 years later, in 1995, Tam Dalyell MP expressed concern in the House of Commons regarding the shortage of taxonomists in the United Kingdom:

HC Deb 13 June 1995 vol. 261 cc431-2W
431W
Mr. Dalyell
To ask the Chancellor of the Duchy of Lancaster what action Her Majesty's Government are taking to train sufficient scientists in methods relating to the identification of species to enable the

Figure 1.3 Professor Ernst Haeckel.

Table 1.1 Selected definitions of ecology

Author	Definition – Ecology may be defined as…
Haeckel (1866)	…the entire science of the relations of the organism to the surrounding exterior world, to which its relations we can count in the broader sense all the conditions of existence. These are partly of organic, partly of inorganic nature. OR …household of nature.
King and Russell (1909)	…the study of the relations between animals and their environment, both animate and inanimate.
Elton (1927)	…scientific natural history.
Eggleton (1939)	…the science which treats of the interrelationships of organisms with their complete environment.
Andrewartha (1961)	…the scientific study of the distribution and abundance of organisms.
Odum (1963)	…the study of the structure and function of nature.
Krebs (1972)	…the scientific study of the interactions that determine the distribution and abundance of organisms.

United Kingdom to fulfil obligations under the biodiversity convention; and what assessment has been made of the shortage of taxonomists. [26328]

The United Nations Convention on Biological Diversity 1992 requires Parties to the Convention to identify components of biodiversity of conservation interest:

Article 7. Identification and Monitoring

Each Contracting Party shall, as far as possible and as appropriate, in particular for the purposes of Articles 8 to 10:

(a) Identify components of biological diversity important for its conservation and sustainable use having regard to the indicative list of categories set down in Annex I;

(b) Monitor, through sampling and other techniques, the components of biological diversity identified pursuant to subparagraph (a) above, paying particular attention to those requiring urgent conservation measures and those which offer the greatest potential for sustainable use;

(c) Identify processes and categories of activities which have or are likely to have significant adverse impacts on the conservation and sustainable use of biological diversity, and monitor their effects through sampling and other techniques; and

(d) Maintain and organize, by any mechanism, data derived from identification and monitoring activities pursuant to subparagraphs (a), (b) and (c) above.

Q1.1.1 Which of the definitions in Table 1.1 do you consider too vague to explain the term 'ecology' to the lay person?

Q1.1.2 Why are taxonomists essential to the proper study of ecology?

Q1.1.3 Explain why a shortage of taxonomists could have had the effect of preventing the United Kingdom from fulfilling its legal obligations under the UN Convention on Biological Diversity.

Q1.1.4 Suggest the types of 'processes' and 'activities' envisaged by Art. 7(c).

Q1.1.5 Explain why it is important that countries cooperate in the conservation of biodiversity.

References/Further Reading

Andrewartha, H. G. (1961). *Introduction to the Study of Animal Populations*. Chicago: University of Chicago Press.

Eggleton, F. E. (1939). Fresh-water communities. *The American Midland Naturalist, 21*(1), 56–74.

Elton, C. (1927). *Animal ecology*. New York: The MacMillan Company.

Friederichs, K. (1958). A definition of ecology and some thoughts about basic concepts. *Ecology, 39*(1), 154–159.

Haeckel, E. (1866). *Generelle morphologie der organismen*. Berlin: Druck und verlag von Georg Reimer.

King, L. A. L., & Russell, E. S. (1909). A method for the study of the animal ecology of the shore. *Proceedings of the Royal Society of Edinburgh, 17*, 225–253.

Krebs, C. J. (1972). *Ecology. The experimental analysis of distribution and abundance*. New York: Harper and Row, Publishers.

Odum, E. P. (1963). *Ecology*. New York: Holt, Rinehart & Winston.

WHAT'S IN A NAME?
VERNACULAR VERSUS SCIENTIFIC NAMES

Figure 1.4 Barn owl (*Tyto alba*).

The vernacular name of a species is its common name, e.g., lion, blackbird, hippopotamus. Vernacular names vary between countries and languages, and over time. There may be several names used for a single species in a particular region.

At various times and in various places in the United Kingdom the barn owl (*Tyto alba*) has been called the screech owl, silver owl, yellow owl, hobby owl, white owl, hissing owl, church owl, Jenny owl, ullat, oolert, willow owl and many other names (Fig. 1.4). To avoid confusion scientists use the binomial system to assign a single Latin or scientific name to each species.

Table 1.2 Examples of New World blackbirds

Vernacular name	Scientific name
Austral blackbird	*Curaeus curaeus*
Bolivian blackbird	*Agelaioides oreopsar*
Brewer's blackbird	*Euphagus cyanocephalus*
Chestnut-capped blackbird	*Chrysomus ruficapillus*
Chopi blackbird	*Gnorimopsar chopi*
Cuban blackbird	*Dives atroviolaceus*
Forbes's blackbird	*Curaeus forbesi*
Jamaican blackbird	*Nesopsar nigerrimus*
Melodious blackbird	*Dives dives*
Oriole blackbird	*Gymnomystax mexicanus*
Pale-eyed blackbird	*Agelasticus xanthophthalmus*

The name 'blackbird' is used for many bird species in different regions of the world. Table 1.2

Figure 1.5 Ocelot (*Leopardus pardalis*).

Table 1.3 The vernacular names of the ocelot (*Leopardus pardalis*)

Country/language	Vernacular names
Argentina	chivi-guazu, cuanguaro, gato onza, tirica
Bolivia	gato bueno, gato onza, tigrezillo
Brazil	gato-maracajá, jaguatirica, maracajá-açu, maracajá-verdadeiro
Colombia	maracaya, maracaja
Costa Rica	manigordo
French Guiana	chat tigre
Guarani	agua-tirica
Mayan	zac-xicin
Nicaragua	mandigordo
Panama	gato tigre, mandigoldo, tigre chico
Paraguay	chivi-guazu, gato onza
Peru	gato onza, pupillo, tigrillo
Spanish	gato onza, ocelote, tigrillo
Surinam	hétigrikati
Venezuela	cuanguaro, manigordo

Adapted from Anon. (2015).

lists examples of some of the species found in the New World.

Q1.2.1 Using the information provided in this exercise, explain why it is important for scientists to use the scientific (binomial) name of a species rather than a vernacular name.

Q1.2.2 Using the information in Table 1.2, explain why the term 'blackbird' has little meaning from an evolutionary and taxonomic point of view.

Q1.2.3 If you were an ecologist studying the distribution of the ocelot (*Leopardus pardalis*) (Fig. 1.5) in the countries of Central and South America by surveying local people using a questionnaire, what precautions would you need to take to ensure that the interviewees knew which animal you were asking about? (See Table 1.3).

Reference/Further Reading

Anon. (2015). Cat Specialist Group, Species Survival Commission. <http://www.catsg.org/index.php?id=88>. Accessed 20.8.15.

EXERCISE 1.3

THE CLASSIFICATION OF ANIMALS

Organisms are classified according to their evolutionary relationships. Animals are allocated a scientific name which consists of two parts indicating the genus to which it belongs (generic name) and a specific name which indicates the particular species within that genus (Fig. 1.6). These names are written in italics. This naming system is called the binomial system of nomenclature because it assigns two names to each species (the binomial name).

Panthera leo is the lion. This species has several subspecies, each of which is assigned a different subspecific (third) name. For example,

the Asiatic lion is called *Panthera leo persica*. This three-part name is known as the trinomial name. Where subspecies exist, the one which was first described has its specific name repeated as its subspecific name. For example, the first lion described was *Panthera leo leo*.

Species are grouped into genera, genera into families, families into orders, orders into classes and classes into phyla. The classification of the bonobo (*Pan paniscus*) is given in Table 1.4.

Q1.3.1 Fig. 1.7 shows the family tree of an imaginary group of animals. Family P is

Figure 1.6 (A) Western lowland gorilla (*Gorilla gorilla gorilla*); (B) chimpanzee (*Pan troglodytes*); (C) bonobo (*Pan paniscus*); (D) orangutan (*Pongo* sp.).

Table 1.4 Classification of the bonobo

Taxon	
Phylum	Chordata
Class	Mammalia
Order	Primates
Family	Hominidae
Genus	*Pan*
Species	*paniscus*

Note that a full classification would include additional levels such as superfamily and subfamily.

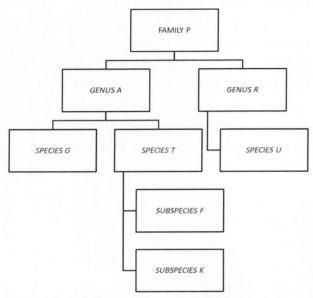

Figure 1.7 Classification of imaginary organisms.

divided into two genera, A and R. Genus A contains species G and T. Species T is divided into two subspecies, F and K. Genus R has a single species, U. Using the information in Table 1.5 and the example layout in Fig. 1.7, draw a family tree showing the evolutionary relationships between the members of the Hominidae.

Table 1.5 Classification of living members of the family Hominidae

Binomial name	Vernacular name
Pongo abelii	Sumatran orangutan
Pan troglodytes troglodytes	Central chimpanzee
Gorilla beringei graueri	Eastern lowland gorilla
Pan paniscus	Bonobo
Homo sapiens	Man
Pan troglodytes schweinfurthii	Eastern chimpanzee
Gorilla gorilla diehli	Cross River gorilla
Pan troglodytes vellerosus	Nigeria-Cameroon chimpanzee
Gorilla beringei beringei	Mountain gorilla
Pongo pygmaeus	Bornean orangutan
Gorilla gorilla gorilla	Western lowland gorilla
Pan troglodytes verus	West African chimpanzee

Q1.3.2 What do the vernacular names of the species listed in Table 1.5 tell us about the likely mechanism by which subspecies have evolved?

Q1.3.3 Which of the subspecies of chimpanzee was the first to be formally described by a scientist?

Q1.3.4 The bonobo is sometimes referred to as the pygmy or dwarf chimpanzee. Why might this be confusing?

Q1.3.5 Is it possible to determine from the family tree that you have produced which species is most closely related to humans?

Reference/Further Reading

Wilson and Reeder's Mammal Species of the World. <http://vertebrates.si.edu/msw/mswcfapp/msw/index.cfm>. Accessed 15.11.16.

EXERCISE 1.4

CONSTRUCTING A DICHOTOMOUS KEY

Field ecologists use dichotomous keys to assist in the identification of organisms. Such keys consist of a series of questions to which there may only be two possible answers. Each answer leads to another question until the end of the sequence is reached and the organism is identified. The following key distinguishes between four types of bird (Fig. 1.8):

Q1	Can it fly?	If YES go to Q2.
		If NO go to Q3.
Q2	Does it have a short downward curving bill?	If YES – Flamingo.
		If NO – Pelican.
Q3	Does it have a very long neck?	If YES – Ostrich
		If NO – Penguin

This is not the only key which could be devised to distinguish between these four types of bird. Note that the number of questions required is one fewer than the number of types of organisms to be distinguished.

Q1.4.1 Why are keys like this referred to as 'dichotomous'?

Q1.4.2 Explain why questions relating to the relative size of a structure are not really appropriate in a key (e.g., 'Does it have a very long neck?')

Q1.4.3 Construct a dichotomous key to distinguish between the six imaginary invertebrate species in Fig. 1.9.

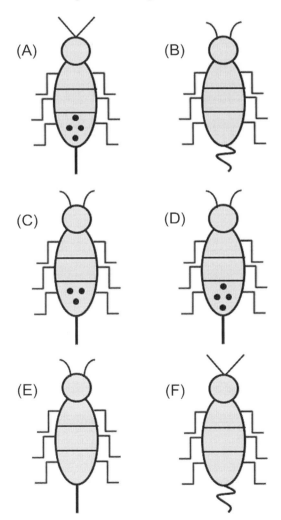

Figure 1.9 Six imaginary invertebrate species.

Figure 1.8 (A) Penguin; (B) ostrich; (C) pelican; (D) flamingo.

Q1.4.4 A dichotomous key would normally lead to an image of the organism that has been identified. This may be a detailed drawing or a photograph. Why is this image important in the identification process?

Reference/Further Reading

Many dichotomous keys exist in a variety of field guides, for example:

Falk, S. (2015). *Field guide to the bees of Great Britain and Ireland*. London: Bloomsbury.

EXERCISE 1.5

GLOBAL BIODIVERSITY: THE NUMBERS OF RECOGNISED SPECIES

The International Union for Conservation of Nature and Natural Resources (IUCN) was founded in 1948 and is now the world's largest and most diverse environmental network. It is based in Geneva, Switzerland, and has 1300 member organisations and access to 16,000 experts. The IUCN is internationally recognised as the global authority on the status of the natural world and maintains a list of estimates of the numbers of described species listed by taxon (Table 1.6).

Table 1.6 Estimated numbers of described species of animals recognised by the IUCN in 2016

Taxon	Species
Vertebrates	
Mammals	5536
Birds	10,424
Reptiles	10,450
Amphibians	7538
Fishes	33,300
Subtotal	**67,248**
Invertebrates	
Insects	1,000,000
Molluscs	85,000
Crustaceans	47,000
Corals	2175
Arachnids	102,248
Velvet Worms	165
Horseshoe Crabs	4
Others	68,658
Subtotal	**1,305,250**
Total	

Adapted from IUCN (2016).

Table 1.7 The relative abundance of selected taxa

	%
Arachnids as a percentage of all invertebrate species	
Invertebrates as a percentage of all animal species	
Fishes as a percentage of all vertebrate species	
Homiotherms (mammals and birds) as a percentage of all vertebrate species	
Insects as a percentage of all animal species	
Mammals as a percentage of all vertebrate species	
Homiotherms as a percentage of all animal species	

Q1.5.1 Calculate the total number of identified animal species from the information in Table 1.6.

Q1.5.2 Draw a pie chart showing the proportion of all animal species that are mammals, birds, reptiles, amphibians, fishes and invertebrates using the data in Table 1.6.

Q1.5.3 Complete Table 1.7 using data from Table 1.6.

Q1.5.4 Most high profile conservation projects tend to be focussed on vertebrates, especially mammals and birds. Use the information in Table 1.6 and the completed Table 1.7 to write a short paragraph justifying expending more effort on the conservation of invertebrates.

Reference/Further Reading

IUCN, (2016). IUCN Red List Version 2016-2. Table 1: Numbers of threatened species by major groups of organisms (1996–2016). <http://cmsdocs.s3.amazonaws.com/summarystats/2016-2_Summary_Stats_Page_Documents/2016_2_RL_Stats_Table_1.pdf>. Accessed 10.10.16.

EXERCISE 1.6

DIVERSITY IN CHALCID WASPS

Stork (1991) collected insects from 10 trees in lowland rainforest in Borneo using knock-down insecticide fogging. Fogging involves spraying a fine mist of droplets from a portable machine. Table 1.8 shows the number of species of chalcid wasps (Chalcidoidea; Hymentoptera) obtained represented by 1, 2, 3…19 individuals.

Q1.6.1 Draw a bar chart of the data in Table 1.8.

Q1.6.2 Calculate the total number of different species of chalcid wasps collected.

Q1.6.3 Calculate the number of individual wasps collected.

Q1.6.4 How many individual wasps were collected from the four commonest species?

Q1.6.5 Suggest reasons why some species of chalcid wasps that may have been present in the forest were not collected in this study.

Q1.6.6 It has been said that 'Common species are rare in nature.' Does your graph support this statement?

Q1.6.7 Suggest two desirable characteristics of the insecticide used to collect these insects.

Table 1.8 Diversity in Chalcidoidea

Number of individuals	Number of species
1	437
2	160
3	54
4	31
5	18
6	10
7	8
8	4
9	6
10	3
11	2
12	1
13	1
14	0
15	0
16	0
17	1
18	0
19	1

Adapted from Stork (1991).

Reference/Further Reading

Stork, N. E. (1991). The composition of the arthropod fauna of Bornean lowland rainforest trees. *Journal of Tropical Ecology*, 7, 161–180.

LOGNORMAL DISTRIBUTION OF SPECIES RELATIVE ABUNDANCE

In communities that consist of a relatively large number of species, the distribution of species relative abundance is almost always lognormal. This is a bell-shaped (Gaussian) distribution and it has been observed in a wide range of groups from moths and diatoms to birds and fishes. It arises because species relative abundance is influenced by a large number of random variables. Other distributions affected by many random variables also exhibit a lognormal distribution, for example, the distribution of wealth in the United States.

Table 1.9 shows the $\log_2$ of the numbers 2, 4, 8 and 16. The $\log_2$ of a number is the power to which you must raise 2 to obtain the number. So, to obtain the value 4 we must raise 2 to the power of 2 (i.e., 2^2) so $\log_2$ of 4 is 2. Similarly, $\log_2$ of 8 is 3 because 8 is $2 \times 2 \times 2$ or 2^3.

The data in Table 1.10 describe the pattern of species abundance in a diatom community.

Q1.7.1 Calculate the $\log_2$ of the ranges of values in column B in Table 1.10 and add them to column C. The first three values have been calculated for you and some of the other values you need are in Table 1.9.

Q1.7.2 Draw a bar chart of the number of species (y-axis) against abundance, i.e., $\log_2$ number of individuals (x-axis) using intervals 1 (rare) to 10 (common). In this graph, interval 1 represents the number of species with 1 or 2 individuals, interval 6 represents the number of species with between 32 and 64 individuals etc.

Table 1.9 Calculation of $\log_2$

Number			$\log_2$
2	2	2^1	1
4	2×2	2^2	2
8	$2 \times 2 \times 2$	2^3	3
16	$2 \times 2 \times 2 \times 2$	2^4	4

Table 1.10 The pattern of species abundance in a diatom community

A	B	C	D
Interval	Number of individuals	$\log_2$ number of individuals	Number of species
1	1–2	1	16
2	2–4	1–2	21
3	4–8	2–3	25
4	8–16		23
5	16–32		20
6	32–64		15
7	64–128		12
8	128–256		8
9	256–512		5
10	512–1024		3

Q1.7.3 If the diatom sample taken was too small would rare or common species be most likely to be overlooked?

Reference/Further Reading

Edden, A. C. (1971). A measure of species diversity related to the lognormal distribution of individuals among species. *Journal of Experimental Marine Biology and Ecology, 6*, 199–209.

THE DISCOVERY OF NEW SPECIES

New species may be discovered in a number of different ways. In the early days of exploration, new species were found when scientists travelled in remote areas and found animals and plants that, although they were well known to indigenous people, nevertheless had not been previously described by scientists. Some species that are rare, either because they are naturally rare or because their populations have been depleted by human activity, may not have been recorded in the wild for many years before being rediscovered, often as isolated relict populations. Occasionally, populations of known species are split and reclassified to create two or more distinct species based on new scientific evidence. This means that some new species are hiding in

Table 1.11 Examples of animal species discovered since 1865

Species	Year of discovery	Discoverer	New (N)/ rediscovered (R)/ reclassified (C)	Location
Santa Cruz giant tortoise (*Chelonoidis donfaustoi*)	2015	Poulakakis et al. (2015)	C	Santa Cruz, Galapagos Islands.
Hog-nosed rat (*Hyorhinomys stuempkei*)	2013	Scientists from Museum Victoria, Indonesia and USA	N	Sulawesi
Black dwarf lowland tapir (*Tapirus pygmaeus*)	2008	Dr Marc van Roosmalen	N	Lowland Amazonia, Brazil
Kipunji (*Rungwecebus kipunji*) – a new monkey genus	2004	Wildlife Conservation Society scientists	N	Montane forest in S.W. Tanzania
Giant muntjac (*Megamuntiacus vuquangensis*)	1994	MacKinnon et al.	N	Northern Vietnam and Laos
Mbaiso tree kangaroo (*Dendrolagus mbaiso*)	1994	Dr Tim Flannery	N	Irian Jaya, Indonesia
Soprano pipistrelle (*Pipistrellus pygmaeus*)	1993	Jones and Van Parjis (1993)	C	Europe
Vu quang ox (*Pseudoryx nghetinhensis*)	1992	Dung, Giao, Chinh, Tuoc, Arctander and MacKinnon	N	Vietnam/Laos
Hairy-eared lemur (*Allocebus trichotis*)	1989	B. Meier	R	N. Madagascar
Fijian crested iguana (*Brachylophus vitiensis*)	1979	Dr John Gibbons	N	Yadua Taba, Fiji
Long-whiskered owlet (*Xenoglaux loweryi*)	1976	Drs J. O'Neill and Gary Graves	N	Peru
Greater bamboo lemur (*Hapalemur simus*)	1972	Dr A. Peyriéras	R	S.E. Madagascar
Iriomote cat (*Felis iriomotensis*)	1967	Yokio Togawa	N	Iriomote, Japan
Mountain pygmy possum (*Burramys parvus*)	1966	Dr K. Shortman	R	Victoria and N.S.W., Australia
Coelacanth (*Latimeria chalumnae*)	1938	Marjorie Courtenay-Latimer	N	South Africa
Kouprey (*Bos sauveli*)	1937	Prof. Achille Urbain	N	Cambodia
Congo peacock (*Afropavo congensis*)	1936	Dr James Chapin	N	Congo
Bonobo (*Pan paniscus*)	1929[a]	Dr Ernst Schwarz	C	Congo
Scaly-tailed possum (*Wyulda squamicaudata*)	1917	W. B. Alexander	N	Australia
Komodo dragon (*Varanus komodoensis*)	1912	Major P.A. Ouwens	N	Komodo, Indonesia
Okapi (*Okapia johnstoni*)	1901	Sir Harry Johnston	N	Ituri Forest, Zaire
Giant panda (*Ailuropoda melanoleuca*)	1869	Père Armand David	N	China
Père David's deer (*Elaphurus davidianus*)	1865	Père Armand David	N	China

Adapted from Rees (2011).
[a]Originally classified as a subspecies, *Pan troglodytes paniscus*, in 1929 but elevated to a new species five years later.

plain sight in natural populations or in museum collections.

Q1.8.1 What type of new evidence could lead to the discovery of a new species due to the reclassification of animals that are already known to science?

Q1.8.2 What circumstances could lead to the rediscovery of a species thought to be extinct in the wild?

Q1.8.3 In what types of habitats have the new species (indicated by 'N') listed in Table 1.11 been found?

Q1.8.4 List those new species in Table 1.11 that have been 'hiding in plain sight' and have been created by a reexamination of existing species (indicated by 'C').

Q1.8.5 The soprano pipistrelle (*Pipistrellus pygmaeus*) was identified when two phonotypes were discovered within the bat populations previously considered to be made up entirely of the common pipistrelle (*P. pipistrellus*). Bats use echolocation to find prey, avoid obstacles and for other purposes. What characteristic of echolocation distinguishes the two phonotypes?

Q1.8.6 Why are new species of mammals and birds unlikely to be found in the United States?

Q1.8.7 Suggest the most likely origin of the second part of the scientific name of the okapi (Fig. 1.10)?

References/Further Reading

Fuller, E. (2014). *Lost animals: Extinction and the photographic record*. Princeton, NJ: Princeton University Press.

Jones, G., & Van Parijs, S. M. (1993). Bimodal echolocation in pipistrelle bats: are cryptic species present?. *Proceedings of the Royal Society, B, 251B*, 119–125.

Poulakakis, N., Edwards, D. L., Chiari, Y., Garrick, R. C., Russello, M. A., Benavides, E., et al. (2015). Description of a new galapagos giant tortoise species (Chelonoidis; Testudines: Testudinidae) from Cerro Fatal on Santa Cruz Island. *PLoS ONE, 10*(10), e0138779. http://dx.doi.org/10.1371/journal.pone.0138779.

Rees, P. A. (2011). *An introduction to zoo biology and management*. Chichester: Wiley-Blackwell.

Figure 1.10 Okapi (*Okapia johnstoni*).

EXTINCTIONS VERSUS DISCOVERY OF NEW SPECIES

Figure 1.11 Nēnē or Hawaiian goose (*Branta sandvicensis*).

By 1952 only 52 individuals of the Hawaiian goose or nēnē (*Branta sandvicensis*) survived in the wild (Fig. 1.11). The species had been decimated by hunting, habitat loss and predation particularly by introduced mongooses, rats and feral cats. The species was saved from extinction by conservationists, notably Sir Peter Scott and the Wildfowl Trust (now the Wildfowl and Wetlands Trust), who established a captive breeding programme.

Since 1500, 61 mammal species and 128 bird species have become extinct, including the dodo (*Raphus cucullatus*) (Fig. 1.12) which disappeared from the island of Mauritius around the end of the 17th century. Table 1.12 shows extinction rates since 1500 for mammal and bird species per million square kilometres.

Figure 1.12 The dodo (*Raphus cucullatus*) was a flightless bird that was endemic to the island of Mauritius and became extinct around 1662. It was predated by introduced cats, dogs, pigs and rats and was taken by sailors for food.

Table 1.12 Extinctions since 1500

	Mammals		Birds	
	Species	Per 10^6 km²	Species	Per 10^6 km²
Continents	3	0.025	6	0.05
Islands	58	4.43	122	9.38

Adapted from Loehle & Eschenbach (2012).

Table 1.13 Number of new species of birds discovered per year

Period	Number of species/year
1938–1941	6.0
1941–1955	2.6
1956–1965	3.5
1966–1975	3.1
1976–1980	2.4
1981–1990	2.4

Based on Vuilleumier, LeCroy, & Mayr (1992).

Q1.9.1 Are island species of mammals and birds more or less susceptible to extinction than continental species?

Q1.9.2 Which of the classes of animals for which data are presented in Table 1.12 has the highest extinction rate per unit area?

Q1.9.3 What ecological factors are likely to be responsible for extinctions on islands?

Q1.9.4 Why are ground-nesting birds particularly susceptible to extinction on some islands?

Q1.9.5 How many new species of birds were discovered in the decade 1981–90 (Table 1.13)?

References/Further Reading

Loehle, C., & Eschenbach, W. (2012). Historical bird and terrestrial mammal extinction rates and causes. *Biodiversity and Distributions, 18*, 84–91.

Vuilleumier, F., LeCroy, M., & Mayr, E. (1992). New species of birds described from 1981 to 1990. *Bulletin of the British Ornithologists' Club, Centenary Supplement A, 122*, 267–309.

THE HISTORICAL ECOLOGY OF THE LARGE MAMMALS OF NGORONGORO CRATER

Figure 1.13 Wildebeest (*Connochaetes taurinus*) herd in Ngorongoro Crater, Tanzania.

The extract below has been taken from a paper by Oates and Rees (2013) which discusses the historical ecology of the large mammal populations in Ngorongoro Crater, Tanzania (Fig. 1.13). The crater is approximately 19 km in diameter and has sides that extend some 600 m from the floor.

The first description of the crater was written in German by Baumann (Organ & Fosbrooke, 1963) who visited in 1892, and the first English language account was by Barns (1921). Baumann referred to the presence of only four large mammal species: wildebeest, zebras Equus burchellii, hippopotamuses Hippopotamus amphibius

and black rhinoceros Diceros bicornis (Organ & Fosbrooke, 1963).

Historical accounts probably overestimated the number of animals living within the crater. In the early 1900s, FR Holmes (Holmes, 1929) and Captain GHR Hurst (Fosbrooke, 1972) estimated that the large mammal population numbered 100,000 and 500,000, respectively. Dr Hans Poeschel, who visited the Siedentopf brothers before the onset of the First World War, stated that, according to several experienced game hunters, the crater held over 20,000 wildebeest (Grzimek & Grzimek, 1960). In 1921, Major AR Dugmore estimated that there were over 75,000 large mammals, while Barns suggested that wildebeest alone numbered 50,000 (Fosbrooke, 1972). Molloy and Harvey in 1959 estimated that the crater contained just 7800 wildebeest (Grzimek & Grzimek, 1960). The first aerial census by Grzimek and Grzimek (1960) counted 8500 large mammals in 1958. A second aerial survey by Turner and Watson (1964) counted over 22,000. Recent counts (1999–2005) show that the total number of wildebeest, buffalo Syncerus caffer, *zebras, Grant's gazelles* Nanger granti *and Thomson's gazelles* Eudorcas thomsonii *alone fluctuates between 15,000 and 28,000 (Estes et al., 2006).*

Early researchers debated whether migration occurred between the crater and the SNP [Serengeti National Park] (e.g., Grzimek & Grzimek, 1960). Estes (1966) established that some wildebeest left the crater during the rains and it is now accepted that a number of species migrate out regularly (Estes & Small, 1981). Several species found in the SNP are absent from the crater, including giraffes Giraffa camelopardalis, *impalas* Aepyceros melampus, *topis* Damaliscus korrigum *and oribis* Ourebia ourebi *(Estes et al. 2006). Giraffes may be absent due to a lack of browse species (e.g.,* Acacia tortilis, *A. melifera and A. seyal) and the open woodland preferred by impalas is absent (Fosbrooke, 1972).*

Q1.10.1 What does Baumann's account of his visit to the crater in 1892 tell us about the diversity and size of the large mammal populations present at the time?

Q1.10.2 What evidence is there in the literature that the sizes of large mammal populations have been overestimated in the past?

Q1.10.3 Is it possible to determine which differences in estimates were affected by losses due to migration and which were not?

Q1.10.4 Explain why, although the Serengeti National Park (SNP) is located near the crater, their large mammal faunas differ.

Q1.10.5 Why would it be difficult to use the estimates given here to make any clear inferences about changes to the large mammal populations of the crater with time?

Q1.10.6 Turner and Watson's 1964 survey involved counting animals from a light aircraft flying at a height of 1000 feet. Grant's gazelle and Thompson's gazelle are both small antelope species and the authors stated that, although it was possible to distinguish between them from 1000 feet, nevertheless, they grouped them together as gazelles. Why do you think they did this?

References/Further Reading

Oates, L., & Rees, P. A. (2013). The historical ecology of the large mammal populations of Ngorongoro Crater, Tanzania, east Africa. *Mammal Review, 43*, 124–141.

Turner, M., & Watson, M. (1964). A census of game in Ngorongoro Crater. *African Journal of Ecology, 2*, 165–168.

PEST ERADICATION IN NEW ZEALAND

In July 2016 the Prime Minister of New Zealand, John Key, announced an ambitious project intended to eradicate all nonnative predators from the country by 2050. The following list of facts and objectives is based on the Prime Minister's press release (Key, 2016):

- In the past New Zealand has seen some of its native species forced into extinction.

- There was once a time when the greatest threat was deforestation and poaching, today it is introduced predators.

- Their impact cannot be overstated – rats, stoats and possums kill around 25 million of our native birds every year.

- They threaten New Zealand's economy with a total economic cost estimated at around $3.3 billion a year.

- By 2050 every single part of New Zealand will be completely free of rats, stoats and possums.

- The Crown will initially invest $28 million over four years to establish a new joint venture company called Predator Free New Zealand Ltd to drive the programme, alongside the private sector.

- The following steps will be taken initially:

  - Increase the amount of New Zealand covered by predator control;

  - Establish regional partnerships and support community led initiatives;

  - Improve the tools available to do the job;

  - Establish more areas of complete elimination as a base to build from; and

  - Invest in long-term scientific breakthroughs to harness the enterprise and ingenuity of the science and business communities.

- By 2025 the Predator Free New Zealand project will see:

  - All introduced predators eradicated from all New Zealand's offshore island nature reserves;

  - 1 million more hectares of mainland New Zealand where predators are suppressed;

  - A demonstration that complete predator eradication can be achieved in areas of at least 20,000 hectares on the New Zealand mainland;

  - The development of a breakthrough science solution capable of removing at least one small mammal predator from the mainland entirely.

- All of this will help with the ultimate goal – to make New Zealand predator free by 2050.

Q1.11.1 Why are nonnative predators a particular problem in New Zealand?

Q1.11.2 Which common predatory domestic pets are likely to have a damaging effect on New Zealand's native fauna?

Q1.11.3 What can owners of these pets do to minimise their effects on native species?

Q1.11.4 What is the economic justification for the eradication project?

The release of animals into the wild in New Zealand is regulated by the Hazardous Substances and New Organisms Act (1996). Sections 2A and 25 are reproduced below.

Hazardous Substances and New Organisms Act 1996

…2A Meaning of term new organism

(1) A **new organism** *is –*

(a) an organism belonging to a species that was not present in New Zealand immediately before 29 July 1998:

(b) an organism belonging to a species, subspecies, infrasubspecies, variety, strain, or cultivar prescribed as a risk species, where that organism was not present in New Zealand at the time of promulgation of the relevant regulation:

(c) an organism for which a containment approval has been given under this Act:

(ca) an organism for which a conditional release approval has been given:

(cb) a qualifying organism approved for release with controls:

(d) a genetically modified organism:

(e) an organism that belongs to a species, subspecies, infrasubspecies, variety, strain, or cultivar that has been eradicated from New Zealand.

…25 Restriction of import, manufacture, development, field testing, or release

(1) No –

(a) hazardous substance shall be imported, or manufactured:

(b) new organism shall be imported, developed, field tested, or released – otherwise than in accordance with an approval issued under this Act or in accordance with Parts 11 to 16.

(1A) Subsection (1) (b) does not apply to –

(a) the importation of an incidentally imported new organism, if it is imported in or on goods lawfully imported under the Biosecurity Act 1993; or

(b) the movement or use of those goods, together with any new organisms incidentally imported while they remain in or on those goods, after their importation.

…(2) No approval shall be issued to import, develop, field test, or release any new organism specified in Schedule 2.

…Schedule 2 Prohibited new organisms*

1. Any snake of any species whatever.

2. Any venomous reptile, venomous amphibian, venomous fish, or venomous invertebrate. (In this item, **venomous** *means capable of inflicting poisonous wounds harmful to human health.)*

3. Any American grey squirrel (Sciurus carolinensis gmelini).

4. Any red squirrel (Sciurus vulgaris).

5. Any musquash (or muskrat) (Ondatra zibethica).

6. Any coypu or nutria (Myocastor coypus).

7. Any beaver (Castor canadensis).

8. Any gerbil (Meriones unguiculatus).

9. Any prairie dog (Cyonomys spp.).

10. Any pocket gopher (Geomys spp. and Thomomys spp.).

11. Any red or silver fox (Vulpes vulpes).

12. Any Arctic fox (Alopex lagopus).

13. Any mongoose (family Herpestidae) other than Suricata suricatta.

14. Any member of the family Mustelidae, subfamily Mustelinae, other than ferrets (Mustela furo), weasels (Mustela nivalis) and stoats (Mustela erminea), and subfamily Lutrinae, other than oriental small clawed otter (Aonyx cineria).

15. *Any mole (family Talpidae).*
16. *Any member of the family Esocidae (e.g., pikes, muskellunge).*
17. *Any member of the families Phalangeridae and Petauridae, other than the Australian brushtail possum (Trichosurus vulpecula).*
18. *Any stickleback (Gasterosteus spp.).*
19. *Any giant African snail (Achatina spp.).*
20. *Any predatory snail (Euglandina rosea).*
21. *Any cane toad (Bufo marinus).*

*Plant species have been excluded.

Q1.11.5 Sarah is considering applying for a permit to import a prairie dog to keep as a pet. What advice would you give her about the likely outcome of her application?

Q1.11.6 Stephen is a scientist interested in the effects of nonnative species on New Zealand forests. He would like to study the ecological effects of releasing red squirrels (*Sciurus vulgaris*) into a remote area of forest on South Island. Is Stephen likely to be given permission for his experiment? Explain your answer.

References/Further Reading

Hazardous Substances and New Organisms Act. (1996). New Zealand.

Key, J., (2016). New Zealand to be Predator Free by 2050. New Zealand Government press release 25 July 2016. <https://www.beehive.govt.nz/release/new-zealand-be-predator-free-2050?utm_source=feedburner&utm_medium=feed&utm_campaign=Feed%3A+beehive-govt-nz%2Fminister%2FJohnKey+%28John+Key+-+beehive.govt.nz%29>. Accessed 7.9.2016.

CITIZEN SCIENCE: BIODIVERSITY AND ENVIRONMENTAL STUDIES

Scientists are increasingly turning to the general public to assist with large-scale research projects ranging from searching for exoplanets in space to monitoring air pollution on Earth. Table 1.14 lists some examples of citizen science projects concerned with the environment and biodiversity.

Q1.12.1 What characteristics of the projects listed in Table 1.14 make them particularly suitable for a citizen science approach?

Q1.12.2 What are the advantages of citizen science projects?

Q1.12.3 What are the disadvantages of using members of the public to collect data for scientific projects?

Q1.12.4 A citizen science project at the Natural History Museum, London, required the participants to learn to distinguish between the native common bluebell (*Hyacinthoides non-scripta*) and the Spanish bluebell (*H. hispanica*). Using the information in Table 1.15 assign each of the bluebell specimens in Fig. 1.14 to the correct species.

Q1.12.5 The projects listed in Table 1.14 have been undertaken in the United Kingdom and the United States where people are relatively affluent. What challenges would a resource-poor country face in attempting to engage citizens in a project to assess the biodiversity

Table 1.15 Morphological differences between common and Spanish bluebells

Common bluebell (*Hyacinthoides non-scripta*)	Spanish bluebell (*Hyacinthoides hispanica*)
Flowers – deep blue, tubular, narrow, tips curled back, mostly arranged on one side of a drooping stem	Flowers – light blue, conical or bell-shaped, spread out tips, arranged on all sides of an upright stem
Leaves – narrow	Leaves – broad

Table 1.14 Examples of citizen science projects concerned with the environment

Project	Activity	Organisation
Effect of climate change on flowering time in orchids	Photographing wild orchids and examining museum specimens around the United Kingdom	Natural History Museum, London
Effect of invasive species and climate change on seaweeds	Recording seaweeds present on beaches	
Recording of cetacean strandings	Reporting beach strandings of whales and dolphins	
Effect of climate change on flowering time in bluebells	Recording locations of native bluebells and Spanish bluebells and their flowering times	
Diversity of microorganisms living in urban environments	Collection of samples of microorganisms growing on the outside of buildings	
Survey of flood events	Taking photos of flooding events	British Geological Survey
Survey of landslides	Taking photos of landslides	
Monarch butterfly migration study	Tracking the migration and recording breeding condition of butterflies	Monarch Joint Venture[a]
Monitoring California condor populations	Reading tag numbers from camera trap photos	Condor Watch

[a]A partnership between a large number of organisations in the USA including the US Fish and Wildlife Service, US Forest Service, US Geological Survey, National Park Service, Bureau of Land Management and the Natural Resources Conservation Service.

Figure 1.14 Native common bluebell (*Hyacinthoides non-scripta*) and the Spanish bluebell (*H. hispanica*): which is which?

of birds in which GPS locations have to be uploaded onto a computer database via the Internet?

Q1.12.6 How does the migration of people to the cities in resource-poor countries affect indigenous biodiversity knowledge and engagement with biodiversity conservation?

Q1.12.7 Governments tend to focus on activities concerned with monitoring biodiversity while conservation NGOs focus on providing biodiversity education. Suggest why citizen science projects might be more successful in engaging the public than the traditional educational approach.

References/Further reading

Braschler, B. (2009). Successfully implementing a citizen-scientist approach to insect monitoring in a resource-poor country. *BioScience, 59,* 103–104.

EXERCISE 1.13

THE ECOLOGICAL VALUE OF ANCIENT WOODLAND

Figure 1.15 Raveden Wood, Bolton, Greater Manchester, United Kingdom: an ancient seminatural woodland.

Raveden Wood is part of the Smithills Estate, located on the northern fringe of Bolton (Greater Manchester, England) and covers an area of about 12.0 ha (29.6 acres). The area was probably planted with trees shortly after the establishment of Smithills Hall in the early 14th century and has remained woodland ever since. The site is classified as ancient seminatural woodland and today it is part of Smithills Country Park (Fig. 1.15). Ancient woodland is land that has been continuously wooded since 1600.

Table 1.16 is a list of species living in the wood identified by staff at Bolton Museum.

Q1.13.1 Why is ancient woodland ecologically important?

Q1.13.2 Which groups of species are likely to have been omitted from this list?

The United Kingdom government is planning to build a new high speed railway line (HS2) linking London with cities in the north of England. The Woodland Trust is the UK's largest woodland conservation charity. It has identified 63 ancient woodlands that will be affected by Phase 1 of the project (London to Birmingham). Of these, 34 woodlands will be directly damaged. The remainder will suffer indirect impacts.

Table 1.16 Species identified in Raveden Wood, Bolton

Taxon	Number of species
Birds	42
Mammals	9
Fishes	2
Amphibians	2
Insects	
Butterflies & Moths	25
Alderflies	1
Bees	1
Beetles	7
Dragonflies	1
Stoneflies	4
True bugs	1
Lacewings	3
True flies	61
Leaf hoppers	1
Centipedes	2
Millipedes	1
Woodlice	2
Molluscs	8
Flowering plants	95
Ferns	7
Horsetails	2
Mosses	15
Liverworts	7
Fungi	51

Adapted from Rees (2002).

Q1.13.3 Measures that would mitigate the loss of this ancient woodland include:

a. the translocation of soil from ancient woodland to new sites; and

b. the planting of new woodlands

Suggest reasons why these are likely to be unacceptable to conservationists.

Reference/Further Reading

Rees, P. A. (2002). *Urban environments and wildlife law: A manual for sustainable development.* Oxford: Blackwell Science.

CHAPTER 2

ABIOTIC FACTORS AND ECOPHYSIOLOGY

This chapter contains exercises concerned with physical factors in the environment (such as air or water temperature and oxygen and carbon dioxide concentration) and the effects these factors have on the physiological functioning of organisms.

Forestry Commission plantation, England

Examining Ecology. DOI: http://dx.doi.org/10.1016/B978-0-12-809354-2.00002-6

INTENDED LEARNING OUTCOMES

On completion of this chapter you should be able to:

- Draw graphs of seasonal changes in the physical environment.

- Draw graphs showing the relationship between changes in the physical environment and the physiological responses of plants and animals.

- Explain the adaptation camels have evolved to survive desert conditions.

- Explain how seabirds remove excess salt from their bodies.

- Interpret the behavioural responses of animals to changes in temperature.

- Analyse seasonal changes in the pattern of temperature change with depth in a lake.

- Describe the effects of changes in oxygen level on the survival of invertebrates.

- Analyse data on the effect of water availability on seed germination.

- Describe and explain the physiological responses that occur during diving in seals.

- Explain the benefits of diapause to an insect.

- Describe and explain the physiological changes that occur when a mammal hibernates.

INTRODUCTION

LIMITS FOR LIFE

Life on Earth exists within a band around the Earth extending down to the bottom of the oceans and up to the highest mountains and beyond to altitudes of over 10,000 m. The highest recorded altitude for a bird is 11,277 m (37,000 ft) for a Ruppell's griffon vulture (*Gyps rueppellii*) which collided with a commercial aircraft over Ivory Coast in 1973 (Carwadine, 1995). Fujii, Jamieson, Solan, Bagley, and Priede (2010) reported a hadal snailfish (*Pseudoliparis amblystomopsis*) from a depth of 7703 m

(25,272 ft) in the Japan Trench in the Pacific Ocean: the deepest recording for a living fish.

Evolution has allowed life to respond to the variety of environmental conditions found on Earth and created life-forms that are able to survive in the most extreme conditions. The study of the relationships between the physiological functioning of an organism and its physical environment is called ecophysiology.

CLIMATIC VARIATION

As a general rule, temperature falls with altitude in the atmosphere and with depth in water. The quantity of sunlight received by the Earth reduces from the equator to the poles due to the angle of the axis of the planet's rotation to the sun. Light levels determine the amount of plant growth that can be sustained and so too does carbon dioxide concentration in the atmosphere. The differential heating of the surface creates air movements and drives the water cycle creating rainfall patterns that support a great variety of life-forms in tropical forests while depriving animals and plants in hot deserts of the water that is essential to their survival. Climatic differences have caused the development of animal and plant communities that vary spatially with distance from the equator and vertically as we ascend mountains or descend into the oceans.

TEMPERATURE REGULATION

Most animal species are poikilothermic and their body temperature is largely determined by the ambient environmental temperature (Fig. 2.1). When the environment becomes too hot or too cold they struggle physiologically and become less active or completely inactive. Behavioural adaptations may allow them to move into the shade to cool down or into the sun if they are too cold, and to increase their movements when

Figure 2.1 Evergreen toad (*Incilius coniferus*).

they enter a cold environment in the hope of finding somewhere warmer.

Birds and mammals are homiothermic and are able to maintain a more or less constant body temperature. This allows them to function even when the environmental temperature varies considerably on a daily or seasonal basis and has allowed some species to disperse into extreme environments.

ADAPTATION TO HARSH ENVIRONMENTS

Some animal species have evolved mechanisms that allow them to avoid adverse conditions. Many bear species hibernate to avoid the cold conditions of winter when food is difficult to find. Some insects enter a period of diapause which allows them to delay their development in anticipation of a period of adverse conditions.

Oxygen is essential to most forms of life but difficult to obtain in some environments. Some animal species are specially adapted to function under water where oxygen levels are low or in water that is contaminated with pollution. Marine species need to remain in osmotic balance while living in a salty environment, and some achieve this by excreting salt through special glands.

The following exercises provide examples of how physical factors vary between habitats and illustrate the adaptations that organisms have evolved which enable them to survive.

References

Carwadine, M. (1995). *The Guinness book of animal records.* Enfield, Middlesex: Guinness Publishing Ltd.

Fujii, T., Jamieson, A. J., Solan, M., Bagley, P. M., & Priede, I. G. (2010). Large aggregation of liparids at 7703 meters and a reappraisal of the abundance and diversity of hadal fish. *Bioscience, 60,* 506–515.

CLIMATIC VARIATIONS

Figure 2.2 Tropical forest, Costa Rica.

Climatic data give the ecologist information about two of the environmental variables that are most important in determining the nature of the ecosystem which will develop in any particular part of the world: rainfall and temperature. Table 2.1 presents such data for two widely separated parts of the Earth (Figs. 2.2 and 2.3).

Q2.1.1 Draw separate climate graphs for each of the two biotopes (A and B) using the data in Table 2.1. Each graph should show temperature on the vertical axis on the left side and rainfall on the vertical axis on the right side. Rainfall should be indicated using a bar chart and temperature using a line graph. All vertical scales should start at zero.

Q2.1.2 For each biotope, calculate:

(a) the mean monthly temperature;

(b) the mean monthly rainfall; and

(c) the total annual rainfall received by each area.

Figure 2.3 Savannah, Tanzania.

Table 2.1 Mean monthly temperatures and rainfall totals for two biotopes

	Biotope A		Biotope B	
Month	Mean monthly temperature (°C)	Rainfall (mm)	Mean monthly temperature (°C)	Rainfall (mm)
January	25	259	25	0
February	25	249	27	0
March	24	310	32	0
April	25	165	34	0
May	24	254	35	15
June	23	188	33	99
July	23	168	29	211
August	24	117	28	211
September	24	221	28	142
October	25	183	29	48
November	25	213	28	8
December	25	292	25	5

Q2.1.3 Describe the main differences in the rainfall pattern in these two biotopes.

Q2.1.4 Why does the mean monthly rainfall data for these two biotopes give an incomplete picture of the rainfall differences between them?

Q2.1.5 Identify which set of data is from a tropical forest and which is from a savannah and explain why.

Q2.1.6 What effect would you expect the rainfall pattern in biotope B to have on the growth of plants?

Reference/Further Reading

Woodward, F. I. (1987). *Climate and plant distribution.* Cambridge: Cambridge University Press.

LAKE STRATIFICATION

Shallow water bodies exhibit little variation in temperature with depth. However, this is not the case in deep lakes. In such lakes the temperature varies with depth and with season. Lake Ekoln in Sweden is a dimictic lake. This means that the water mixes from the bottom to the top during two mixing periods each year. In the winter the surface is frozen; in the summer the lake is thermally stratified, with an epilimnion at the top and a hypolimnion at the bottom.

Q2.2.1 Draw a graph of the data in Table 2.2 with the depth of the lake on the vertical axis (zero, i.e., the surface of the water, should be at the top of the graph). Draw both lines against the same axes.

Table 2.2 The vertical distribution of summer and winter temperatures in a dimictic lake, Lake Ekoln, Sweden

| Depth (m) | Temperature (°C) | |
	Summer (June 1969)	Winter (March 1970)
0	19.00	0
2	18.75	1.30
4	18.50	2.20
6	12.00	2.40
8	10.00	2.50
10	9.00	2.30
12	8.00	2.10
14	7.00	2.00
16	6.50	1.80
18	6.25	1.70
20	6.15	1.80
22	6.05	1.90
24	6.00	2.20
26	6.00	2.60
28	6.00	–
30	6.00	–
32	6.00	–
34	6.00	–

Data extracted from Kvarnäs & Lindell (1970).

Q2.2.2 Give two reasons why the scales on this graph are unusual.

Q2.2.3 Mark the following zones on the summer graph line:

(a) epilimnion;

(b) hypolimnion;

(c) thermocline.

Q2.2.4 What happens to the thermocline as the winter approaches? Explain why.

Q2.2.5 What would you expect to happen to oxygen levels below the thermocline in summer.

Reference/Further Reading

Kvarnäs, H., & Lindell, T. (1970). *Hydrologiska studier i Ekoln*. UNGI Rapport 3.

CARBON DIOXIDE AND PHOTOSYNTHESIS

Plants need light, water and a source of carbon to produce carbohydrates by photosynthesis (Fig. 2.4). The carbon source is carbon dioxide in the atmosphere. The data in Table 2.3 show the effects of changing light levels and the availability of carbon dioxide on cucumber plants (*Cucumis sativus*) grown in greenhouse conditions.

Q2.3.1 Draw a single graph showing changes in the rate of photosynthesis with increasing light intensity (highest intensity = 7) in atmospheres of 0.03% and 0.13% carbon dioxide (i.e., two separate lines).

Q2.3.2 What evidence is there from the graph that carbon dioxide availability is limiting the rate of photosynthesis when present at a concentration of 0.03%?

Q2.3.3 Using the evidence from your graph, suggest the likely effect of a global increase in carbon dioxide concentration in the atmosphere on plant production.

Table 2.3 The effect of different light intensities on the rate of photosynthesis of cucumber plants grown in 0.03% and 0.13% carbon dioxide at 20°C

Light intensity (arbitrary units)	Rate of photosynthesis (mm^3 CO_2 cm^{-2} h^{-1})	
	0.03% CO_2	0.13% CO_2
1	45	60
2	58	110
3	70	150
4	75	177
5	79	190
6	79	193
7	79	198

Q2.3.4 What human activities have led to an increase in carbon dioxide concentration in the atmosphere?

Q2.3.5 Vast quantities of carbon dioxide are stored in the oceans, mainly as hydrogen carbonate ions. Which group of organisms utilises this?

Q2.3.6 How would the planting of large areas of forest affect the concentration of carbon dioxide in the atmosphere? Explain your answer

Reference/Further Reading

Kramer, P. J. (1980). Carbon dioxide concentration, photosynthesis, and dry matter production. *Bioscience, 31*, 29–33.

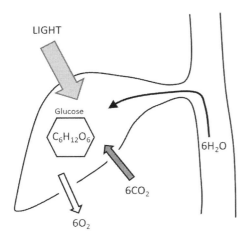

LIGHT

Glucose

$C_6H_{12}O_6$

$6H_2O$

$6CO_2$

$6O_2$

Figure 2.4 Photosynthesis in the leaf of a green plant.

THE EFFECT OF WATER TABLE DEPTH ON SEED GERMINATION

Juncus effusus is a species of rush normally found growing in association with water and waterlogged soils. The effect of the depth of the water table on the germination of seeds of this species is shown in Table 2.4.

Q2.4.1 Calculate the percentage germination for each depth of the water table.

Table 2.4 The effect of water table depth on germination in *Juncus effusus*

Water table depth (cm below soil surface)	Number of seedlings germinated from 6000 seeds	Germination (%)
0	540	
5	151	
10	12	
20	0	

Data modified from Lazenby (1955).

Q2.4.2 Draw a scatter diagram showing the relationship between the depth of the water table (*x*-axis) and the percentage of seeds that germinated (*y*-axis).

Q2.4.3 What type of correlation appears to exist between the depth of the water table and the number of seeds that germinated?

Q2.4.4 Why is it particularly important for the seeds of this species to respond in this way to the presence of water?

Reference/Further Reading

Lazenby, A. (1955). Germination and establishment of *Juncus effusus* L.: II. The interaction effects of moisture and competition. *Journal of Ecology*, 43, 595–605.

EXERCISE 2.5

THE EFFECT OF OXYGEN LEVEL ON MIDGE LARVAE SURVIVAL

Midges (*Chironomus*) lay their eggs in freshwater. Chironomid larvae require oxygen in order to survive and this is transported to the tissues by the blood pigment haemoglobin. Walshe (1948) examined the survival of chironomid larvae from four different species under anaerobic conditions. Some of her results are summarised in Table 2.5.

Table 2.5 Survival of chironomid larvae from four species in the absence of oxygen

Hours without oxygen	Percentage of larvae of each species alive			
	A	B	C	D
0	100	100	100	100
3	100	100	100	100
10	22	100	100	100
22	0	30	–	100
30	–	5	90	100
46	–	0	75	100
53	–	–	–	100
70	–	–	48	92
94	–	–	20	78
105	–	–	–	30
120	–	–	0	0

Based on Walshe (1948).
Note: – = no recording.

Q2.5.1 Draw a single graph of the data presented in Table 2.5 showing the effect of oxygen deprivation on each species.

Q2.5.2 Describe briefly the effect of oxygen deficiency on the four species.

Q2.5.3 Which two species are likely to be stream dwellers and which two species are likely to dwell in lakes or stagnant water? Give reasons for your choices.

Q2.5.4 Using your graph, determine how long species D would have to be deprived of oxygen in order for 50% of the larvae to die.

Reference/ Further Reading

Walshe, B. M. (1948). The oxygen requirements and thermal resistance of chironomid larvae from flowing and still waters. *Journal of Experimental Biology, 25,* 35–44.

TEMPERATURE SELECTION IN TWO FISH SPECIES

Many species of animals have been shown to prefer a particular temperature range when placed in a temperature gradient. They are less active at their preferred temperature and thus arrive there by a type of behaviour known as thermokinesis. Fisher and Elson (1950) examined temperature preference in the speckled trout (*Cynoscion nebulosus*) and the Atlantic salmon (*Salmo salar*) when placed in a tank with a temperature gradient from 0°C to 22°C. Some of their observations are presented in Table 2.6.

Table 2.6 Temperature preferences in trout and salmon

Temperature (°C)	Frequency (% of total observations)	
	Trout	**Salmon**
0	0	0
2	0.5	0
4	2.5	0
6	7.0	0.5
8	20.0	4.5
10	32.0	10.0
12	25.0	23.0
14	10.0	32.0
16	2.5	21.0
18	0.5	7.0
20	0	2.0
22	0	0

Based on Fisher & Elson (1950).

Q2.6.1 Draw a single graph of the data in Table 2.6.

Q2.6.2 What range of temperature was each species able to tolerate?

Q2.6.3 What is the preferred temperature of each species?

Q2.6.4 At what temperatures would you expect each species to be most active?

Q2.6.5 Why is it important for fish to respond behaviourally to a temperature gradient?

Reference/ Further Reading

Fisher, K. C., & Elson, P. F. (1950). The selected temperature of Atlantic Salmon and Speckled Trout and the effect of temperature on the response to an electrical stimulus. *Physiological Zoology, 23*, 27–34.

ADAPTATION TO LIFE UNDERWATER: DIVING IN SEALS

Aquatic mammals exhibit a variety of physiological adaptations to diving. Seals, sea lions (Fig. 2.5) and many other mammals, exhibit a 'diving' reflex which allows them to stay underwater for an extended period of time. This response is triggered when the face is immersed in water. Changes to the circulatory system during diving in seals are illustrated by the data in Tables 2.7 and 2.8.

Q2.7.1 Draw a graph showing the change in heart rate of a harbour seal while diving.

Q2.7.2 Draw a vertical line on the graph indicating:

(a) the point at which the dive began;

(b) the point at which the seal surfaced.

Q2.7.3 Describe what happens to blood flow to the heart during diving and explain why.

Q2.7.4 Why is blood flow to the diaphragm so low during diving?

Q2.7.5 Identify the organs to which blood is preferentially diverted during diving and explain why.

Reference/Further Reading

Kooyman, G. L., Castellini, M. A., & Davis, R. W. (1981). Physiology of diving in marine mammals. *Annual Review of Physiology*, 43, 343–356.

Figure 2.5 California sea lion (*Zalophus californianus*) diving.

Table 2.8 Changes in blood flow to major organs in the Weddell seal (*Leptonychotes weddellii*) during submersion

Organ	Blood flow as % of predive flow
Lungs	58
Retina	61
Pancreas	5
Psoas muscle*	6
Heart, ventricles	17
Brain cortex	108
Diaphragm	5
Cerebellum	93
Heart, atria	17
Liver	5

* Located in the lumbar region of the back

Table 2.7 The effect of diving on the heart rate of a harbour seal (*Phoca vitulina*)

Time from beginning of study (mins)	0	0.5	1.0	1.5	2.0	2.5	3.0	3.5	4.0	4.5	5.0	5.5	6.0	6.5	7.0
Heart rate (beats/min)	122	158	140	37	30	28	20	18	20	19	19	17	170	159	161

SURVIVING SALTY ENVIRONMENTS: MAINTAINING OSMOTIC BALANCE

Figure 2.6 Northern fulmar (*Fulmaris glacialis*).

Marine animals take in a great deal of salt from the sea and from their food. Some birds, such as gulls, petrels and fulmars (Fig. 2.6), are able to stay in osmotic balance by excreting some salt through a nasal gland and some in urine via the cloaca (Table 2.9).

Q2.8.1 For nasal and coacal secretions in the black-backed gull calculate:

(i) the total volume;

(ii) the total quantity of sodium (mmol).

Table 2.9 Nasal and cloacal excretion by a black-backed gull following ingestion of 134cm^3 of sea water

Time (min)	Nasal secretion		Cloacal secretion	
	Volume (cm^3)	Sodium (mmol)	Volume (cm^3)	Sodium (mmol)
15	2.2	1.7	5.8	0.28
40	10.9	8.2	14.6	1.04
70	14.2	11.1	25.0	2.00
100	16.1	12.5	12.5	0.76
130	6.8	5.4	6.2	0.21
160	4.1	3.3	7.3	0.07
175	2.0	1.5	3.8	0.05
Total				

Adapted from Schmidt-Nielsen (1960).

Q2.8.2 Which is the more important mechanism in removing salt?

Q2.8.3 Calculate the sodium concentration (mmol L^{-1}) at each point in time for the nasal secretion and the cloacal secretion, e.g., at 15 minutes, sodium concentration in the nasal secretion is:

$$1.7 \times (1000/2.2) = 773 \, \text{mmol} \, L^{-1}.$$

Q2.8.4 Draw a graph showing changes in sodium concentration with time for nasal secretion and cloacal excretion.

Q2.8.5 Draw a second graph of this data using a log scale on the vertical axis (sodium concentration).

Q2.8.6 Why is your second graph more useful than your first in identifying differences in the pattern of release of sodium excretion from the two sources?

Q2.8.7 Table 2.10 shows differences in the concentration of the nasal secretion of various birds. Account for these differences with reference to their ecology.

Table 2.10 Variation in the sodium concentration of the nasal secretion of birds

Species	mmol L^{-1}	Habitat	Diet
Mallard (*Anas platyrhynchos*)	400–600	Inland/coastal	Mainly plants
Cormorant, double crested (*Phalacrocorax auritus*)	500–600	Coastal	Mainly fishes
Herring gull (*Larus argentatus*)	600–800	Coastal	Invertebrates, fishes, eggs, chicks, offal
Leach's petrel (*Oceanodroma leucorhoa*)	900–1100	Oceanic	Marine plankton, fish, marine algae

Adapted from Schmidt-Nielsen (1960).

Reference/Further Reading

Schmidt-Nielsen, K. (1960). The salt-secreting gland of marine birds. *Circulation, 11*, 955–967.

LIVING IN THE DESERT: TEMPERATURE TOLERANCE IN THE CAMEL

Figure 2.7 Bactrian camel (*Camelus ferus*).

Hot deserts are among the most extreme environments on Earth. Camels have evolved a number of physiological adaptations that allow them to survive such conditions (Fig. 2.7). Table 2.11 compares the rectal temperature of camels with that of other mammalian species. The range of body temperatures experienced by camels living under desert conditions and the amount of water they lose are indicated in Tables 2.12 and 2.13.

Q2.9.1 How do camels differ from the other animals listed in Table 2.11 with regard to the range of body temperature experienced?

Q2.9.2 In what circumstances does the camel store excess heat in its body?

Table 2.11 Rectal temperatures of selected species

	Average		Range	
	°C	°F	°C	°F
Goat	39.1	102.3	38.5–39.7	101.3–103.5
Rabbit	39.5	103.1	38.6–40.1	101.5–104.2
Stallion	37.6	99.7	37.2–38.1	99.0–100.6
Camel	37.5	99.5	34.2–40.7	93.6–105.3

Data based on Anderson (1984).

Q2.9.3 How does this help the camel to conserve water?

Q2.9.4 How does the maintenance of a high body temperature reduce heat gain from the environment?

Q2.9.5 How is heat lost during the night without the use of water?

Table 2.12 Physiological measurements on watered and water deprived camels

	Normal (control) camel (watered each day and fully hydrated)	Camel deprived of drinking water
Maximum rectal temperature (late afternoon) (°C)	38	41
Minimal rectal temperature (at night) (°C)	36	34
Evaporation during the 10 hottest hours of the day (L)	9.1	2.8

Table 2.13 Effect of fur on water loss in a camel

Coat condition	Fur intact	Fur shorn
Water loss (L per hour)	0.45	0.9

Q2.9.6 What role does fur play in temperature regulation in the camel?

Q2.9.7 It takes 2426 kJ of energy to evaporate 1 litre of water. How much energy has the camel that was deprived of water stored in the 10 hottest hours of the day by allowing its body temperature to rise compared with the fully hydrated camel?

References/Further Reading

Anderson, B. E. (1984). Temperature regulation and environmental physiology. In M. J. Swenson (Ed.), *Dukes' Physiology of Domestic Animals* (10th ed). Ithaca, New York: Cornell University Press.

Schmidt-Nielsen, K. (1964). *Desert animals: physiological problems of heat and water*. Oxford: Clarendon.

EXERCISE 2.10

DIAPAUSE IN *GRAPHOLITA*

Diapause is a delay in the development of an animal in response to a recurring period of adverse environmental conditions. The oriental fruit moth (*Grapholita molesta*) is native to China, but has been introduced to North America, Europe and many other places around the world. It is an agricultural pest and its larvae feed on peaches, apples, plums and many other fruits. *Grapholita* larvae (caterpillars) exhibit diapause in response to the approach of winter.

Q2.10.1 Draw a line graph of the data in Table 2.14.

Table 2.14 The percentage of *Grapholita* larvae in diapause after exposure to different periods of daylight in the laboratory

Hours of light per day	Percentage of larvae in diapause
0	1
3	12
6	12
9	70
11	90
12	99
13	98
14	3
18	0
21	0
24	0

Based on Dickson (1949).

Q2.10.2 What is the critical minimum number of hours of daylight after which most larvae enter diapause?

Q2.10.3 The arrival of which season does this signal in southern California?

Q2.10.4 Dickson (1949) found that, in orchards in southern California, virtually all of the larvae that hatched from eggs after about 25 August entered diapause. Using the data in Table 2.15 and your graph, explain how *Grapholita* is effectively predicting future adverse conditions.

References/Further Reading

Dickson, R. C. (1949). Factors affecting the induction of diapause in the oriental fruit moth. *Annals of the Entomological Society of America*, *42*, 511–537.

USNO. (2016). US Naval Observatory, Astronomical Applications Department, Washington DC. <http://aa.usno.navy.mil/cgi-bin/aa_durtablew.pl?form=1&year=2015&task=1&state=CA&place=Los+Angeles> Accessed 22.11.16.

Table 2.15 Approximate number of hours of daylight on the first day of each month in Los Angeles California, 2015

Jan	Feb	Mar	Apr	May	Jun	Jul	Aug	Sep	Oct	Nov	Dec
10.0	10.5	11.5	12.5	13.5	14.3	14.4	13.8	12.8	11.8	10.8	10.0

Data adapted from USNO (2016).

EXERCISE 2.11

THE EFFECT OF TEMPERATURE ON DUSTING BEHAVIOUR IN AN ASIAN ELEPHANT

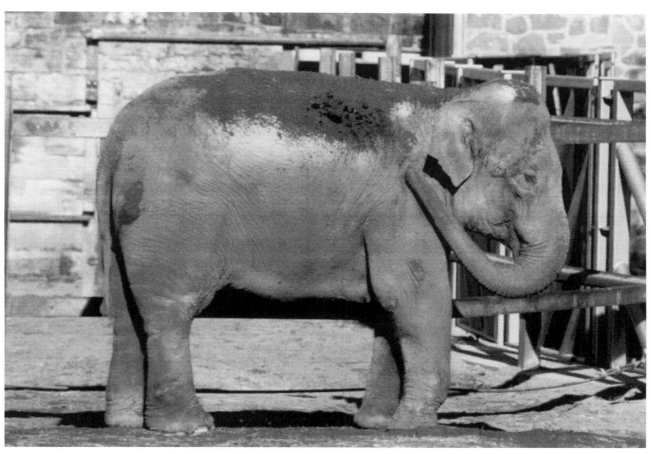

Figure 2.8 An Asian elephant (*Elephas maximus*) dusting.

In hot conditions elephants throw dust (soil) over their bodies (Fig. 2.8). The reason for this is unclear. The data in Table 2.16 show the frequency of dusting behaviour in an adult female Asian elephant (*Elephas maximus*) kept in a zoo (where 1.0 = 100% of the time). Recordings were made over 29 days spread over a period of almost eight months (between February and September).

Q2.11.1 Draw a scatter diagram of the data in Table 2.16, plotting temperature on the *x*-axis and dusting frequency on the *y*-axis.

Q2.11.2 Add a line-of-best fit through the points.

Q2.11.3 Calculate a correlation coefficient for the relationship between maximum daily temperature and dusting frequency. (See Chapter 10, Statistics).

Q2.11.4 Describe the relationship between temperature and dusting frequency.

Q2.11.5 What is the lowest temperature at which dusting was observed?

Table 2.16 The frequency of dusting behaviour in an adult female Asian elephant (*Elephas maximus*) in different temperatures

Study day	Frequency of dusting behaviour	Max daily temperature (°C)
1	0.07	18.4
2	0.02	21.3
3	0.23	24.2
4	0.20	19.7
5	0.02	18.1
6	0.00	18.6
7	0.50	28.0
8	0.23	20.7
9	0.08	23.6
10	0.14	24.4
11	0.48	25.1
12	0.33	23.4
13	0.00	16.8
14	0.15	21.3
15	0.10	19.3
16	0.00	13.6
17	0.02	14.8
18	0.04	15.1
19	0.02	16.1
20	0.07	12.6
21	0.02	14.4
22	0.17	17.9
23	0.00	13.9
24	0.00	7.8
25	0.00	6.1
26	0.00	6.0
27	0.00	11.6
28	0.04	12.7
29	0.00	4.3

Data based on Rees (2002).

Q2.11.6 Suggest possible reasons why elephants exhibit dusting behaviour.

Q2.11.7 Is it possible to determine which of the possible reasons is likely to be true from this study?

Reference/Further Reading

Rees, P. A. (2002). Asian elephants (*Elephas maximus*) dust bath in response to an increase in environmental temperature. *Journal of Thermal Biology, 27,* 353–358.

EXERCISE 2.12

HIBERNATION IN THE BLACK BEAR

Black bears (*Ursus americanus*) hibernate during the winter for up to seven months of the year. During this time they do not drink, eat, urinate or defecate, they lower their heart rate and body temperature, and reduce their metabolic rate. A reduction in metabolic rate reduces the quantity of energy required to keep the body alive. This is important to the bears' survival when they are unable to search for food.

Øivind et al. (2011) recorded various physiological parameters in hibernating black bears in order to examine the relationship between body temperature and the suppression of metabolic rate.

Q2.12.1 Why is oxygen consumption rate used as a measure of metabolic rate?

Q2.12.2 The data in Table 2.17 indicate the core body temperature and oxygen consumption of a black bear on the 1st and 15th of the month from January to May of a single year.

(a) Draw a graph showing the change in core temperature between 1st January and 15th May. Use a minimum value of 30°C and a maximum value of 38°C on the y-axis.

(b) Draw a graph showing the change in oxygen consumption between 1st January and 15th May. Use a minimum value of $0\,\mathrm{mL\,g^{-1}\,h^{-1}}$ and a maximum value of $0.3\,\mathrm{mL\,g^{-1}\,h^{-1}}$ on the y-axis.

Q2.12.3

(a) On each graph, add a vertical line at a position representing 10th April. This indicates the date of emergence from hibernation.

(b) Write 'Hibernation' to the left of each vertical line and 'After emergence' to the right.

Q2.12.4 Calculate the difference between the highest body temperature after emergence and the lowest body temperature during hibernation.

Q2.12.5 Calculate the difference between the highest core body temperature recorded after emergence and the lowest recorded during hibernation.

Q2.12.6 What evidence is there from this data that metabolic suppression during hibernation is independent of lowered body temperature?

Reference/Further Reading

Øivind, T., Blake, J., Edgar, D. M., Grahn, D. A., Heller, H. C., & Barnes, B. M. (2011). Hibernation in black bears: Independence of metabolic suppression from body temperature. *Science, 331*, 906–909.

Table 2.17 Minimum oxygen consumption and corresponding core body temperature in a black bear during hibernation and recovery from hibernation

Month	Day	Core body temperature (°C)	O$_2$ consumption (mL g^{-1} h^{-1})
January	1	33.7	0.07
	15	31.8	0.06
February	1	30.9	0.05
	15	31.0	0.06
March	1	32.3	0.07
	15	33.5	0.08
April	1	35.4	0.10
	15	37.6	0.12
May	1	37.5	0.25
	15	37.3	0.28

Data based on Øivind et al. (2011).

MICROCLIMATE PREFERENCES IN WOODLICE

Many invertebrates exhibit a type of behaviour known as a kinesis. A kinesis is a movement which is proportional to the strength of a stimulus but the response is not directional. Woodlice are small

crustaceans that exhibit kineses which tend to kept them in suitable environments. They become more active when they are in an unfavourable environment. This makes it more likely that they

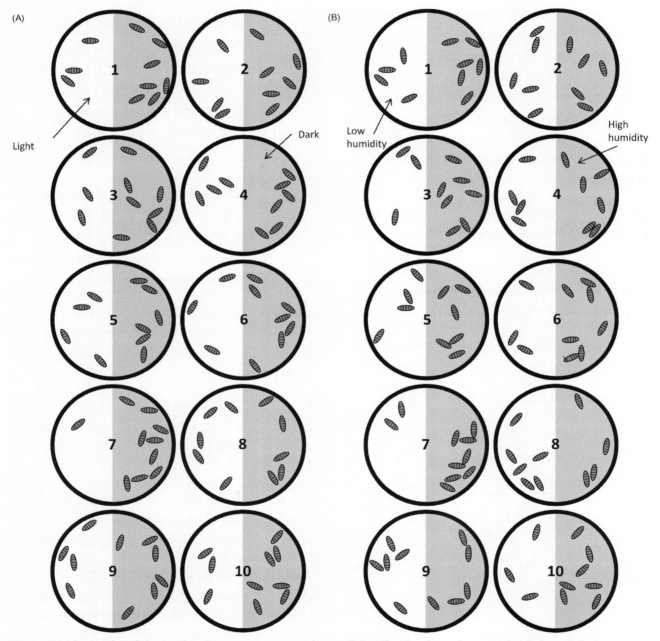

Figure 2.9 The results of choice chamber experiments with woodlice offered a choice between (A) light and dark conditions and (B) humid and dry conditions.

will wander into a favourable environment at which time they become less active so that they are less likely to move away from it.

Habitat preferences in woodlice can be investigated using a choice chamber. This is a simple piece of apparatus that may be used to study the responses of small invertebrates to different environmental conditions. It consists of a circular plastic base with a transparent lid. This experiment was designed to investigate habitat selection in woodlice with respect to light and humidity.

Ten choice chambers were set up offering a choice between light and dark conditions. One half of the lid of the chamber was covered with metal foil to exclude light while the other half was left uncovered and exposed to ambient light levels in the laboratory. Ten woodlice were placed in each choice chamber through the small hole in the centre of the lid. After 20 minutes the locations of the woodlice were recorded.

A second experiment was conducted using the same choice chambers, but this time the lids were left uncovered on both sides. A piece of wet paper towel was placed in one side of the chamber and a small tray of dried silica gel in the other. Silica gel absorbs moisture from the air, so this effectively creates a humid side and a dry side to the choice chamber. Once again ten woodlice were placed in each chamber and the location of each after 20 minutes was recorded. The results of both experiments are shown diagrammatically in Fig. 2.9.

Q2.13.1 If woodlice showed no preference for either the dark side or the light side of the choice chambers in the first experiment, how many individuals would you expect to find in each side of a single chamber at the end of the experiment?

Table 2.18 Calculation of observed and expected values for the calculation of chi-squared

Chamber number	Light	Dark
1		
2		
3		
4		
5		
6		
7		
8		
9		
10		
Total (observed values)	(A)	(B)
Total (expected values)	$\dfrac{A + B}{2}$	$\dfrac{A + B}{2}$

Q2.13.2 A chi-squared test may be used to determine whether or not there is a statistically significant difference between the results obtained in the experiment and those we would have expected if the woodlice showed no preference for either dark or light conditions. Complete Table 2.18 using data from Fig. 2.9 and then use the method shown in Chapter 10, Statistics, to calculate the chi-squared value.

Q2.13.3 Repeat the procedure outlined in Q2.13.2 for the second experiment in which the woodlice chose between dry and humid conditions.

Q2.13.4 Summarise the microclimate preferences of woodlice with respect to light and humidity.

Reference/Further Reading

Morris, M. C. (1999). Using woodlice (Isopoda, Oniscoidea) to demonstrate orientation behaviour. *Journal of Biological Education*, 33, 215–216.

CHAPTER 3

ECOSYSTEMS, ENERGY AND NUTRIENTS

This chapter contains exercises concerned with energy flow, nutrient cycling and the structure and development of ecosystems.

Ants consuming a caterpillar

Examining Ecology. DOI: http://dx.doi.org/10.1016/B978-0-12-809354-2.00003-8

INTENDED LEARNING OUTCOMES

On completion of this chapter you should be able to:

- Identify trophic levels in food chains and pyramids of numbers, energy and biomass.

- Calculate the assimilation efficiency of an animal.

- Construct a pyramid of biomass.

- Calculate the efficiency with which energy is transferred between trophic levels.

- Recognise pyramids of numbers representing different types of food chains (grazing, parasitic, detritus, etc.).

- Identify the organisms responsible for different types of chemotrophic processes.

- Distinguish between grazing and detritus food chains.

- Analyse the energy budget of an organism.

- Distinguish between farming systems that exhibit high productivity and those that have low productivity.

- Recognise alternative terms for each trophic level.

- Construct a food web from information about feeding habits.

- Identify the main processes that occur in the cycles of major nutrients.

- Recognise the chemical symbols for elements and compounds involved in major nutrient cycles.

- Determine the role of humans in nutrient cycles.

- Describe the role of major nutrients in biological systems.

- Explain how nutrients are lost from ecosystems.

- Describe the effect of soil pH on nutrient availability.

- Describe and explain the effects of deforestation on nutrient cycles.

- Recognise the relative importance of various nitrogen sources and sinks in lakes.

- Calculate the benefit–cost ratio and net yield return for a range of fertiliser applications.

- Draw graphs illustrating the benefits of repeated applications of organic fertiliser on soil nutrient status.

- Explain the changes that occur in ecosystems during ecological succession.

INTRODUCTION

The functioning of all ecosystems depends upon the flow of energy from one organism to another and the passing of nutrients in a cycle between the physical environment and living things (DeAngelis,1980).

FOOD CHAINS AND FOOD WEBS

The energy in most organisms has ultimately been derived from the sun. Carnivores eat herbivores; herbivores eat green plants; and the green plants photosynthesise using light from the sun to make carbohydrates from carbon dioxide and water (using chlorophyll):

$$6CO_2 + 6H_2O + \text{light energy} \xrightarrow{\text{chlorophyll}}$$
$$C_6H_{12}O_6 + 6O_2$$

Organisms that obtain their energy from the physical environment rather than by eating other organisms are known as autotrophs. Green plants are a type of autotroph known as a phototroph, because they use light energy. Some bacteria obtain energy from chemical compounds and are known as chemotrophs.

All organisms obtain energy from their food, principally by breaking down carbohydrates during respiration. For aerobic organisms – those that use oxygen to break down food – the equation is the opposite of that for photosynthesis (although the two processes are completely different):

$$C_6H_{12}O_6 + 6O_2 \xrightarrow{\text{metabolic enzymes}}$$
$$6CO_2 + 6H_2O + \text{energy}$$

The energy released is used for growth, reproduction and to keep the body tissues alive.

However, some of the energy is lost as heat. It is the heat generated by respiration that ultimately restricts the length of food chains.

When a herbivore eats a plant, some of the energy contained in the tissues of the plant is used by the herbivore, but some of it is lost as the heat of respiration. When the herbivore is eaten by a carnivore, once again, some of the energy is used by the carnivore but some is lost when the carnivore respires. If we consider a parcel of energy captured by a green plant and then passed along a food chain to a herbivore and then a carnivore, at each stage in this transfer of energy some is lost as the organisms respire. This means that the parcel of energy is much smaller by the time it reaches the carnivore.

This helps to explain why carnivores are less numerous than their herbivorous prey and why herbivores are sustained by very large numbers of green plants. Although the energy lost as heat is lost from the food chain it serves a useful purpose in homiothermic animals (mammals and birds) by providing a source of heat to help maintain their body temperature above that of their surroundings.

Apart from the losses due to respiration, there are other reasons why the energy in one trophic level does not all pass to the next in any food chain. Some plant material is not eaten by herbivores but dies and decomposes. Animals lose energy in their urine and faeces, and when they shed antlers, fur, feathers or skin (Fig. 3.1). When an animal dies it may be eaten by other animals (scavengers, such as vultures) or its body may simply decompose.

The energy in the body parts of animals and plants that are not eaten is not lost from the living things in the ecosystem. The type of food chain discussed so far is called a grazing food chain. However, the process of decomposition involves other types of organisms that form

Figure 3.1 Fallow deer (*Dama dama*) antlers. These are discarded each year and contain both energy and nutrients that may be used by other organisms.

Figure 3.2 A bracket fungus growing on a damaged tree.

decomposer food chains. These include detritivores – earthworms, insects and other invertebrates that breakdown the dead material – and bacteria and fungi that break it down further into molecules small enough for them to absorb as an energy source (Wetzel, 1995) (Fig. 3.2). The process of decomposition also produces a heat loss because the organisms involved also have to respire to stay alive. When garden and food waste is composted the process produces heat because of the activity of the decomposer organisms breaking down the organic matter. In some ecosystems the decomposer food chains account for more of the energy flow than do the grazing food chains.

In real ecosystems individual food chains do not occur in isolation. They interconnect because most organisms eat more than one food, although some are very specialised feeders, such as the koala (*Phascolarctos cinereus*) that feeds on eucalyptus leaves and the giant panda (*Ailuropoda melanoleuca*) that feeds almost exclusively on bamboo.

Food webs cannot be considered to be fixed because many species are generalists with respect to food and will switch from one prey species to another depending upon availability. Seasonal variation may occur in feeding behaviour because of the availability of seeds, fruits and other plant materials or because of the seasonal presence of migratory species.

ECOLOGICAL PYRAMIDS

Ecologists analyse patterns in ecosystems in a number of ways. In 1927 the English ecologist Charles Elton published his book *Animal Ecology*, in which, among other things, he described the concept of a 'pyramid of numbers' as a means of representing ecosystem structure in terms of feeding relationships (Elton, 1927). This simply involves counting all of the organisms in each trophic level (Table 3.1) and representing the total at each level as a horizontal bar. This is effectively

Table 3.1 Alternative terms for trophic levels in a food chain

Top carnivores/ super carnivores/ apex predators	T_4	Tertiary consumers
Carnivores	T_3	Secondary consumers
Herbivores	T_2	Primary consumers
Green plants	T_1	Primary producers

a bar chart on its side, but each bar is centred over all of the others creating a symmetrical shape. Since, in most ecosystems, the number of green plants is greater than the number of herbivores and the number of herbivores is greater than the number of carnivores, the shape is similar to that of a pyramid, with a wide base (representing green plants) and a narrow top (representing carnivores or top carnivores). Elton recognised that most food chains only have four or five links.

Food chains may be studied in a number of ways. Simple observation will allow us to determine what each species is feeding on in some ecosystems. An alternative is to examine the stomach contents of animals which, of course, involves killing the animals and depends on the ability to identify partly digested animal and plant remains. A third method uses radioactive isotopes. For example, if phosphorus-32 is injected into grass its presence in other species some time later would allow us to trace its path from one species to another (Marples, 1966).

Nutrient Cycles

Living things are made up of a wide range of nutrients and as one organism feeds on another they are passed along the food chain. Nitrogen in the soil may be absorbed – as nitrate – by a green plant and used to build a protein molecule within a plant cell. When plant tissue is eaten by a herbivore this protein may be digested and the nitrogen could end up in a different protein in an animal cell. If the herbivore is eaten by a carnivore the same process could result in the nitrogen forming part of a different protein in the cell of a carnivore. When a plant or an animal dies its proteins will be broken down during the process of decomposition and will be released into the soil and then possibly into the atmosphere from where it may be washed back into the soil and taken up by a plant in the form of nitrate.

Nutrients move from the physical to the biological components of the ecosystem and back again in a cyclical manner creating nutrient cycles. They are sometimes called biogeochemical cycles, emphasising the interrelationship between geology and biology observed in these cycles. In theory we could study the cycles of all of the nutrients in an ecosystem. In practice, ecologists have concentrated on attempting to describe and understand the cycling of the major elements such as carbon (C), nitrogen (N) and sulphur (S). Some of these nutrients are important in limiting primary production and animal growth (e.g., nitrogen) while others are important in our understanding of how the climate changes with time (e.g., carbon).

Nutrient Cycles and People

In some situations it may become necessary to add nutrients to an ecosystem. Agricultural ecosystems produce animal and plant products that need nutrients in order to grow and reproduce. When the products are removed from the farm to be sold (e.g., cattle or wheat) the nutrients go with them. Consequently, such systems are continually being depleted of nutrients. In order to continue to function these nutrients must be replaced, either by adding organic fertiliser, such as manure, or by adding chemical fertilisers.

Sometimes nutrients may become too abundant. If nitrogen fertiliser is added to farmland it may be washed out by rainfall into the local rivers and lakes and even eventually, make its way to the sea. This may result in an overgrowth of algae, resulting in an algal bloom in a process known as cultural eutrophication (Smith, Joye, and Howarth, 2006). Organic fertilisers are generally more environmentally friendly than chemical fertilisers because they release nutrients more slowly.

As we extract oil from the ground and burn it as fuel we release carbon (as carbon dioxide) which has been locked up in the earth for millions of years into the atmosphere. Here it acts as a greenhouse gas causing the gradual heating of the atmosphere and consequent changes in the global climate. This is affecting the geographical patterns of agricultural production, the distribution of pests and disease vectors, and the distributions of rare species of plants and animals. Planting trees helps to sequester this carbon as it is absorbed by leaves to be used in photosynthesis. Cutting down forests has the opposite effect, especially if their wood is then burned as fuel.

Ecological Succession

Ecosystems develop and change over time. This process is known as ecological succession. A primary succession is one which takes place on newly exposed rock where no soil has previously developed. Mosses and lichens will colonise the rock. They will break it down and add humus to it when they die and decompose. As the succession progresses one biological community will be replaced by another; the soil will deepen and larger animals and plants will appear. The organisms present at any point in the succession will modify the environment making it suitable

for other species which may then colonise it. Succession may end with the establishment of a climax community. The nature of this climax community varies from ecosystem to ecosystem. Beech-maple climax forest occurs in some parts of the United States dominated by tall trees. The concept of a climax community has fallen out of favour with many ecologists because a climax community will take a long time to develop and vegetation dynamics are more complex than climax theory suggests.

A secondary succession is one which takes place where a soil already exists, e.g., after a forest has been felled or damaged by fire. Secondary successions generally progress faster than primary successions because plant seeds and spores are already present in the soil and animal species will move into the area from nearby having become established during the primary succession.

The following exercises examine the structure and functioning of ecosystems, the manner and efficiency with which energy flows through food chains, nutrient cycles and the effect of adding nutrients to ecosystems, and ecosystem development.

References

DeAngelis, D. L. (1980). Energy flow, nutrient cycling, and ecosystem resilience. *Ecology, 61,* 764–771.

Elton, C. (1927). *Animal Ecology.* London: Sidgwick and Jackson.

Marples, T. G. (1966). A radionuclide tracer study of arthropod food chains in a Spartina salt marsh ecosystem. *Ecology, 47,* 270–277.

Smith, V. H., Joye, S. B., & Howarth, R. W. (2006). Eutrophication of freshwater and marine ecosystems. *Limnology and Oceanography, 51,* 351–355.

Wetzel, R. G. (1995). Death, detritus, and energy flow in aquatic ecosystems. *Freshwater Biology, 33,* 83–89.

EXERCISE 3.1

A STEPPE ECOSYSTEM: BIOTIC AND ABIOTIC FACTORS

Figure 3.3 Kiang (*Equus kiang*): a wild ass found in the steppes of northern Asia.

Table 3.2 Biotic and abiotic components of a steppe ecosystem

Lichens	Wild boar	Polecat	Sedges
Fungi	Fire	Roe deer	Ice
Geese	Hares	Soil bacteria	Wolf
Wind	Sand lizards	Eagles	Starlings
Finches	Daisy	Snow	Tulip
Stoat	Temperature	Locusts	Ground squirrel
Crows	Falcons	Thistles	Rain
Iris	Frost	Fox	Fescue grass
Ducks	Kiang	Rivers	Sunlight
Humidity	Rabbits	Heron	Marmot (a rodent)
Sage brush	Feather grass	Wild cats	Kestrels
Buzzards	Air	Lily	Saiga (antelope)
Gazelles	Hyacinth	Rock	Przewalski's horse

A steppe is a mid-latitude grassland consisting of a flat, generally treeless plain extending across Eurasia. Characteristically, steppes have a semi-arid continental climate with extremes of temperature occurring in summer and winter, high temperatures during the day and low temperatures at night.

Table 3.2 contains some of the major components of this type of ecosystem, including a wild equid known as the kiang (*Equus kiang*) (Fig. 3.3).

Q3.1.1 From an ecological point of view these components may be divided into two fundamentally different categories. Give a name to each of these two categories and three examples of each.

Q3.1.2 Group the components of the table into five distinct categories of ecological significance.

Give a name to each of these categories; four of the categories should be based on positions in the food chain (counting carnivores and top carnivores as a single group).

Q3.1.3 List the features that the steppe possesses which make it an ecosystem.

Q3.1.4 A taxonomist might classify the organisms in Table 3.2 into mammals, birds, reptiles, insects, grasses, etc. Why is this type of classification of less interest to an ecologist than that produced in answer to Question 3.1.2?

Reference/Further Reading

Bai, Y., Wu, J., Xing, Q., Pan, Q., Huang, J., Yang, D., et al. (2008). Primary production and rain use efficiency across a precipitation gradient on the Mongolia plateau. *Ecology*, 89, 2140–2153.

EXERCISE 3.2

FOOD WEBS IN AN ENGLISH WOODLAND

The feeding relationships between organisms in an ecosystem may be illustrated by the construction of food chains. A food chain is simply a linear sequence showing, for example, green plants collecting energy from light and then, for each animal species, what it eats and what eats it. A simple food chain might look like this:

light → green plant → rabbit → fox → eagle

Each organism in this food chain represents a different trophic level (Fig. 3.4)

In a real ecosystem, most animals feed on more than one other species. If we draw a diagram of all of these interrelated food chains we construct a food web. For complex ecosystems it is almost impossible to construct a complete food web. Apart from the difficulty of establishing all of the feeding relationships at any particular point in time, the situation is made even more complex by seasonal differences (e.g., in ecosystems where migratory species are present for only part of the year).

Table 3.3 is a very simplified representation of the food web in Wytham Woods in Oxfordshire, England. This woodland has been widely studied by ecologists because of its proximity to the University of Oxford and includes a variety of common species such as titmice (e.g., blue tits (*Cyanistes caeruleus*) and great tits (*Parus major*) (Fig. 3.5)) and small mammals including the wood mouse (*Apodemus sylvaticus*) (Fig. 3.6).

Q3.2.1 Draw a food web for Wytham Woods using the information on feeding relationships in Table 3.3.

Q3.2.2 Complete Table 3.4, indicating the trophic level to which each organism belongs: primary producer, primary consumer, secondary consumer, tertiary consumer.

Q3.2.3 Populations of which taxa are likely to benefit if the numbers of voles and mice increased?

Table 3.3 Feeding relationships in a simplified food web for Wytham Woods, England

Energy flows in the food web	
Energy passes from: →	**→ Energy passes to:**
Light	Herbs/Trees & bushes/Oak trees
Herbs	Insects/Voles & mice
Trees & bushes	Voles & mice/Winter moth/Titmice/Litter
Oak trees	Winter moth/Other leaf feeders/Litter
Insects	Spiders/Titmice/Voles & mice
Spiders	Parasites/Titmice
Winter moth	Titmice
Other leaf feeders	Titmice
Litter	Earthworms
Earthworms	Owls
Titmice	Weasels
Voles & mice	Owls/Weasels

Simplified from Varley (1970).

Figure 3.4 Trophic levels in a food chain.

Figure 3.5 Great tit (*Parus major*).

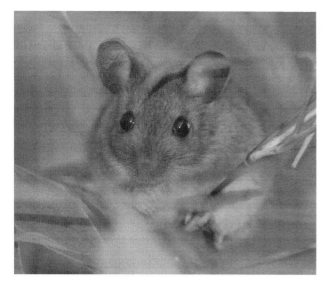

Figure 3.6 Wood mouse (*Apodemus sylvaticus*).

Table 3.4 The trophic level of selected species from a food web in Wytham Woods

Organism	Trophic level
Oak tree	
Parasites	
Owls	
Weasels	
Winter moth	

Q3.2.4 Suggest methods that could be used to determine the feeding relationships in the woodland.

Q3.2.5 Which major group of organisms is missing from your food web?

Reference/Further Reading

Varley, G. C. (1970). The concept of energy flow applied to a woodland community. In A. Watson (Ed.), *Animal populations in relation to their food resources* (pp. 389–405). Oxford: Blackwell.

EXERCISE 3.3

ECOLOGICAL PYRAMIDS

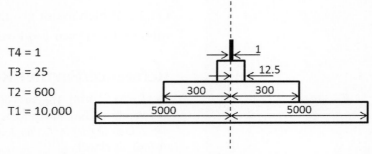

T4 = 1
T3 = 25
T2 = 600
T1 = 10,000

1
12.5
300   300
5000   5000

Figure 3.7 How to draw an ecological pyramid. The value for each trophic level should be divided by two and this value is the width of the bar on either side of the centre line. The top bar (T_4) is often so narrow that it has to be drawn as single line. Not drawn to scale.

Pyramids of numbers, biomass and energy are used to examine standing crop and energy flow in ecosystems. Silver Springs is a freshwater ecosystem in Florida, United States. Data for this system are presented in Table 3.5.

Q3.3.1 Draw a pyramid of biomass to scale on graph paper using the data in Table 3.5. You may find it useful to draw a vertical line down the centre of the page before you begin as this will help you to keep the pyramid symmetrical (Fig. 3.7).

Q3.3.2 Which major group of organisms has been omitted from this pyramid?

Q3.3.3 Explain why it would be difficult to draw a pyramid of numbers for Silver Springs to scale.

Q3.3.4 Calculate the efficiency with which energy is transferred from one trophic level to the next by completing Table 3.6.

Q3.3.5 Explain why there are differences in the efficiency of energy transfer between trophic levels.

Q3.3.6 Explain what happens to the energy which is not transferred to the next trophic level in a food chain when one organism feeds on another.

Pyramids of numbers representing parasitic food chains have unusual characteristics. The data in Table 3.7 relate to a food chain in which bacteria are parasitising nematode worms which in turn are fish parasites.

Table 3.5 The number of organisms, biomass and energy in each trophic level of Silver Springs, Florida

Trophic level	Number of organisms	Biomass (g m^{-2})	Energy (kJ m^{-2} year^{-1})
Tertiary consumers T_4	0.01	1.5	67
Secondary consumers T_3	24×10^3	11.0	1602
Primary consumers T_2	41×10^4	37.0	14×10^3
Primary producers T_1	27×10^{10}	809.0	87×10^3

Based on Odum (1957).

Table 3.6 The efficiency of energy transfer between trophic levels in Silver Springs

Efficiency of transfer from T_3 to T_4	$(67/1602) \times 100 =$	4.18%
Efficiency of transfer from T_2 to T_3		
Efficiency of transfer from T_1 to T_2		

Table 3.7 The number and biomass of organisms in the trophic levels of a parasitic food chain

Organism	(A) Number of individuals	(B) Mass of each individual (g)	(C) Total biomass (g)
Bacteria	10^9	10^{-15}	
Nematodes	10^3	0.1	
Fish	1	1000	

Q3.3.7 Draw a food chain showing the relationships between these organisms, indicating the direction of the flow of energy.

Q3.3.8 Which organism is at the top of this food chain?

Q3.3.9 Calculate the standing crop biomass in each trophic level (C = A × B).

Q3.3.10 Draw:

a. a pyramid of numbers; and

b. a pyramid of biomass,

for this food chain (not to scale).

Q3.3.11 Would a pyramid of energy for this food chain be similar in shape to the pyramid of numbers or the pyramid of biomass? Explain your answer.

References/Further Reading

Fath, B. D., & Killian, M. C. (2007). The relevance of ecological pyramids in community assemblages. *Ecological Modelling, 208*, 286–294.

Odum, H. T. (1957). Trophic structure and productivity of Silver Springs, Florida. *Ecological Monographs, 30*, 187–206.

ENERGY FLOW IN ECOSYSTEMS

Most of the energy entering ecosystems on Earth comes from sunlight (insolation). This energy then passes from one organism to another along food chains from green plants to carnivores. Fig. 3.8 shows the fate of 100 units of light energy falling on a hypothetical ecosystem. In this diagram:

gross production - respiration = net production.

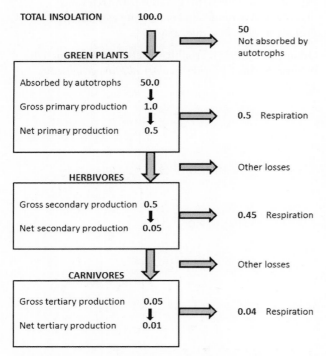

Figure 3.8 The fate of 100 units of light energy falling on a hypothetical ecosystem. Note—Other losses are small and have not been quantified.

Q3.4.1 What happens to the light energy that is not absorbed by the green plants?

Q3.4.2 What percentage of the total insolation is:

    a. absorbed by green plants;

    b. converted into net primary production;

    c. converted into net secondary production;

    d. converted into net tertiary production?

Q3.4.3 What percentage of the gross production is used in respiration in

    a. green plants;

    b. herbivores;

    c. carnivores?

Q3.4.4 In what form is the energy lost in respiration?

Q3.4.5 Apart from respiration, what other energy losses occur in the food chain between:

    a. the green plants and the herbivores;

    b. the herbivores and the carnivores?

Reference/Further Reading

Kozlovsky, D. G. (1968). A critical evaluation of the trophic level concept. I. Ecological efficiencies. *Ecology*, *49*, 48–60.

EXERCISE 3.5

A COMPARISON OF ENERGY BUDGETS IN TWO ECOSYSTEMS

Energy budgets for two contrasting ecosystems in the United States are shown in Fig. 3.9: an old field community in Michigan, and Silver Springs, a freshwater ecosystem in Florida. The diagram shows the amount of energy entering the ecosystems as light from the sun (total insolation) and its fate up to the point where it is used to produce carbohydrates in the green plants. The units for the values shown are kJ m^{-2} y^{-1}.

Q3.5.1 Account for the differences in the insolation at the two sites.

Q3.5.2 For each ecosystem, calculate:

a. the percentage of the total insolation which is used by the green plants (to produce the gross primary production);

b. the percentage of the total insolation that is not used by the green plants;

c. the percentage of the total insolation that becomes the net primary production;

d. the percentage of the gross primary production that is used in respiration.

Q3.5.3 Comment on the efficiency of ecosystems that trap energy from the sun using evidence from your calculations.

References/Further Reading

Golley, F. B. (1960). Energy dynamics of a food chain of an old-field community. *Ecological Monographs*, 30, 187–206.

Odum, H. T. (1957). Trophic structure and productivity of Silver Springs, Florida. *Ecological Monographs*, 27, 55–112.

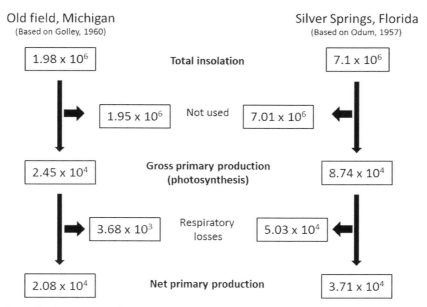

Figure 3.9 Energy budgets for two contrasting ecosystems.

EXERCISE 3.6

CHEMOTROPHISM

Chemotrophs are organisms that obtain their energy from chemicals found in the physical environment. The chemical reactions shown in Table 3.8 represent the processes used by three different types of chemotrophs.

Q3.6.1 Are these organisms autotrophs or heterotrophs? Explain your answer.

Q3.6.2 Name the process from which these types of organisms obtain their energy.

Q3.6.3 Equations in Table 3.8 used by three groups of bacteria:

  a. colourless sulphur bacteria (e.g., *Thiobacillus*);

  b. nitrifying bacteria (e.g., *Nitrosomonas*);

  c. iron bacteria (e.g., *Leptothrix*).

  Pair each of these reactions with the appropriate chemotroph.

  Biogas is a mixture of gases released from the breakdown of organic matter such as food waste, manure, sewage and decomposing plant material. It can be compressed so that it can be used to power internal combustion engines, which in turn may be used to generate electricity. Unfortunately, biogas contains hydrogen

Table 3.8 Chemical reactions used by selected chemotrophs to obtain energy

i	$4FeCO_3 + O_2 + 6H_2O \rightarrow 4Fe(OH)_3 + 4CO_2 + Energy$
ii	$2S + 3O_2 + 2H_2O \rightarrow 2H_2SO_4 + Energy$
iii	$2NH_3 + 3O_2 \rightarrow 2HNO_2 + 2H_2O + Energy$

sulphide (H_2S) which is highly corrosive to internal combustion engines and chemical removal is expensive.

  H_2S is a toxic compound produced by the anaerobic digestion of organic materials. The bacterium *Thiobacillus thioparus* is capable of breaking down H_2S using the following reaction:

$$H_2S + 2O_2 \rightarrow SO_4^{2-} + 2H^+$$

Q3.6.4 What chemical process does the bacterium use to remove the H_2S from the biogas?

Q3.6.5 How could chemotrophic bacteria be used to make the generation of electricity from biogas more cost-effective?

Reference/Further Reading

Syed, M., Soreanu, G., Falletta, P., & Béland, M. (2006). Removal of hydrogen sulphide from gas streams using biological processes - A review. *Canadian Biosystems Engineering, 48*, 2.1–2.14.

EXERCISE 3.7

GRAZING AND DETRITUS FOOD CHAINS

The energy trapped by primary producers in an ecosystem is either consumed by grazing animals or used by decomposers (bacteria and fungi) during the decay of detritus. The proportion of this energy which is used by grazers varies from one ecosystem to another (Fig. 3.10). Some systems have predominantly grazing food chains while others favour detritus food chains. Table 3.9 compares the fate of energy in a variety of ecosystems.

Q3.7.1 Why is almost all of the net primary production in plankton and algae consumed by herbivores?

Q3.7.2 Are the herbivores in forests generally larger or smaller than the plants on which they feed?

Q3.7.3 Herbivores are unable to digest the lignin present in woody tissues. How does this affect the proportion of energy in forests which goes into the detritus food chains?

Q3.7.4 In a grazed meadow, the cattle may only take about 15% of the net primary production.

   a. What percentage of the total grazed net primary production is taken by other herbivores?

   b. Assuming no other farm animals are present, what types of organisms would these other herbivores be?

Q3.7.5 In which of the ecosystems listed in Table 3.9 are detritus food chains using more of the net primary production than grazing food chains?

Q3.7.6 What role do soil animals such as earthworms and insects play in the decomposition process?

Table 3.9 The fate of net primary production in selected ecosystems

Primary producer	Percentage of net primary production	
	Eaten	Decomposed
Plankton	~100.0	<0.01
Algae	~100.0	<0.01
Spartina marsh	6.5	93.5
Grazed meadow	37.0	63.0
Beech forest	71.0	29.0
Tropical rainforest	67.0	33.0

Figure 3.10 In a grazed meadow less that 40% of the net primary production may be eaten by herbivores.

Reference/Further Reading

Wetzel, R. G. (1995). Death, detritus, and energy flow in aquatic ecosystems. *Freshwater Biology, 33*, 83–88.

ENERGY SOURCES IN AQUATIC ECOSYSTEMS

The character of a watershed determines whether a stream's energy sources are mostly autochthonous (produced within) or allochthonous (produced outside) (Fig. 3.11). Table 3.10 contains data collected by Likens (1975) for deciduous forest and by Wissmar, Richey, and Spyridakis (1977) for coniferous forest.

Figure 3.11 Leaves and twigs are allochthonous sources of energy in a stream.

Table 3.10 Comparison of energy sources in two ecosystems (gC m^{-2} y^{-1})

Energy source	Deciduous forest			Coniferous forest-lake
	Forest	Streams	Lake	
Autochthonous	941	1	88	4.8
Allochthonous	3	615	18	9.4

Data on deciduous forest adapted from Likens (1975); and coniferous forest adapted from Wissmar, Richey, & Spyridakis (1977).

Q3.8.1 What percentage of the energy in the lake in the deciduous forest studied by Likens (1975) came from the lake itself?

Q3.8.2 Describe in simple terms the differences in the origin of the energy within the streams and the lake studied by Likens (1975).

Q3.8.3 Was the deciduous forest itself heavily dependent upon outside sources of energy?

Q3.8.4 Compare the relative importance of the two energy sources in the lake in the deciduous forest studied by Likens (1975) and the lake in the coniferous forest studied by Wissmar et al. (1977).

Q3.8.5 How is knowledge of the source of energy in ecosystems important in ecosystem management?

References/Further Reading

Likens, G. E. (1975). Primary production in inland aquatic ecosystems. In H. Leith & R. H. Whittaker (Eds.), *The primary productivity of the biosphere* (pp. 185–202). New York Inc: Springer-Verlag.

Wissmar, R. C., Richey, J. E., & Spyridakis, D. E. (1977). The importance of allochthonous particulate carbon pathways in a subalpine lake. *Journal of the Fisheries Board of Canada, 34*, 1410–1418.

EXERCISE 3.9

ENERGY BUDGET OF A BANK VOLE

All organisms require energy in order to function. Some of this energy is used to maintain normal body functions and growth, and some is lost in respiration. An energy budget is an attempt to account for all of the energy taken in as food in terms of tissue growth, respiratory losses and so on. This will be affected by a variety of factors including activity and ambient temperature. Table 3.11 shows such a budget for a bank vole (*Myodes glareolus*) kept under laboratory conditions during summer (Fig. 3.12).

Table 3.11 Daily energy budget for a bank vole in summer

	Energy ($\times 10^3$ Joules)
Food consumed	61.9
Assimilated	54.4
New tissue production	1.3
Respiratory heat loss	53.1
Urine	2.5
Faeces	5.0

Based on Gorecki (1968).

Figure 3.12 Bank vole (*Myodes glareolus*).

Table 3.12 Annual production in a population of bank voles

Number of individuals produced per year	50
Mean weight at death (g)	23
Calorific value of dried mammal tissue (Jg^{-1})	20,900

Q3.9.1 Draw a diagram illustrating the source and fate of all energy used by a bank vole.

Q3.9.2 Calculate the amount of energy assimilated as a percentage of the energy consumed.

Q3.9.3 Calculate the percentage of the assimilated energy that is:

a. lost as the heat produced by respiration;

b. used to produce new tissue.

Q3.9.4 Calculate the percentage of the consumed energy that is lost in:

a. urine;

b. faeces.

Q3.9.5 Calculate the amount of energy a pregnant female bank vole would need to consume per day assuming that her energy requirements were increased by approximately 60% during pregnancy.

Q3.9.6 The annual production of a small mammal population may be calculated if the number of animals born per year is known, along with their mean weight at death and the energy value of their tissues.

Calculate the annual production in the population of bank voles shown in Table 3.12 using the following equation:

$$\text{Annual production (J)} = number \times mean\,weight\,(g) \times calorific\,value\,(Jg^{-1})$$

Reference/Further Reading

Gorecki, I. (1968). Metabolic rate and energy budget in a bank vole. *Acta Theriologica*, 8, 341–365.

EXERCISE 3.10

WILDLIFE BIOMASS IN THE SERENGETI

The Serengeti Ecological Unit in Tanzania is one of the most important wildlife areas in Africa. The dominant animal species are large mammals (Fig. 3.13). The biomass of vegetation in the ecological unit was estimated at 2×10^7 g km^{-2} (wet weight). Table 3.13 shows the large mammal populations in the area along with the mean weight of each species.

Q3.10.1 List the seven species of predators (carnivores) in Table 3.13.

Q3.10.2 Complete Table 3.13 by calculating the biomass of each species.

Q3.10.3 Calculate:

a. the total number of predators;

b. the total number of herbivores.

Q3.10.4 Account for the difference in the total number of carnivores and herbivores in this ecosystem.

Q3.10.5 Calculate:

a. the total biomass of predators;

b. the total biomass of herbivores.

Table 3.13 The number and biomass of large mammal species in the Serengeti Ecological Unit, Tanzania (Area 25,500 km^2)

Species	Number	Mean weight (kg)	Total biomass (kg)
Wildebeest	370,000	128	47,360,000
Plains zebra	193,000	164	
Lion	1650	120	
Thomson's gazelle	980,000	12	
Grant's gazelle	3100	32	
Eland	7200	225	
Leopard	500	35	
Cheetah	500	40	
Topi	26,000	82	
Hartebeest	20,000	95	
Spotted hyena	6000	55	
Impala	75,000	32	
Hunting dog	1100	18	
Waterbuck	3200	131	
Giraffe	8000	716	
Black-backed jackal	13,500	6	
Warthog	17,000	40	
Buffalo	38,000	420	
Golden jackal	5000	6	

Based on data in Schaller (1972).

Figure 3.13 Serengeti wildlife: (A) a female lion (*Panthera leo*) guarding a kill; (B) spotted hyenas (*Crocuta crocuta*); (C) a plains zebra (*Equus quagga*); (D) a male impala (*Aepyceros melampus*).

Q3.10.6 Draw a pyramid of biomass (showing the wet weights) for the carnivores, herbivores and the vegetation in 1 km² of the Serengeti (not to scale).

Q3.10.7 Explain how the vegetation in this ecosystem is able to support the great variety of herbivores present.

Q3.10.8 Why do very large numbers of plains zebra and wildebeest migrate each year between the Serengeti and the Masai Mara in Kenya?

Reference/Further Reading

Schaller, G. B. (1972). *The Serengeti lion. A Study of Predator-Prey Relations*. Chicago: University of Chicago Press.

ASSIMILATION EFFICIENCY IN THE AFRICAN ELEPHANT

Gross assimilation efficiency is a measure of the efficiency with which an animal converts food into its own body cells. The relationship between ingestion, assimilation and egestion is illustrated in Fig. 3.14. The amount of food eaten and dung voided by an eleven-year-old captive African elephant (*Loxodonta africana*) over a seven-day period is shown in Table 3.14. Note that some food was wasted (i.e., not eaten).

Food eaten $(I) = O - W$

Assimilation efficiency (%) $= \dfrac{I - E}{I} \times 100$

where,

I = dry weight of food ingested;
E = dry weight of dung egested.

Q3.11.1 Complete Table 3.14 by calculating the missing values.

Table 3.14 Food eaten and dung voided by an African elephant in seven days

Food offered	Total wet weight (kg)	Dry matter %	Total dry weight (kg)	Variable
Hay	112	83		
Straw	30	81		
Carrots	293	11		
Sprouts	94	13		
Beetroot	3	9		
Total food offered			=	O
Waste (food not eaten)	89	21	=	W
Food eaten (offered – waste)			=	I
Dung voided	579	19	=	E

Based on Rees (1982).

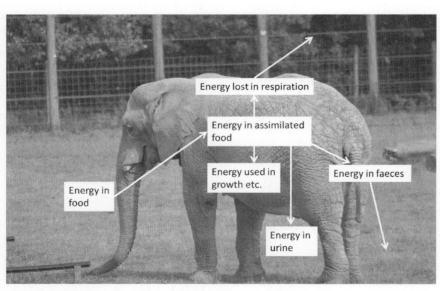

Figure 3.14 The fate of energy in food eaten by an elephant.

Q3.11.2 Calculate the gross assimilation efficiency, taking care to use dry weights and to subtract the dry weight of the waste from the dry weight of the total food offered.

Q3.11.3 Explain why it is important to use dry weight in these calculations.

Q3.11.4 Determine the ratio of the dry weight of food consumed to the dry weight of the dung voided. (Express this as *x* units of food: 1 unit of dung, in terms of dry weight).

Q3.11.5 A wild African elephant produces 3 kg of dung per hour (wet weight) at a constant rate throughout the day. The dung is 20% dry matter. Using your answers to questions 3.11.2 and 3.11.4 above, calculate the dry weight of food eaten in 24 hours.

Reference/Further Reading

Rees, P. A. (1982). Gross assimilation efficiency and food passage time in the African elephant. *African Journal of Ecology, 20*, 193–198.

CAPYBARA FARMING IN VENEZUELA

Figure 3.15 A capybara (*Hydrochaerus hydrochaeris*).

The capybara (*Hydrochaerus hydrochaeris*) is a large rodent that forms an important part of the diet of the indigenous people of Venezuela (Fig. 3.15). The Venezuelan government licences ranchers to crop capybara on their ranches. Table 3.15 compares meat production data from capybara with those from beef cattle.

Q3.12.1 Why does this study compare meat production from one cow with that from 18 capybara?

Q3.12.2 Calculate the mean live weight of a capybara at slaughter.

Q3.12.3 Which of the two animals produces meat at the faster rate?

Q3.12.4 From the information in Table 3.15 suggest two reasons for this.

Q3.12.5 Express the quantity of meat produced by beef cattle in one year as a percentage of that produced by capybara in the same period.

Table 3.15 A comparison of meat production from capybara and beef cattle in Venezuela

	Capybara	Beef cattle
Number of offspring per year	6	1
Age at slaughter weight (years)	1.5	4.5
Number of individuals slaughtered during 4.5 years	18	1
Sum total of live weights at 4.5 years (kg)	540	360
Carcase at 50% of live weight (kg) produced in 4.5 years	270	180
Carcase meat produced in 1 year (kg)	60	40

Based on Kyle (1987).

Q3.12.6 How may the use of capybara as food help to assure its future survival in the wild?

References/Further Reading

Kyle, R. (1987). Rodents under the carving knife. *New Scientist, 1566*, 58–62.

Moreira, J. R., & Macdonald, D. W. (1996). Capybara use and conservation in South America. In V. J. Taylor & N. Dunstone (Eds.), *The exploitation of mammal populations* (pp. 88–101). Netherlands: Springer.

THE NITROGEN CYCLE

Figure 3.16 A decomposing mustelid. This animal's body will provide nitrogen for scavengers, decomposers and for the soil.

Nitrogen is an important element in many of the biochemicals found in living things (Fig. 3.16). It is found in the structure of proteins and enzymes which are essential to the structure and functioning of cells. The atmosphere is 78% nitrogen but few organisms can utilise this. Terrestrial plants obtain their nitrogen from the soil while animals obtain it by eating other organisms.

The conversions shown in Table 3.16 are some of the processes which occur in the nitrogen cycle. In particular they show the importance of two genera of bacteria: *Nitrosomonas* and *Nitrobacter*.

Q3.13.1 Use the information in Table 3.16 to construct part of the nitrogen cycle as a single flow diagram.

Q3.13.2 Which two sources of nitrogen for plants are shown by your diagram?

Q3.13.3 Add another component to your diagram showing a manmade source of nitrogen for plants.

Table 3.16 Chemical conversions that occur in the nitrogen cycle

A	Ammonium NH_4^+	$\xrightarrow{\textit{Nitrosomonas}}$	Nitrite NO_2^-
B	N in inorganic matter and humus	$\longrightarrow$	Ammonium NH_4^+
C	Nitrite NO_2^-	$\xrightarrow{\textit{Nitrobacter}}$	Nitrate NO_3^-
D	Ammonium NH_4^+	$\xrightarrow{\text{soil water}}$	N in plants (e.g., in protein)
E	Nitrate NO_3^-	$\xrightarrow{\text{soil water}}$	N in plants (e.g., in protein)
F	N in plants (e.g., in protein)	$\xrightarrow{\text{death}}$	N in organic matter and humus

Q3.13.4 The atmosphere is 78% nitrogen. Lightning (electrification) can fix this nitrogen, putting nitrates back into the soil. Add a component to your flow diagram showing this process.

Q3.13.5 Nitrates are highly soluble in water. What are the likely consequences of this for:

a. the crop farmer; and

b. freshwater ecosystems?

Q3.13.6 If land is contaminated with chemicals which inhibit the growth of *Nitrosomonas* and *Nitrobacter* what would be the likely consequences for plant growth?

Reference/Further Reading

Gruber, N., & Galloway, J. N. (2008). An Earth-system perspective of the global nitrogen cycle. *Nature, 451,* 293–296.

NITROGEN BALANCE FOR LAKE MENDOTA, WISCONSIN

Lake Mendota is located in Dane County, Wisconsin, in the United States. It has an area of approximately 9800 acres (3966 ha) and a maximum depth of around 80 feet (24 m). Table 3.17 shows the nitrogen inputs into (sources) and losses from the lake (sinks), based on data published in 1968.

Q3.14.1 Using the data in Table 3.17 draw a flow diagram illustrating the nitrogen balance in Lake Mendota using arrows whose thickness is drawn approximately to scale. For example:

Wastewater

This arrow should be 8.1 units thick (e.g., approximately 8 mm).

Table 3.17 Nitrogen inputs and losses from Lake Mendota

Input		Loss	
Source	Per cent	Sink	Per cent
Wastewater	8.1	Outflow	16.4
Surface water	14.7	Denitrification	11.1
Precipitation	17.5	Fish catch	4.5
Groundwater	45.0	Weeds	1.3
Fixation	14.4	Sedimentation	66.7
Total	100.0	Total	100.0

Modified from Brezonik & Lee (1968).

Q3.14.2 What is the most important source of nitrogen in the lake?

Q3.14.3 What is the most important sink for nitrogen in the lake?

Q3.14.4 Suggest a possible major source of the nitrogen in:
   a. wastewater;
   b. surface water.

Q3.14.5 Explain how nitrogen is lost from the lake by denitrification.

Q3.14.6 What would be the likely outcome of excessive inputs of nitrogen into a lake?

Reference/Further Reading

Brezonik, P. L., & Lee, G. F. (1968). Dentrification as a nitrogen sink in Lake Mendota, Wisconsin. *Environmental Science & Technology, 2*, 120–125.

THE ECONOMICS OF FERTILISER APPLICATION

Farmers apply nitrogen fertiliser to the soil to increase crop yields. However, there is a limit beyond which further application of fertiliser does not provide any economic benefit. Table 3.18 shows the effect of different levels of nitrogen fertiliser application on yield, benefit–cost ratio and net yield return in a cereal crop. Net yield return is a measure of the value of the increased yield by adding fertiliser.

$$\text{Value of crop} = £200.00\,t^{-1}$$

$$\text{Cost of fertiliser} = £1.00\,kg^{-1}$$

$$\text{Benefit–cost ratio} = \frac{(Y_f - Y_u) \times V}{C}$$

$$\text{Net yield return} = (Y_f - Y_u) \times V - C$$

Where,

Y_f = Yield of fertilised crop (t ha^{-1})

Y_u = Yield of unfertilised crop (t ha^{-1})

V = Crop value (£ t^{-1})

C = Total cost of fertiliser (£)
   = *kg applied × price kg^{-1}*

Q3.15.1 Complete Table 3.18 by calculating values for benefit–cost ratio and net yield return for all nitrogen application rates using the formulae provided.

Q3.15.2 Draw graphs showing the effect of increasing nitrogen application rate on:

 a. yield;

 b. benefit–cost ratio; and

 c. net yield return.

Table 3.18 The effect of nitrogen application on yield in a cereal crop

Nitrogen applied (kg ha^{-1})	Yield (t ha^{-1})	Benefit–cost ratio	Net yield return (£)
0	2.9	–	–
50	3.4	2.0	50
100	4.3		
150	4.7		
200	4.8		

Q3.15.3 What would be the implications of a benefit–cost ratio of less than 1.0?

Q3.15.4 Repeat the calculations assuming a crop value of £150.00 t^{-1} and a nitrogen fertiliser cost of £1.50 kg^{-1}.

Q3.15.5 Explain what happens when nitrogen is applied at a rate of 200 kg ha^{-1} with these new values.

Reference/Further Reading

Scobie, G. M., & St-Pierre, N. R. (1987). Economics of phosphorus fertiliser use on pastures 1. Long-run maintenance requirements. *New Zealand Journal of Experimental Agriculture, 15,* 435–443.

SEWAGE SLUDGE CAKE AS A FERTILISER

Some farmers apply dried sewage sludge (cake) to their land as a fertiliser. If sludge cake is applied to a field at an application rate of 25 t ha^{-1} only 15% of the nitrogen is available in the first year, approximately 38 kg ha^{-1}. In the second and subsequent years this is reduced by approximately 50% each year, as illustrated in Table 3.19.

Q3.16.1 Calculate the total available nitrogen each year by completing Table 3.19, assuming additional applications of cake for a total of six years.

Q3.16.2 Using the format shown in Table 3.20, construct a table showing the effect of applying this cake over a period of ten years by completing the empty cells and calculating the totals for each column.

Q3.16.3 Draw a graph showing the effect of applying cake in six successive years on the total amount of available nitrogen.

Q3.16.4 Use evidence from your graph to describe the benefits which may be derived from using sludge cake rather than an inorganic nitrogen fertiliser.

Table 3.19 Available N kg ha^{-1} from sewage sludge cake in the first and subsequent years after application

Year	1	2	3	4	5	6
Total N kg ha^{-1}	38	19	10	5	2	1

Table 3.20 Total available nitrogen (kg ha^{-1}) for six successive years

Year	Year after first application					
	1	2	3	4	5	6
1	38	19	10	5	2	1
2	–	38	19	10	5	2
3	–	–				
4	–	–	–			
5	–	–	–	–		
6	–	–	–	–	–	
Total N kg ha^{-1}	38	57				

Q3.16.5 What reservations might a farmer have about using sludge cake to fertilise agricultural land?

Reference/Further Reading

Bhogal, A., Nicholson, F. A., Chambers, B. J., & Shepherd, M. A. (2003). Effects of past sewage sludge additions on heavy metal availability in light textured soils: Implications for crop yields and metal uptakes. *Environmental Pollution, 121*, 413–423.

EXERCISE 3.17

THE EFFECT OF pH ON CROP GROWTH

The availability of plant nutrients is in part determined by soil pH. Soil pH describes the relative acidity or alkalinity of the soil. Pure water has a close to neutral pH (7.0). A pH value below 7.0 is acidic (i.e., the hydrogen ion concentration is higher than at neutral pH). A pH value above 7.0 is alkaline (i.e., the hydrogen ion concentration is lower than at neutral pH). The pH scale is logarithmic (See Fig. 3.17). Most plant nutrients are optimally available within a pH range of 6.5 to 7.5 (See Table 3.21).

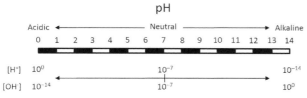

Figure 3.17 The pH scale.

Q3.17.1 Draw a graph of the effect of soil pH on yield in corn and oats using data from Table 3.22. Comment on the differences between the two graph lines.

Q3.17.2 Identify the range of pH values which produces the highest yields in all seven of these crops.

Q3.17.3 Which of these crops produces the lowest yield at each of the following pH values?:

a. 4.7;

b. 5.7;

c. 7.5.

Q3.17.4 Which pH produces the optimum yield in Timothy?

Table 3.21 Effect of pH on the availability of selected plant nutrients

Nutrient	Symbol	Effect of pH
Phosphorus	P	When pH >7.5 phosphate ions react quickly with calcium (Ca) and magnesium (Mg) forming less soluble compounds. In acidic conditions (pH < 7.0) phosphate ions react with (Al) and iron (Fe) forming less soluble compounds
Molybdenum	Mo	Less available under acidic pH; more available at slightly alkaline pH values
Nitrogen Potassium Sulphur	N, K, S	These and most other nutrients tend to be less available when soil pH is above 7.5, and are optimally available at a slightly acidic pH (6.5 to 6.8)

Table 3.22 Relative yield of selected crops grown in a corn, small grain, legumes or timothy rotation at different pH levels

Crop	pH				
	4.7	5.0	5.7	6.8	7.5
Corn	34	73	83	100	85
Wheat	68	78	89	100	99
Oats	77	93	99	98	100
Barley	0	23	80	95	100
Alfalfa	2	9	42	100	100
Soybean	65	79	80	100	93
Timothy	31	47	66	100	95

Adapted from Smith & Doran (1996).

Q3.17.5 Explain why liming an acidic soil may increase crop yield.

Q3.17.6 If the soil pH was 6.8 which crops would not benefit from liming?

Reference/Further Reading

Smith, J. L., & Doran, J. W. (1996). Measurement and use of pH and electrical conductivity for soil quality analysis. *Methods for assessing soil quality. Soil Science Society of America Special Publication, 49,* 169–182.

TIMANFAYA NATIONAL PARK, LANZAROTE: AN OPPORTUNITY TO STUDY SUCCESSION

Timanfaya National Park in Lanzarote (Canary Islands) was created in 1974 to protect a large area which had recently been affected by volcanic activity. At the park's western boundary larva flowed into the Atlantic Ocean and this coastal zone is also protected. The most recent eruption took place in 1834 and scientists have had a rare opportunity to study the process of succession on the larva fields which this created.

Q3.18.1 With reference to Table 3.23, explain why Timanfaya has an unusual flora.

Q3.18.2 How would you expect the composition of this flora to change with time?

Q3.18.3 What would you expect to happen to the larva as a result of the presence of plants listed in Table 3.23.

Q3.18.4 What adaptations would you expect the vascular plants to have evolved in order to survive in this inhospitable environment?

Q3.18.5 Using the information in Tables 3.23, 3.24 and 3.25, write a paragraph justifying:

a. the existence of Timanfaya National Park.

b. the local regulations forbidding visitors from walking on the larva.

c. locating the park's visitor centre below ground.

Table 3.23 Terrestrial plant groups in Timanfaya National Park

Group of organisms	Number of species
Lichens	200
Mosses	15
Algae	5
Vascular plants	177

Table 3.24 Species identified from the coastal zone of Timanfaya National Park

Group of organisms	Number of species
Marine plants	105
Invertebrates	120
Fishes	59

Table 3.25 Climate data for Lanzarote

Annual rainfall	125 mm
Mean annual temperature	21°C
Maximum summer temperature	28°C
Mean water temperature	20°C
Trade winds blowing for much of the year. Occasional warm winds from Africa.	

d. restricting visitor access to a small part of the park. Most of the park may only be toured by coach.

References/Further Reading

Bramwell, D. (1990). Conserving biodiversity in the Canary Islands. *Annals of the Missouri Botanical Garden*, 28–37.

Nedelcu, A. (2010). A model for sustainable development of tourism in the Canary Islands, Spain. *Journal of EcoAgriTourism*, 6, 131–138.

EXERCISE 3.19

SUCCESSION IN BIRD SPECIES

Succession in ecosystems is most easily observed in plant species because they tend to give the ecosystem its most readily identifiable characteristics. For example, a tropical forest contains a great variety of large tree species, a savannah contains many grass species and so on.

If a grassland develops into a forest during the process of succession the ecologist could not fail to notice this, but changes in the fauna may be more difficult to identify, especially in the smaller species, those that live in soil or water, and those that are highly mobile. Nevertheless, these changes are also important in determining the character of the ecosystem and are inextricably linked to changes in the flora.

Tables 3.26 and 3.27 show changes which occurred within the avifauna during a secondary succession in an upland area in the Piedmont Region of Georgia in the United States.

Q3.19.1 Draw a graph showing the changes which have occurred with time to:

a. the number of bird species;

b. the total number of individual birds.

Q3.19.2 Account for these changes with reference to changes in the flora.

Q3.19.3 What is a secondary succession and how does it differ from a primary succession?

Table 3.26 The effect of succession on the number of breeding passerine birds in an upland ecosystem in the Piedmont Region of Georgia, United States

Age (years)	Number of bird species	Number of individual birds
2	2	15
3	2	40
15	7	110
20	9	136
25	7	87
35	10	93
60	20	158
100	18	239
200	19	228

Data derived from Johnston & Odum (1956).

Table 3.27 The dominant plant taxa present at different times during the succession

Age (years)	Plant taxa
2–20	Grasses, shrubs
25–60	Pine forest
100–200	Oak–hickory climax forest

Q3.19.4 Explain what is mean by 'oak–hickory climax forest'.

Reference/Further Reading

Johnston, D. W., & Odum, E. P. (1956). Breeding bird populations in relation to plant succession on the Piedmont of Georgia. *Ecology, 37*, 50–62.

CHAPTER 4

DETERMINING ABUNDANCE AND DISTRIBUTION

This chapter contains exercises concerned with the methods used by ecologists to determine where a species occurs – its distribution – and how many individuals of that species are present in a particular area – its abundance.

Spotted hyena (*Crocuta crocuta*)

Examining Ecology. DOI: http://dx.doi.org/10.1016/B978-0-12-809354-2.00004-X

INTENDED LEARNING OUTCOMES

On completion of this chapter you should be able to:

- Discuss the difficulties of producing maps of the distribution of organisms.

- Identify and explain patterns in the spatial distribution of organisms.

- Determine an appropriate method of estimating the size of a population of organisms.

- Calculate the density of a population.

- List factors which may affect the accuracy of population estimates.

- Construct a calendar of catches and use it to determine changes in the size of an animal population.

- Estimate the size of a population of mobile organisms using a variety of methods: mark-release-recapture, removal trapping, quadrat sampling, transect counts, net sampling.

- Estimate population size by determining the relationship between population size and the number of breeding sites used by animals.

- Identify types of spatial distributions of organisms using statistics: clumped, uniform and random.

- Recognise the effects of sample size on population estimates.

- Analyse data on plant distribution and habitat selection in animals.

- Use simple simulations to study the abundance and distribution of organisms.

- Determine the position of an organism using the UTM coordinate system.

- Analyse population data produced by computer simulation.

INTRODUCTION

The two most fundamental things that we can know about the ecology of any species are its abundance and its distribution. However, for many species in many parts of the world we have little information about either of these aspects of their ecology. This is especially true for remote areas of the world and extreme habitats, and for those species that are small or present at relatively low densities. Some historical records of animal and plant distributions relied heavily upon information and specimens collected by amateur naturalists, often for museums. Although many of these amateurs were self-taught experts, others were less proficient at identification and some historical records undoubtedly contain errors.

DISCOVERING NEW SPECIES

Many species still remain hidden from scientists. Even quite large animals have remained undetected until recent times because of the inaccessibility of their habitats. The okapi (*Okapia johnstoni*) may be 1.8 m tall yet was unknown to science until 1901 and the Komodo dragon (*Varanus komodoensis*) was unknown to the Western world until 1910 (Fig. 4.1). New mammals are still occasionally discovered; a new rodent species, the hog-nosed rat (*Hyorhinomys stuempkei*), was discovered on Sulawesi Island in 2015 by scientists from Louisiana State University Museum of Natural Science.

MAPPING DISTRIBUTIONS

Many distribution maps consist of little more than shaded areas within an outline of an entire country or continent. For example, *The Kingdon Field Guide to African Mammals* (Kingdon, 1997) shows the distribution of the leopard (*Panthera pardus*) as most of Africa south of the Sahara. In reality, leopards are very difficult to find, and restricted to particular habitats.

Some countries have produced detailed maps of the distributions of their native and naturalised species. In the United Kingdom these consist of maps where the country is divided into 10-km squares and the presence of a species in any particular square is indicated by shading it in or filling it with a dot. The mapping of animal and plants species can now be done very precisely by using Global Positioning System (GPS) equipment to record locations.

Figure 4.1 The Komodo dragon (*Varanus komodoensis*) was unknown to the Western world until 1910.

Estimating Population Size

Many methods have been devised to determine or estimate population size (Henderson and Southwood, 2016). At the simplest level, the size of a population may be determined by counting all of the animals or plants. This may be easy if the organisms are large and easy to find. However, most populations are not like this and ecologists must resort to the use of sampling methods to calculate a population estimate.

In most cases it is not possible to count all of the individuals in a population of animals or plants directly. Calculating a population estimate usually depends upon taking samples which are representative of the population as a whole. If the density of organisms can be established – the number of individuals per unit area – the total population size for a specified area can be calculated (Fig. 4.2). Other methods involve marking a proportion of the animals in a population and assuming that the proportion marked will remain constant, thereby allowing a population estimate to be calculated from any sample where some individuals bear marks and others do not.

Great care must be taken in devising sampling methods. Any method which produces biased samples – samples which are not representative of the population as a whole – will produce

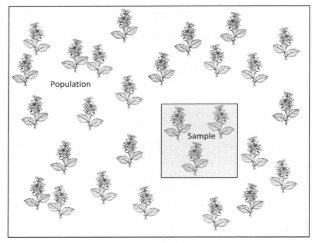

Figure 4.2 The relationship between a sample and a population.

population estimates which are either too high or too low. As a general rule, samples should be taken at random. That is to say, each individual in the population should be equally likely to be selected in the sampling process. In practice, in the field, this is often very difficult to achieve.

The following exercises examine a variety of methods of mapping organisms and measuring their abundance.

References

Henderson, P., & Southwood, T. R. E. (2016). *Ecological methods* (4th ed.). Chichester: Wiley.

Kingdon, J. (1997). *The Kingdon Field Guide to African Mammals*. London: A & C Black Publishers Ltd.

EXERCISE 4.1

RECORDING THE DISTRIBUTION OF ORGANISMS

The United Kingdom is one of the best biologically recorded countries in the world. The Biological Records Centre records the distribution of organisms in 10 km × 10 km squares. It was created in 1964 as a national focus for recording the distribution of terrestrial and freshwater species. A separate map is produced for each species. Records of sightings of each species are collected by amateur naturalists and professional biologists and submitted to the recorders for the county where the sighting was made. Country recorders send their data to the Biological Records Centre which then collates the data and produces maps.

(A)

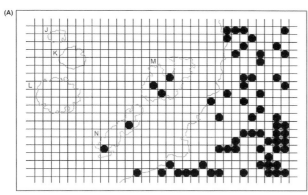

(B)

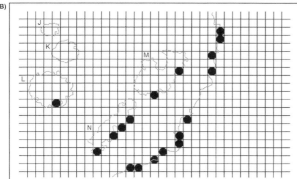

Figure 4.3 Maps of the distributions of two hypothetical mammal populations.

Each map shows the distribution of a particular species in the United Kingdom by dividing the map into 10-km squares based on the Ordnance Survey. If the species is present in a particular square the square is shaded or filled with a dot. Maps generally represent the distribution of a species as indicated by data collected over a relatively long period of time. Sometimes dots of different colours are used to indicate past and present distributions, e.g., pre-1930, 1930–69, 1970–86 and so forth.

Q4.1.1 A dot-map of the hypothetical distribution of a small mammal species is shown in Fig. 4.3A. Suggest reasons why there are no dots on islands J, K and L.

Q4.1.2 Fig. 4.3B shows the hypothetical distribution of a large mammal species.

a. What type of mammal is this likely to be?

b. Why is this distribution map misleading for this type of animal?

Q4.1.3 Suggest three reasons why a species may appear to be absent from a particular 10-km square.

Q4.1.4 Why are museum specimens of use in establishing the distributions of organisms?

Q4.1.5 Using the information in Fig. 4.4 explain why mapping a distribution in 10-km squares may be misleading.

Q4.1.6 Using a single dot to represent the presence of a species in a 10-km square

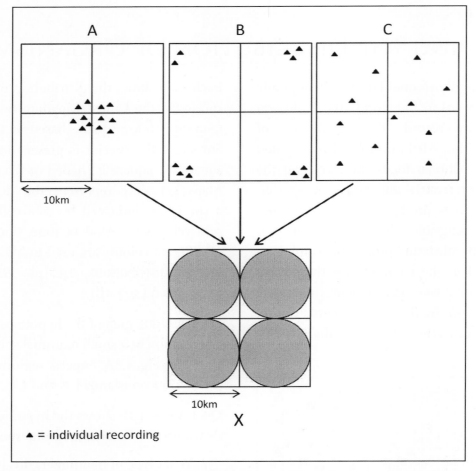

Figure 4.4 Three hypothetical distributions of recordings (A, B and C), each of 12 organisms which would be recorded as identical in the 10-km square dot-map mapping system used by the Biological Records Centre (X).

results in the loss of detailed information about location. What other information is not provided by a single dot?

Q4.1.7 What types of events and processes may be investigated using dot-maps?

Reference/Further Reading

The Biological Records Centre. <http://www.brc.ac.uk/home>.

PROBLEMS IN DETERMINING THE HISTORICAL DISTRIBUTION OF ORGANISMS

Engaging in the study of natural history was an important pastime in 19th century Britain. In the early days of botanical study simply recording the locations where particular species of plants were to be found was an important contribution to science. Knowing where plants used to grow is essential if we are to assess the effects of recent changes in the environment such as climate change or the effects of pollution. However, this presupposes that these early records were accurate. Historical documents show that sometimes they were not.

Over 200 years ago, in 1831, Rev. Hugh Davies published the first flora of a Welsh county; his catalogue of the plants of Anglesey was entitled *Welsh Botanology*. Davies was known by other British botanists as 'the Welsh Linnaeus'. In 1825 the botanist Dr John Roberts wrote a letter to Prof. John Stevens Henslow of Cambridge, listing about 40 plants that he believed Davies added to his *Welsh Botanology* which could not be justified (Jones, 2007).

Q4.2.1 Using information in Table 4.1, suggest two possible sources of error in historical records of the natural distribution of plants.

Q4.2.2 What does Roberts' assertion that he did not find any specimens of *Pyrola minor* in Anglesey prove?

Q4.2.3 Why were the gardens of stately homes a significant source of exotic species in the past?

Table 4.1 Some comments by Dr John Roberts and Mr J. Griffith on species listed in *Welsh Botanology* (Edwards, 1831)

Species	Comments by Roberts (in a letter of 1825)	Comments by Griffith (1895)	Current status in the United Kingdom*
Salvia pratensis	Inserted upon the authority of no Botanist and probably came from a garden.	Does not grow wild in Anglesey… the specimens were *S. verbenaca*.	Considered native but may have been introduced. Not recorded as present in Anglesey.
Campanula rapunculus	Probably came from Baron Hill gardens [the gardens of a stately home in Beaumaris, Anglesey].	Not mentioned.	Naturalised; rare in wild and in cultivation. Not recorded as present in Anglesey.
Pyrola minor	Taken upon hearsay. I never saw any of them in this country.	…nor have I seen it growing in [Anglesey]…	Native; recorded in Anglesey.
Cucubalis baccifer	A complete error – see Smith's *English Botany*.	I have carefully searched the locality…but in vain.	Probably an alien. Not recorded as present in Anglesey.
Cerastium humile	I believe of his own creation.	Not listed as present by Griffith.	Identification unclear so impossible to determine current status.

*Based on Online Atlas of the British and Irish Flora, Biological Records Centre, UK. http://www.brc.ac.uk/plantatlas/ Accessed 26.11.2015.

References/Further Reading

Edwards, H. (1831). *Welsh botanology*. London: Printed for the author by W. Marchant.

Griffith, J. E. (1895). *The Flora of Anglesey and Carnarvonshire*. Bangor: Nixon and Jarvis.

Jones, D. (2007). *The Botanists and Mountain Guides of Snowdonia*. Gwynedd, Wales: Llygad Gwalch, Pwllheli.

COMMUNITY SCIENCE: MOORS FOR THE FUTURE

The *Moors for the Future* Partnership has been working to preserve 8000 years of moorland heritage in the Peak District and South Pennines, in northern England, since 2003. This landscape has been subjected to 200 years of damage, leaving large areas of upland devoid of vegetation. The lead body is the Peak District National Park Authority but it works in partnership with statutory agencies, water companies and charities including the Environment Agency, Natural England, the National Trust, the Royal Society for the Protection of Birds (RSPB), Severn Trent Water, United Utilities, Yorkshire Water and Pennine Prospects. The project also involves representatives of the moorland owner and farming community.

Moors for the Future uses volunteers in some of its ecological survey projects. One of these projects is attempting to determine the distribution of *Sphagnum* moss in the Peak District and South Pennines with a view to monitoring changes in this over time. Fig. 4.5 is a copy of the 'Community Science *Sphagnum* Survey Guide' provided to volunteers that participate in the survey work.

Q4.3.1 What global environmental change may be reflected in the future pattern of distribution of *Sphagnum* moss and why?

Q4.3.2 Why is *Sphagnum* particularly sensitive to changes in water quality?

Q4.3.3 Why are volunteers asked not to try to identify *Sphagnum* moss to species level?

Q4.3.4 Fig. 4.6 is a map of eight patches of *Sphagnum* moss seen from above (labelled A–H). Using the instructions in the survey guide, determine how many patches of *Sphagnum* should be recorded and identify them from their labels. Remember, you must stay on the path.

Q4.3.5 List three pieces of information that volunteers are asked to collect in order to determine the location of *Sphagnum* patches.

Q4.3.6 Explain the role of *Sphagnum* in providing wildlife habitat.

Reference/Further Reading

Moors for the Future <http://www.moorsforthefuture.org.uk/>.

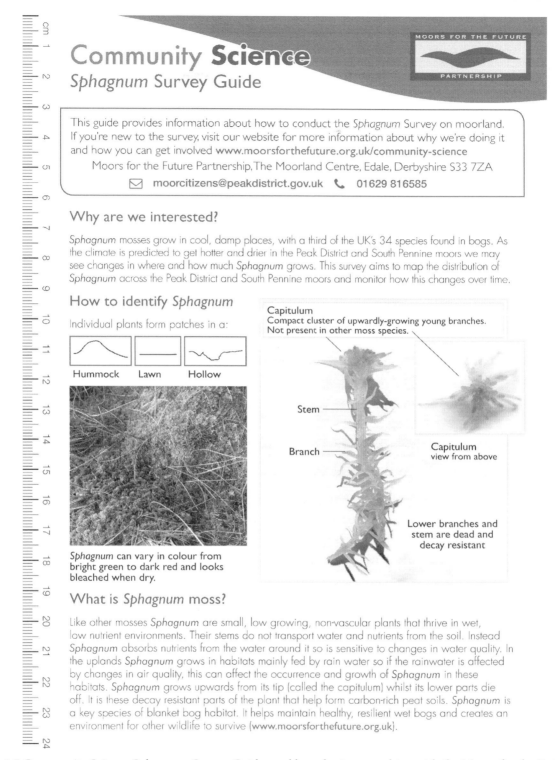

Community Science
Sphagnum Survey Guide

MOORS FOR THE FUTURE
PARTNERSHIP

This guide provides information about how to conduct the *Sphagnum* Survey on moorland. If you're new to the survey, visit our website for more information about why we're doing it and how you can get involved **www.moorsforthefuture.org.uk/community-science**

Moors for the Future Partnership, The Moorland Centre, Edale, Derbyshire S33 7ZA

✉ **moorcitizens@peakdistrict.gov.uk** ☎ **01629 816585**

Why are we interested?

Sphagnum mosses grow in cool, damp places, with a third of the UK's 34 species found in bogs. As the climate is predicted to get hotter and drier in the Peak District and South Pennine moors we may see changes in where and how much *Sphagnum* grows. This survey aims to map the distribution of *Sphagnum* across the Peak District and South Pennine moors and monitor how this changes over time.

How to identify *Sphagnum*

Individual plants form patches in a:

Hummock   Lawn   Hollow

Sphagnum can vary in colour from bright green to dark red and looks bleached when dry.

Capitulum
Compact cluster of upwardly-growing young branches. Not present in other moss species.

Stem

Branch

Capitulum view from above

Lower branches and stem are dead and decay resistant

What is *Sphagnum* moss?

Like other mosses *Sphagnum* are small, low growing, non-vascular plants that thrive in wet, low nutrient environments. Their stems do not transport water and nutrients from the soil. Instead *Sphagnum* absorbs nutrients from the water around it so is sensitive to changes in water quality. In the uplands *Sphagnum* grows in habitats mainly fed by rain water so if the rainwater is affected by changes in air quality, this can affect the occurrence and growth of *Sphagnum* in these habitats. *Sphagnum* grows upwards from its tip (called the capitulum) whilst its lower parts die off. It is these decay resistant parts of the plant that help form carbon-rich peat soils. *Sphagnum* is a key species of blanket bog habitat. It helps maintain healthy, resilient wet bogs and creates an environment for other wildlife to survive (www.moorsforthefuture.org.uk).

Figure 4.5 Community Science Sphagnum Survey Guide used by volunteers working with the Moors for the Future Partnership in northern England. Reproduced with permission, *Moors for the Future* Partnership, Heritage Lottery Funded Community Science.

Where and when to survey

We are surveying Public Rights of Way across the Peak District and South Pennine moorlands every year. Please visit our website to see which routes have been surveyed and which still need visiting this year. You can carry out a survey at any time of year, weather permitting.

What to take with you

- A Recording Form (available from our website).
- A map covering the survey area – either an OS Map or a print out of a satellite image.
- A camera, if available, to send us photos of your findings.
- A GPS unit, or freely available smartphone app such as GridPoint GB or ViewRanger, to record the exact location of *Sphagnum* patches.

Conducting the survey

- Fill in the information at the top of the recording form. **When**, **where** and **who** are important.
- Walk the route, scanning 2 metres (2m) either side of the path for *Sphagnum*. If the path is wide you may need to walk up one side and down the other to ensure an even search.
- Record each patch of *Sphagnum* you see. Please don't worry about identifying *Sphagnum* moss to species level as this can be tricky even for experts. If you would like information on how to do this please contact us and we'd be happy to help. To start with we are looking to map where *Sphagnum* is and where it isn't.
- Some patches may be clearly continuous whereas others are made up of a number of small, less distinct patches. If patches are less than 2m apart, treat them as a single patch.

For each patch of *Sphagnum* you find:

- Mark the location of its centre point on your map as accurately as you can or take a GPS reading. On the Recording Form, note down any features that will help you identify the patch location when you enter your results (taking a photo of obvious landmarks might help).
- Record how wide the patch is at its widest point (up to 2m from the edge of the path) and how long the patch is at its longest point to the nearest 25cm (approximately the length of this card).
- Record whether the *Sphagnum* is growing as a hummock, a lawn or in a wet hollow (overleaf).
- If you have a camera (photos from phones are fine) take photos of the patch: one up close to help verify your findings, and one of the whole patch to compare with findings in future years.

Health and Safety

Before setting out, ensure that you have read and understood the health and safety guidelines provided at training and available on our website **www.moorsforthefuture.org.uk/community-science**

> Submit your results through **www.moorsforthefuture.org.uk/community-science/submit-results** or by post as soon as possible, even if you did not find any *Sphagnum* – knowing where there isn't *Sphagnum* is as important as knowing where there is!

Thank you for your involvement in the Moors for the Future Partnership's Community Science Project, an important initiative to collect long-term data to help us understand how moorlands and the species they support are responding to climate change. For information about other surveys you can get involved in visit **www.moorsforthefuture.org.uk/community-science**

LOTTERY FUNDED

 www.facebook.com/MoorCitizens   @MoorCitizens

 moorcitizens@peakdistrict.gov.uk   01629 816585

Materials designed by

Figure 4.5 (Continued)

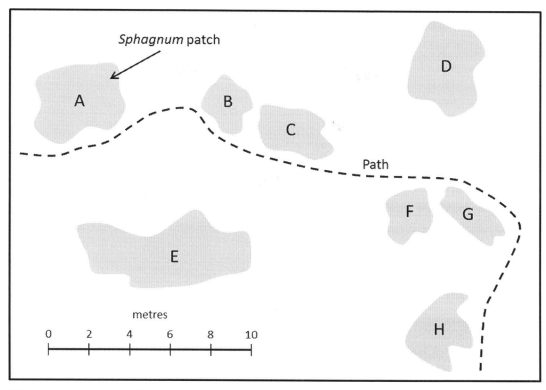

Figure 4.6 Hypothetical map of *Sphagnum* moss distribution.

BIRD RINGING

In many ecosystems around the world wild birds are an easily observable element of the fauna. Bird watching is a popular pastime among large numbers of people and many join organisations, such as the Royal Society for the Protection of Birds (RSPB) in the United Kingdom, the American Birding Association (ABA) in North America and *Birds New Zealand*, as a focus for their interest. An army of scientists and volunteers around the world monitors populations of a large numbers of bird species by attaching identification marks to them in the form of metal or plastic leg rings, plastic collars or wing tags. This process is collectively known as 'bird ringing' or 'bird banding'.

Figs. 4.7 (A–D) illustrate some of the equipment and techniques used in ringing and handling small songbirds. Fig. 4.8 is a greylag goose (*Anser anser*) that is identifiable from a bright orange neck collar.

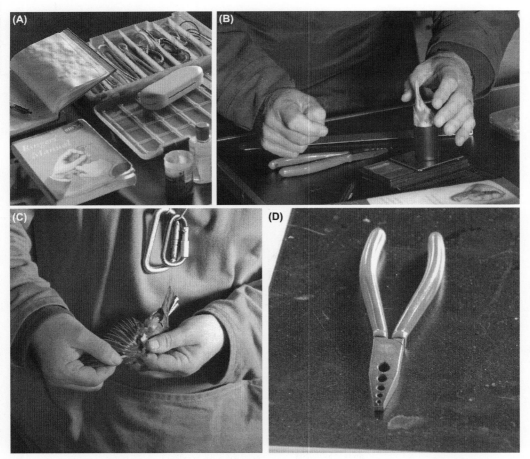

Figure 4.7 Bird ringing. (A) Equipment including rings and record book; (B) weighing a small songbird; (C) ageing a chaffinch (*Fringilla coelebs*); (D) bird ringing pliers used to close metal leg rings (note the different size rings that can be fitted).

Figure 4.8 A greylag goose (*Anser anser*) with a coloured neck collar.

Q4.4.1 Which aspects of the biology of birds may be studied by ringing?

Q4.4.2 What behaviour of greylag geese (*Anser anser*) makes a collar a more useful means of identification than a leg ring?

Q4.4.3 Some ringed birds may be found dead by members of the public. What information should a bird ring contain, other than an identification number, to ensure that information about birds that are found dead is not lost?

Q4.4.4 How could bird ringing be used to study the effects of climate change on bird populations?

Reference/Further Reading

Bub, H. (1995). *Bird trapping and bird banding: A handbook for trapping methods all over the world*. Ithaca, New York: Cornell University Press.

Exercise 4.5

ZONATION ON A ROCKY SHORE

Figure 4.9 Algal species exhibit zonation on rocky shores.

Animals and plants often exhibit a zonation in their distribution when influenced by a gradient in an environmental variable. On a rocky shore, the diversity of organisms at the high water mark differs markedly from that at the low water mark due to differences in tidal exposure (Fig. 4.9).

Table 4.2 shows the species of algae present at 10 sampling stations at a range of positions from the high water mark to the low water mark on a rocky shore in South Wales.

Q4.5.1 Draw a diagram illustrating the distribution of algae along the shore using a single horizontal line to represent each of the taxa listed in Table 4.3 as illustrated in Fig. 4.10.

Table 4.2 The algal species present at 10 stations located on a rocky shore in Bracelet Bay, near Swansea, South Wales

Position	Station	Algal species present
High water	1	A, B, C, D, E, F
↑	2	B, C, D, E, F
	3	D, E, F, G, H, I, J, K
	4	E, H, I, J, K, L, M, N, O
Mid shore	5	E, F, H, I, J, L, M, N, P
	6	H, I, K, L, M, O, P, Q
↓	7	I, K, L, P, Q, R
	8	I, J, K, P, Q, R, S
	9	I, J, K, M, N, O, P, R, T, U, V, W
Low water	10	K, O, P, R, T, U, V, W, X, Y

Based on a Fig. 12.2 in Bennett & Humphries (1974).

Table 4.3 Key to codes used to identify organisms in Table 4.2

Code	Alga
A	*Diatoma* sp.
B	*Pelvetia canaliculata*
C	*Porphyra umbilicalis*
D	*Chaetomorpha*
E	*Fucus spiralis*
F	*Enteromorpha intestinalis*
G	*Plocamium coccineum*
H	*Ascophyllum nodosum*
I	*Ulva lactuca*
J	*Polysiphonia* sp.
K	*Corallina officinalis*
L	*Ectocarpus* sp.
M	*Ceramium* sp.
N	*Gracilaria* sp.
O	*Polyides* sp.
P	*Fucus serratus*
Q	*Laurencia* sp.
R	*Gigartina* sp.
S	*Lomentaria articulate*
T	*Chondrus crispus*
U	*Lithothamnion* sp.
V	*Rhodochorton* sp.
W	*Scytosiphon* sp.
X	*Laminaria digitata*
Y	*Rhodymenia palmate*

Q4.5.2 Which station experiences the shortest period of time exposed to the air?

Q4.5.3 Which two algal species appear to be the least tolerant to drying out, based on their location on the shore?

Q4.5.4 What percentage of all of the algal taxa on this shore occur at:

a. the high water mark (station 1);

b. the low water mark (station 10)?

Q4.5.5 Draw a bar chart showing the total number of taxa present at each station along the shore.

Q4.5.6 Identify the two stations which were occupied by the largest number of taxa and suggest a reason why this should be.

Q4.5.7 How does this type of zonation differ from a plant succession?

Reference/Further Reading

Bennett, D. P., & Humphries, D. A. (1974). *Introduction to field biology* (2nd ed.). London: Edward Arnold.

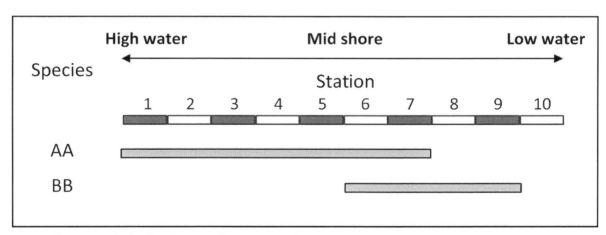

Figure 4.10 How to graph the distribution of species on a shore.

THE EFFECT OF SAMPLE SIZE ON POPULATION ESTIMATES OBTAINED USING QUADRAT SAMPLING

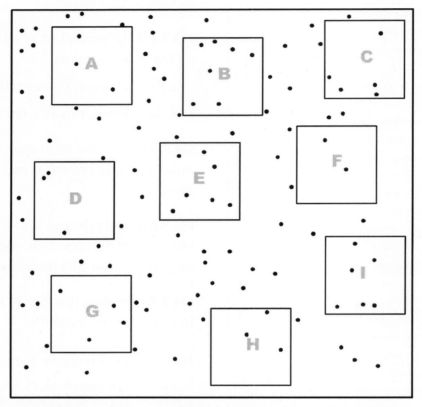

Figure 4.11 Nine quadrat samples taken from a hypothetical population of immobile organisms.

The size of a population of immobile organisms (e.g., plants or sessile animals such as limpets) may be estimated using quadrat sampling. A quadrat is simply a square. It may be a metal frame with sides of one metre or a larger area marked out on the ground. Regardless of the size of the quadrat the method of calculating the population estimate is the same. It involves counting the number of organisms per unit area (density) and then using this density to estimate the number in the total area under study. For example, if the density is 5 individuals per m² and the study area is 5 m × 2 m (i.e., 10 square metres) the estimated population size is 5 × 10 = 50 individuals.

Sampling using a portable quadrat (e.g., a metre square) should be done at random, for example, by throwing it backwards over the shoulder so that the sampler cannot control where it lands.

If only one quadrat is sampled in order to determine the density, the area sampled could, by chance, contain an unusually high density of organisms, an unusually low density of organisms, or no organisms at all. If more than one quadrat sample is taken a mean quadrat

Table 4.4 Population estimates calculated from quadrat samples

Sample	Number of plants in quadrat	Mean number of plants per quadrat	Population estimate
A	3	3	$3 \times 25 = 75$
B	7	$(3 + 7)/2 = 5$	$5 \times 25 = 125$
C	5	$(3 + 7 + 5)/3 = 5$	$5 \times 25 = 125$
D			
E			
F			
G			
H			
I			

density may be used to calculate the population estimate:

$$\text{Population estimate } N = \frac{\text{total study area}}{\text{quadrat area}} \times \text{mean number of individuals per quadrat}$$

The purpose of this exercise is to examine the effect of increasing the sample size – the number of quadrats sampled – on the population estimate.

Fig. 4.11 represents the distribution of a plant species. Nine $1\,\text{m} \times 1\,\text{m}$ quadrat samples (A–I) were taken in order to estimate the population size within the total area of $5\,\text{m} \times 5\,\text{m}$ ($25\,\text{m}^2$).

Q4.6.1 Construct a table similar to Table 4.4 and complete the missing cells using the information in Fig. 4.11.

Q4.6.2 Draw a line graph showing the effect of sample size on the population estimate.

Q4.6.3 The actual number of plants in the population is 112. Draw a horizontal line on your graph from 112 on the y-axis.

Q4.6.4 Construct a second table similar to Table 4.4. This time reverse the order in which the quadrats are sampled (I to A). Now, quadrat I is the first quadrat sampled; quadrat H is the second quadrat sampled; quadrat G is the third, and so on.

Q4.6.5 Draw a second graph similar to that produced in answer to Q4.6.2 making sure that you plot the newly assigned quadrat letter on the x-axis (e.g., I = 1st, H = 2nd, etc.).

Q4.6.6 Why do both lines end at the same value even though they have different shapes?

Q4.6.7 What can you conclude about the effect of sample size on population estimate?

Reference/Further Reading

Evans, C. D., Troyer, W. A., & Lensink, C. J. (1966). Aerial census of moose by quadrat sampling units. *The Journal of Wildlife Management, 30,* 767–776.

ESTIMATING THE SIZE OF A POPULATION OF MOBILE ANIMALS

Figure 4.12 Red squirrel (*Sciurus vulgaris*).

Animals that move quickly, or are rarely seen, are difficult to count. This problem may be overcome by the use of a mark-release-recapture method of counting. These methods require animals to be caught, marked and then released back into the population. When animals are captured on a subsequent occasion the proportion of previously marked animals (recaptures) may be used to estimate the population size. The simplest of these methods involves the use of the Lincoln Index:

$$N = \frac{n_1 \times n_2}{r}$$

Where,

N = the population estimate;

n_1 = the total number of animals caught on the first trapping occasion (and marked);

n_2 = the total number caught on the second trapping occasion;

r = the number caught on both occasions or recaptures (i.e., the number of marked animals in the second sample).

An ecologist trapped and marked red squirrels (*Sciurus vulgaris*) (Fig. 4.12) in an area of woodland in order to estimate their population size using the Lincoln Index. She caught 21 animals and attached a small plastic collar to each animal as a mark before returning it to the woodland. One week later she returned to the woodland and trapped 15 squirrels, some of which were marked (as indicated with ' × ' in Table 4.5).

Q4.7.1 Calculate the population estimate using the Lincoln index.

Q4.7.2 If animals 3, 11 and 14 had lost their marks but had still been captured on the second trapping occasion, what would have been the population estimate?

Q4.7.3 If animals 3, 11 and 14 had lost their marks and animals 1, 4 and 13 had died before the second trapping occasion, what would have been the population estimate?

Q4.7.4 Calculate the population estimate if five unmarked animals had migrated into the population between the two trapping occasions and all of them had been caught on the second trapping occasion in addition to the 15 animals shown in Table 4.5 (and assuming none of these animals had lost its mark).

Reference/Further Reading

Henderson, P., & Southwood, T. R. E. (2016). *Ecological methods* (4th ed.). Chichester: Wiley.

Table 4.5 Squirrels captured during the second trapping

Squirrel number														
1	2	3	4	5	6	7	8	9	10	11	12	13	14	15
×		×	×			×	×	×		×		×	×	

ESTIMATING POPULATION SIZE BY REMOVAL TRAPPING

Figure 4.13 Snails are suitable subjects for removal trapping studies.

The size of some animal populations may be estimated by a technique called 'removal trapping'. This involves trapping animals in a population on a number of successive occasions (and removing them) and then using a simple graphical method to predict the point at which no further animals would be caught.

Table 4.6 shows the number of snails captured (and then temporarily removed) in a small woodland over an 8-day period (Fig. 4.13).

Q4.8.1 Complete column C in Table 4.6 by calculating the cumulative total of snails

Table 4.6 The numbers of snails removed from a hypothetical population by successive samples

Sample number (day)	Number captured in this sample	Cumulative total previously captured
(A)	(B)	(C)
1	38	0
2	40	38
3	33	78
4	31	
5	27	
6	30	
7	24	
8	22	

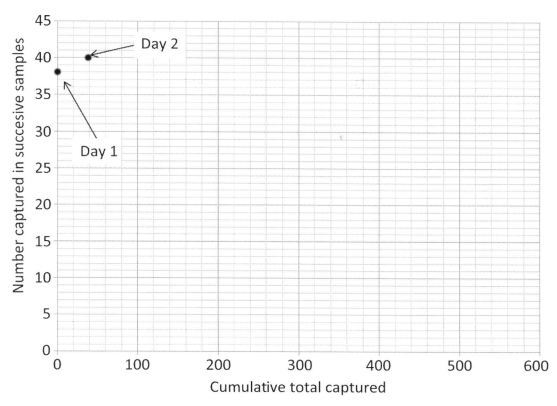

Figure 4.14 The estimation of population size by extrapolation from the numbers removed by successive samples (removal trapping).

captured and removed for each sample day. For example, the total previously captured on sample day 4 is the total captured on all previous days (1, 2 and 3) from column B, but not on day 4 itself.

Q4.8.2 Draw a scatter diagram by plotting the number captured on each sample day (B) on the *y*-axis against the cumulative total previously removed (C) on the *x*-axis. The *x*-axis will need to extend from zero to 600 and the *y*-axis from zero to 45 (Fig. 4.14).

Q4.8.3 Estimate the size of the population by constructing a line-of-best-fit through the points and extrapolating this until it intersects the *x*-axis. The point at which the line crosses the *x*-axis indicates the estimated population size.

Q4.8.4 Explain why is it important that the trapping effort (e.g., the number of traps used, time spent trapping each day, etc.) is kept constant.

Reference/Further Reading

Zippin, C. (1958). The removal method of population estimation. *Journal of Wildlife Management*, *22*, 82–90.

STUDYING ANIMAL POPULATIONS USING A CALENDAR OF CATCHES

A calendar of catches is a tabular representation of the membership of a population between two points in time that indicates which individuals are assumed to be present along with information about when animals enter and leave the population. It is constructed by repeatedly trapping identifiable animals (or recording their presence by direct observation) on a number of occasions through the study and assuming that they are present in the population between consecutive recordings. For example, if a population is trapped over a 6-week period and animal A is only recorded in week 1 and week 6 it is assumed to have been continuously present (Table 4.7).

Table 4.8 indicates the days on which animals were captured during a 10-week study.

Q4.9.1 Draw a calendar of catches for this population based on Table 4.8.

Table 4.7 A simple calendar of catches for three hypothetical animals (A, D and F).

Individual	Week					
	1	2	3	4	5	6
A ♂		•———	———	———	———	——•
D ♂			•———|———•———|———•		
F ♀	•———|———•———|———|———•———|———•			
New (+)	2	0	1	0	0	0
Disappeared (−)	0	0	0	0	0	1
Males	1	1	2	2	2	2
Females	1	1	1	1	1	0
Population size	2	2	3	3	3	2

Table 4.8 The days on which seven hypothetical animals were present in a population studied over a 10-week period

Individual	Weeks when captured
K ♂	1, 3, 6, 10
R ♀	1, 2, 3, 5, 8, 9, 10
L ♂	6, 7, 8
S ♀	1, 2, 5, 8, 10
M ♂	1, 4, 7
T ♀	1, 2, 3, 6, 7, 9, 10
U ♀	2, 4, 5, 6, 8, 10

Q4.9.2 Draw a single graph showing changes in:

a. the number of males;

b. the number of females;

c. the total population size,

from week one to week 10.

Q4.9.3 Suggest two reasons why new animals may appear in the population.

Q4.9.4 Suggest two reasons why animals may disappear from the population.

Q4.9.5 Suggest reasons why the pattern of capture of males appears to be different from that of females.

References/Further Reading

Delany, M. J. (1974). *The ecology of small mammals.* London: Edward Arnold.

Rees, P. A. (2015). *Studying captive animals: A workbook of methods in behaviour, welfare and ecology.* Chichester: Wiley-Blackwell.

POPULATION DYNAMICS OF A ZOO POPULATION OF CHIMPANZEES

The dynamics of a population of animals kept in a zoo may be studied using zoo records. The size of a zoo population may increase due to births or movements of animals into the group from elsewhere. It may decrease due to deaths or because animals are moved to other zoos. Table 4.9 is a summary of records for a hypothetical population of chimpanzees (*Pan troglodytes*) (Fig. 4.15). Assume that a census of the zoo's stock is made on 1st January each year and that each record relates to presence or absence on this date. So, for example, Jerry left the zoo in 2009 but was present on 1st January so is indicated as present in 2009. For simplicity, an animal that died or left in a particular year or was born or arrived in a particular year is recorded here as present for all of that year.

Q4.10.1 Construct a 'calendar of catches' for this population from 1996 to 2014 (showing every year). For males and females separately you should indicate the total number present, the number that arrived from another source and left to join another institution, the number that were born and the number that died. Use the following letters in your table to indicate the year that animals moved into or out of the population: A = arrived; L = left; B = born; D = died.

Q4.10.2 During which period was the highest number of chimpanzees present?

Q4.10.3 How many males were present in 2010?

Q4.10.4 How many females were present in 2012?

Q4.10.5 How could the table be produced to reflect more accurately the changes in the population?

Reference/Further Reading

Rees, P. A. (2015). *Studying captive animals: A workbook of methods in behaviour, welfare and ecology*. Chichester: Wiley-Blackwell.

Table 4.9 Changes in the population of a hypothetical population of chimpanzees living in a zoo

Name	Arrived	Left	Born	Died	Still present in 2014
Mitch ♂	1999				Yes
Jerry♂	1996	2009			No
Simon♂			2001		Yes
Mandy♀	1996			2003	No
Clara♀	1996				Yes
Jenny♀	1997			2011	No
Eve♀	2004				Yes
Tessa♀	1998				Yes

Figure 4.15 Chimpanzees (*Pan troglodytes*) at Chester Zoo, United Kingdom.

ESTIMATING POPULATION SIZE INDIRECTLY: BADGERS AND SETTS

The European badger (*Meles meles*) is a large mustelid with a body length (including tail) of some 90 cm (Fig. 4.16). It is native to most of Europe and parts of the Middle East. The following information about the biology of badgers has been largely abstracted from Rees (2002).

Distribution

Badgers are common and widespread throughout the British Isles. They prefer hilly country with a sandy soil, but also favour deciduous woodland interspersed with fertile grassland which supports high densities of earthworms. They are particularly common in the southwest of England, in the border counties between England and Wales and in the north and northwest of England. They are generally absent from very large conurbations, mountainous areas and lowlands which are wet or liable to flooding (Neal and Cheeseman, 1996).

Behaviour

Badgers are generally nocturnal but may occasionally be encountered above ground in daylight. They are usually found in wooded areas, but also occur in treeless localities, and are often seen in suburban gardens. Badgers build a sett in the slope of a hill, a sunken road, sandpits, quarries and other high banks into which they can burrow. Badgers are highly social animals and up to 15 may live in the same sett. They do not hibernate but in very cold weather they may remain underground for many days. Badgers eat insects, birds, small mammals, roots and other plant material. Their home ranges vary depending upon the habitat but may be up to 1500 ha (3700 acres) in moorland areas.

Badger Setts

Badger setts vary in size and complexity. A sett may have a single entrance leading into a large chamber. Some are just a few metres long and may consist of three or four tunnels with perhaps two entrances. Others are extremely large with a complex network of interconnecting tunnels and chambers, the entire structure sometimes occupying three storeys. One excavated sett was found to cover an area of 704 square metres,

Figure 4.16 A badger (*Meles meles*).

with 80 entrances, 354 m of tunnels and 20 chambers (Roper et al., 1991). Another covering a slightly larger area (740 square metres) but with a similar tunnel length (360 m) had only 38 entrances but a total of 78 chambers (Leeson and Mills, 1977).

Setts are not static structures but change considerably over time. Badgers continue to excavate new tunnels long after the size of the sett has exceeded their needs. They may end up like cities in which some parts are allowed to become derelict while new tunnels and chambers are excavated elsewhere in the complex. Sometimes old entrances that have not been used for some time (perhaps a year or so) may be cleared out and that section of the sett may then be renovated. Some setts are believed to be hundreds of years old and will have been occupied by many generations of badgers.

Signs

Badgers regularly use the same paths and runs. Where they pass through long grass there may be obvious areas of flattened vegetation. It is believed that some of these paths may have been used by badgers for centuries. Such paths often lead to drinking places or latrines that may contain mounds of droppings. The entrance to an occupied sett may show signs of activity such as freshly deposited soil and vegetation that have been removed from the living areas. Piles of ferns and grass (used as bedding) which have been removed from the sett by badgers may have a faint musty smell. Loose badger hairs (black, silver and white) may also be found in the vicinity of the sett on twigs and fences. Badgers may leave deep scour marks on wooden posts after

Table 4.10 Badgers and badgers setts identified in the Cheltenham area, England

Location	Setts	Badgers	Location	Setts	Badgers	Location	Setts	Badgers
1	1	2	18	1	4	35	1	2
2	2	12	19	1	4	36	4	10
3	1	4	20	1	6	37	1	4
4	1	2	21	1	4	38	3	15
5	1	6	22	6	10	39	4	8
6	1	15	23	1	2	40	1	3
7	3	18	24	6	10	41	1	4
8	1	2	25	2	4	42	4	8
9	1	10	26	6	15	43	1	4
10	1	15	27	1	15	44	5	8
11	1	2	28	1	4	45	1	6
12	1	2	29	3	12	46	1	6
13	3	6	30	3	10	47	1	3
14	4	6	31	3	8	48	1	2
15	3	10	32	1	4	49	1	2
16	2	6	33	2	8	50	2	9
17	2	8	34	8	14	51	2	11

Based on Humphries (1958).

sharpening and cleaning their claws. Near the sett badgers may remove the bark from the lower part of sycamore trees to obtain the sweet sap. They may also scrape shallow holes (snaffles) to get at roots. Droppings resemble those of dogs in shape and are always firm. Their colour, size and content vary but they often contain the undigested remains of insects such as beetles.

Breeding

Mating occurs from mid-July to about the end of August. Badgers exhibit delayed implantation so that the fertilised eggs do not implant until the end of the year. Young are born from the end of February to the end of April, usually 2–3 cubs, that first appear above ground at 6–8 weeks old. Exceptionally, delay in implantation may result in young being born as much as a year after mating. Such a delay may occur if a female is psychologically stressed, for example, by being captured.

Humphries (1958) collected data on the number of badgers and badger setts in the Cheltenham area of England (Table 4.10).

Q4.11.1 Calculate the total number of:

a. setts;

b. badgers.

Q4.11.2 What is the ratio of badgers to setts in this area?

Q4.11.3 A total of 76 setts were recorded in a locality with similar ecological characteristics to those of the Cheltenham area. Calculate the approximate size of the badger population.

Q4.11.4 Apart from direct observation of badgers entering and leaving a sett, how might you establish that it is in current use?

Q4.11.5 Why is the badger generally absent from areas liable to flooding?

Q4.11.6 What advantages does the capacity to delay implantation confer on a female badger?

Q4.11.7 Road deaths are common in badgers. What aspects of their biology make them susceptible to road accidents?

References/Further Reading

Humphries, D. A. (1958). Badgers in the Cheltenham area. *School Science Review, 139,* 416–425.

Leeson, R. C., & Mills, B. M. C. (1977). *Survey of excavated badger setts in the county of Avon.* London: Agricultural Development and Advisory Service, Ministry of Agriculture, Fisheries and Food.

Neal, E., & Cheeseman, C. (1996). *Badgers.* London: T & A.D. Poysner Ltd..

Rees, P. A. (2002). *Urban environments and wildlife law.* Oxford: Blackwell Science Ltd.

Roper, T. J., Tait, A. I., & Christian, S. (1991). Internal structure and contents of three badger (*Meles meles*) setts. *Journal of Zoology, 225,* 115–124.

ANALYSIS OF SPATIAL DISTRIBUTIONS: CLUMPED, UNIFORM OR RANDOM?

The spatial distribution of the individuals in a population (e.g., plants or slow moving animals such as snails) may be examined using data collected from quadrat samples. This is useful when looking for differences between various species or for changes in the distribution of a single species with time.

Dispersion may be measured by calculating an index of dispersion (I) using the formula:

$$I = \frac{variance}{mean}$$

There are three basic types of distribution:

(i)	Random	Variance = mean	$I = 1$
(ii)	Uniform (regular)	Variance < mean	$I < 1$
(iii)	Clumped (contiguous)	Variance > mean	$I > 1$.

The data in Table 4.11 show the number of individuals recorded in 10 randomly located quadrats used to sample three different plant species.

Q4.12.1 For each species calculate:

a. the mean number of individuals recorded per quadrat;

b. the variance.

Table 4.11 The number of individual plants from each of three hypothetical species (A, B and C) recorded in 10 quadrats

Quadrat	Species		
	A	B	C
1	18	21	5
2	16	3	10
3	17	0	6
4	16	16	14
5	18	18	8
6	17	2	9
7	15	1	6
8	16	15	11
9	15	19	9
10	16	2	13

Q4.12.2 Calculate the index of dispersion (I) for each species and determine whether it has a random, uniform or clumped distribution.

Q4.12.3 How might the size of the quadrat used affect the calculation of the index of dispersion?

Reference/Further Reading

Henderson, P., & Southwood, T. R. E. (2016). *Ecological Methods* (4th ed.). Chichester: Wiley.

ESTIMATING THE SIZE OF A LARGE MAMMAL POPULATION BY TRANSECT SAMPLING

The size of a population of large animals (e.g., deer) confined within a discrete area may be determined by estimating the density of the animals using transects. A transect is a line of known length (L) along which an observer walks (or drives or flies) while counting animals located within a fixed distance (w) either side of the line. If the total area is 10,000 m^2 and the density is 4 animals per 1000 m^2 the total population size is calculated as (10,000/1000) × 4 = 40 animals. In most circumstances it would be sensible to make several transect counts and calculate the mean number of animals per transect in order to make an estimate of the population. If the population estimate is made using sample counts from several transects the formula that should be used is:

$$\text{Population estimate } N = \frac{\text{total study area}}{\text{transect area}} \times \frac{\text{mean number of individuals per transect}}{}$$

Q4.13.1 The distribution of a herd of deer is shown in Fig. 4.17. Using the values below for the dimensions shown in the figure, estimate the population size of the deer using the number of animals per transect calculated from:

a. the count from transect A only;

b. the mean count from transects A and B;

c. the mean count from all three transects A, B and C.

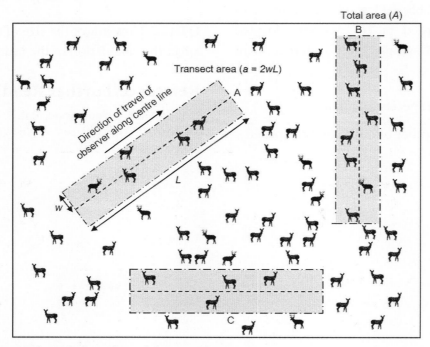

Figure 4.17 Transect sampling of a hypothetical deer population.

Total Area $(A) = 1{,}600{,}000\,\text{m}^2$ (160 ha).

$w = 75\,\text{m}$

$L = 700\,\text{m}$

Area of transect $= 2wL$

Q4.13.2 Count all of the animals in Fig. 4.17 and comment on the accuracy of the three population estimates you have obtained.

Q4.13.3 Suggest two reasons why some animals within the transects may be overlooked in a field study of this type.

Q4.13.4 Suggest some advantages and disadvantages of conducting a transect count from a vehicle confined to travelling on roads (Fig. 4.18).

Reference/Further Reading

Burnham, K. P., Anderson, D. R., & Laake, J. L. (1980). Estimation of density from line transect sampling of biological populations. *Wildlife Monographs, 72,* 3–202.

Figure 4.18 Staff and students from the University of Salford and Manchester Metropolitan University conducting a large mammal census from a Mercedes Unimog 4 × 4 truck in Tanzania.

SAMPLING ZOOPLANKTON POPULATIONS

The accuracy of population estimates is affected by the sampling method used. Zooplankton are often sampled with a plankton net dragged behind a boat, using a flow meter to measure the volume of water sampled. Table 4.12 shows estimates of the numbers of different types of zooplankton obtained using plankton nets with a range of mesh sizes.

Q4.14.1 If 5 cubic metres of water ($5\,m^3$) pass through a plankton net dragged behind a boat, and 750 copepods are collected, what is the concentration of copepods per cubic metre?

Q4.14.2 Which mesh size gives the most accurate information about the population densities of the copepods and the chaetognaths? Explain why.

Q4.14.3 Calculate the density of copepods obtained with the largest mesh size as a percentage of that obtained with the smallest mesh size.

Q4.14.4 Suggest a reason why mesh size does not appear to affect greatly the estimates of siphonophores.

Q4.14.5 Draw a graph showing the relationship between mesh size and the number of chaetognaths captured per m^3.

Q4.14.6 Some zooplankton, for example, fish larvae and chaetognaths, may only be a fraction of a millimetre in diameter but several millimetres long. Why does this cause a problem when sampling with nets?

Q4.14.7 What problems are likely to occur using a plankton net with a very fine mesh when zooplankton are very abundant?

Q4.14.8 How is this problem likely to affect population estimates?

Reference/Further Reading

Omori, M., & Hamner, W. M. (1982). Patchy distribution of zooplankton: Behaviour, population assessment and sampling problems. *Marine Biology*, *72*, 193–200.

Table 4.12 The number of organisms per m^3 of water, calculated from samples collected using nets of a range of mesh sizes

Taxon	Mesh aperture size (μm)				
	1000	650	300	200	70
Copepods	7	20	450	780	2500
Chaetognaths	4	20	35	90	700
Siphonophores	15	20	20	20	10

Exercise 4.15

Estimating the Size of Whale Populations

The grey whale (*Eschrichtius robustus*) is a baleen whale. The species is split into two distinct populations: the eastern North Pacific (American) and the western North Pacific (Asian) population. These whales undertake long migrations along the coastlines between their summer feeding grounds and their winter breeding grounds. Table 4.13 shows the grey whale population estimates for the eastern North Pacific Ocean.

Q4.15.1 Whale populations are surveyed by counting individuals from a ship. How is the presence of a whale detected (Fig. 4.19 (A and B))?

Q4.15.2 Scientists engaged in whale surveys record the distance to the whale and the bearing (angle) from the ship to the place where the whale was seen. What does this allow the scientists to calculate?

Q4.15.3 A marine biologist sights a minke whale in the north Atlantic Ocean west of Iceland. The UTM position of the boat at the time is 26V 507693mE (east) 7168942mN (north) (Fig. 4.20) and the whale is on a bearing

Figure 4.19 Minke whale (*Balaenoptera acutorostrata*): (A) diving after returning to the surface; (B) water spout (low and barely visible in this species).

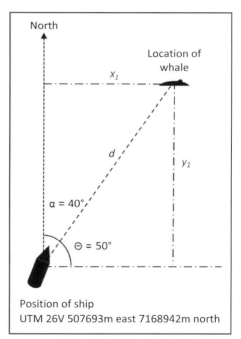

Figure 4.20 Calculating the UTM position of an animal from an observation point. (d = distance from observer to whale; angle α = compass bearing to whale, measured from north; angle $\theta = 90° - \alpha$).

of 40° at a distance of 500metres. The UTM (Universal Transverse Mercator) coordinate system divides the Earth into blocks and each block is divided up by lines one metre apart running north–south and east–west. The 'easting' position is the distance to the east (equivalent to the x-axis on a graph); the 'northing' position is the distance to the north (equivalent to the y-axis on a graph). This is similar to the manner in which traditional grids work on a map.

Calculate the UTM position of the whale as follows:

The easting position of the whale = Easting position of boat + x_1

where,

$$x_1 = d \times \sin \alpha$$

The northing position of the whale = Northing position of boat + y_1

where,

$$y_1 = d \times \sin \theta$$

Note: The distance x_1 is calculated as follows:

$$\sin \alpha = \frac{opposite}{hypotenuse}$$

so, $\sin 40° = \dfrac{x_1}{d}$

Table 4.13 Estimates of grey whale (*Eschrichtius robustus*) populations in the eastern North Pacific Ocean

Year	'Best' estimate	Approximate 95% confidence interval
1997/98	21,000	18,000–24,000
2000/01	16,500	14,000–18,000
2001/02	16,000	14,000–18,000
2006/07	19,000	17,000–22,000

Adapted from IWC (2016).

Rearranged, this becomes:

$$x_1 = d \times \sin 40°$$

Q4.15.4 How would you expect weather conditions to affect surveys?

Q4.15.5 What is meant by the term '95% confidence interval' in relation to the population estimates in Table 4.13?

Q4.15.6 Is there any clear evidence that the grey whale population of the eastern North Pacific Ocean has changed in size between 1997/98 and 2006/07? (Table 4.13). Explain your answer.

References/Further Reading

Barlow, J., & Taylor, B. L. (2005). Estimates of sperm whale abundance in the northeastern temperate Pacific from a combined acoustic and visual survey. *Marine Mammal Science, 21,* 429–445.

Hammond, P. S. (1990). Capturing whales on film–estimating cetacean population parameters from individual recognition data. *Mammal Review, 20,* 17–22.

IWC (2016). International Whaling Commission. Whale Population Estimates. <https://iwc.int/estimate#table>. Accessed 22.11.16.

ESTIMATING POPULATION SIZE USING THE LINCOLN INDEX: A SIMULATION

The Lincoln Index is a simple statistical method used to estimate the size of a population of mobile organisms, generally animals. The method requires that a sample of the population is trapped, marked and returned to the population. The ratio of marked to unmarked animals is now fixed (assuming that no births, deaths, immigration or emigration occur). If a second sample is taken from the population some individuals will be marked and some unmarked (assuming that the sample is random and sufficiently large). The ratio of marked to unmarked animals in the sample should be the same as the ratio in the population as a whole. This fact makes it possible to calculate the population size using the Lincoln Index.

Imagine a theoretical population of 20 animals, i.e., $N = 20$ (Fig. 4.21). Eight animals are captured on the first trapping occasion, marked and returned to the population (n_1). Now, 8 out of 20 individuals are marked, but in a field study, we would not know this; we would know only that 8 individuals were marked. On the

second trapping occasion we catch 10 animals (n_2). Since 40% of the population is marked (i.e., 8/20), if we capture 10 animals 40% should carry marks (i.e., 4 of the 10). The number of marked animals in the second sample is r. So, $n_1 = 8$, $n_2 = 10$ and $r = 4$. Using the Lincoln Index our population estimate is calculated as:

$$N = \frac{n_1 \times n_2}{r}$$

$$N = \frac{8 \times 10}{4} = 20$$

This exercise uses beads to simulate this method and is useful for examining the effect of various factors on the accuracy of the estimates obtained.

Apparatus

200 white beads
100 red beads
Large brown envelope

Method

1. Place 200 white beads in the envelope. The beads represent unmarked animals and the envelope represents the environment.

2. Remove 20 beads. These are captured animals.

3. Replace these 20 white beads with 20 red beads. The animals are now marked. Instead of painting a red mark on each of the captured white beads we have replaced them with red beads. This is the first trapping occasion.

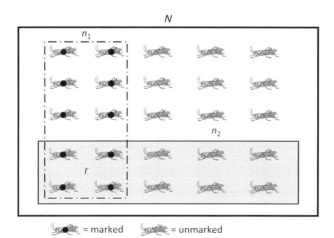

= marked   = unmarked

Figure 4.21 The statistical basis of the Lincoln Index.

The total number of beads is still 200, but now consists of 180 white beads (unmarked animals) and 20 red beads (marked animals).

4. Mix the beads so that the red beads are randomly distributed in the envelope.

5. Remove 40 beads from the tray at random without looking as you select them. Remove all 40 at the same time without replacing any. This is the second trapping occasion.

6. Count up the number of beads that are red (marked) and the number of white beads (unmarked) and record the values.

7. Calculate the population estimate as follows:

$$N = \frac{n_1 \times n_2}{r}$$

where,

n_1 = Number of individuals marked on the first trapping occasion (red);

n_2 = Number of animals trapped on the second occasion (red and white);

r = Number of recaptures (i.e., animals marked on the first occasion and also caught on the second occasion i.e., the red beads in the second sample).

8. Repeat the exercise using a different number of marked animal, e.g., 30, 40 ,50, 60, 70, etc. In other words use 30 red and 170 white, 40 red and 160 white, and so on. Note that the total number of beads is always 200.

Results

Draw a graph showing the effect of changing the number of marked animals in the population on the population estimate.

If you have sufficient time, it is best to take an average estimate for each number marked, e.g., perform the experiment say 4 times with 20 marked before increasing this to 30.

Discussion

Q4.16.1 Describe the effect of increasing the proportion of animals marked on the population estimate.

Q4.16.2 Suggest an optimal number that should be marked in order to obtain an accurate estimate of the population size.

Additional experiments

1. Calculate the population size when 20 of the 200 beads are red. Your second sample should be 40 beads taken at random. Now simulate a large number of births occurring between the first and second trapping occasions by replacing all of the beads in the envelope (20 red, 180 white) and then adding another 40 white beads (unmarked animals born between the two trapping occasions). You now have a population of 240 beads but only 20 are marked (red). Make another estimate by taking 40 beads at random from the envelope.

Q4.16.3 How do births between trapping occasions affect the estimate?

2. Calculate the population size when 20 of the 200 beads are red. Your second sample should be 40 beads taken at random. Now simulate the loss of marked animals due to deaths. To do this, remove 10 of the red beads from the envelope and discard them. Now take a second sample of 40 beads at random and calculate a population estimate.

Q4.16.4 How does the loss of marks affect the estimate?

Reference/Further Reading

Henderson, P., & Southwood, T. R. E. (2016). *Ecological methods* (4th ed.). Chichester: Wiley.

POPULATION SIZE AND HABITAT SELECTION IN TWO SPECIES OF GULLS

Figure 4.22 Counting gulls on Walney Island, Cumbria, UK. Inset: Left – Lesser black-backed gull (*Larus fuscus*); right – herring gull (*L. argentatus*).

Two species of gulls occur at high densities on the sand dunes at South Walney Nature Reserve in Cumbria, England: the herring gull (*Larus argentatus*) and the lesser black-backed gull (*L. fuscus*) (Fig. 4.22). Both populations were sampled in March 1987, using temporary quadrats with sides of 20 metres, marked out using four surveying poles as the corners. Twenty-two quadrats were located at random within the reserve: 11 on dunes and 11 in the dune slacks (low-lying depressions between the dunes). After leaving the birds to settle, observers moved back to the quadrats and counted the number of each species present from a distance using binoculars.

The total area of the reserve utilised by the gulls was 1,096,300 m². The area sampled by a single quadrat was 400 m² (i.e., 20 m × 20 m). For each species a population estimate may be calculated as follows:

$$N = mean\,density\,per\,quadrat \times \frac{1,096,300}{400}$$

or, alternatively,

$$N = median\,density\,per\,quadrat \times \frac{1,096,300}{400}$$

Fig. 4.23 is a stylised map of the areas sampled by the study, although the quadrats show the actual counts of birds made within the 22

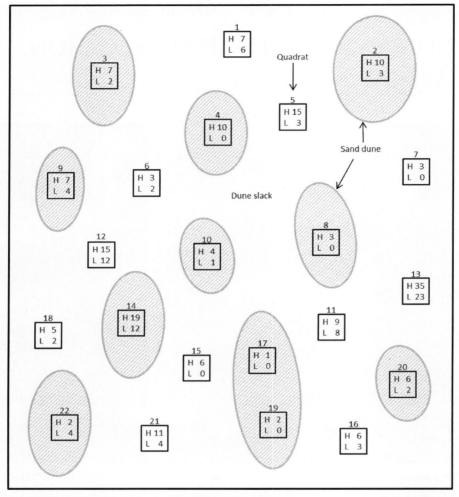

Figure 4.23 A stylised map of the study area showing data collected by the author and students at Bolton College (United Kingdom) in March 1987. Each square represents a 20m x 20m quadrat. The hatched areas are sand dunes.

quadrats. In each quadrat, the number of herring gulls counted is indicated by the letter 'H' and the number of lesser black-backed gulls by the letter 'L'. For example, quadrat 1 was located in a dune slack and contained 7 herring gulls and 6 lesser black-backed gulls.

Q4.17.1 Construct a table similar to Table 4.14 using the data from Fig. 4.23.

Q4.17.2 Calculate:

a. the mean density per quadrat and standard deviation for herring gulls:

   i. in dune slacks;

   ii. in sand dunes.

Table 4.14 The numbers of each gull species counted from each quadrat on sand dunes and in dune slacks.

Quadrat number	Number of gulls	
	Herring	Lesser black-backed
Dune slack samples		
1	7	6
5	15	3
etc…	etc…	etc…
Total from dune slacks		
Dune samples		
2	10	3
etc…	etc…	etc…
Total from dunes		
Total from both habitats		

b. the mean density per quadrat and standard deviation for lesser black-backed gulls:

i. in dune slacks;

ii. in sand dunes.

Q4.17.3 Calculate the mean and median densities per quadrat for each species, regardless of habitat (i.e., for all 22 quadrat samples).

Q4.17.4 Estimate the size of the:

a. herring gull population;

b. lesser black-backed gull population;

c. total gull population (by adding together estimates from a. and b. above), using:

i. the mean density;

ii. the median density.

This is a total of six population estimates in all.

Q4.17.5 Which species appeared to have the highest density, and in which habitat?

Q4.17.6 Which species appeared to have the lowest density, and in which habitat?

Q4.17.7 In which species and in which habitat was the greatest amount of variation in density observed?

Q4.17.8 What is the advantage of using the median density for calculating estimates from these data, rather than the mean density?

Q4.17.9 Which particular quadrat would make the use of mean densities to calculate population estimates inadvisable, and why?

Table 4.15 A chi-squared 2 × 2 contingency table

		Gull species		
		Herring gull	Lesser black-backed gull	Row totals
Habitat	Dune	a	b	$a + b$
	Dune slack	c	d	$c + d$
	Column totals	$a + c$	$b + d$	$a + b + c + d = n$

Q4.17.10 It is possible to test whether or not there is any association between the two gull species and the two habitats they occupy by using a 2 × 2 contingency table (see Chapter 10: Statistics).

a. Copy out and complete Table 4.15. The letters a, b, c, and d represent the total number of each species in each habitat.

b. State the null hypothesis to be tested.

c. Calculate the value of χ^2 using the formula provided in Chapter 10, Statistics, and determine whether or not there is a statistically significant association between habitat type and species.

Reference/Further Reading

Hosey, G. R., & Goodridge, F. (1980). Establishment of territories in two species of gull on Walney Island, Cumbria. *Bird Study*, *27*, 73–80.

EXERCISE 4.18

THE FACTORS AFFECTING PLANT POPULATION ESTIMATES OBTAINED BY QUADRAT SAMPLING: A COMPUTER SIMULATION

The size of a population of plants confined within known boundaries may be estimated by quadrat sampling. If a quadrat is thrown randomly onto the ground a number of times and the number of plants located within it counted, the population size may be estimated as follows:

$$N = \frac{\text{area occupied by population}}{\text{quadrat area}} \times \text{mean number of organisms per quadrat}$$

For this method to be useful it is important to establish the minimum number of quadrat samples that needs to be taken to obtain an accurate population estimate. If too few quadrats are counted the population may be inadequately sampled. Alternatively, if a large number of quadrats is thrown but only produces the same estimate as may be obtained with fewer quadrats, unnecessary effort is being expended.

A computer simulation was used to investigate the factors affecting the size of population estimates obtained by quadrat sampling. A population of 100 randomly located points – representing plants – was established with an area of 150 × 150 units (i.e., 22,500 units^2) representing a field (Fig. 4.24). Table 4.16 contains data collected using 20 quadrats of six different sizes (5, 10, 20, 30, 40 and 50 units).

− = No estimate possible because mean density = 0.

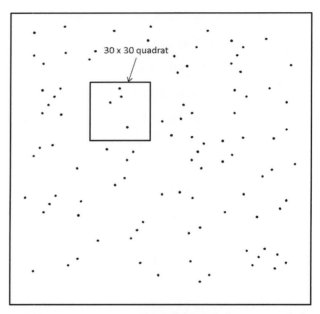

Figure 4.24 A representation of the population sampled by the computer simulation using quadrats. The population is located in an area of size 150 × 150 units.

x = Number of plants counted in quadrat.

N = Estimate of population size calculated using the mean density calculated from all quadrats examined thus far.

When a quadrat of side 5 units was used only one plant was recorded from a total of 20 quadrats. The first quadrat 'thrown' with sides of 10 units contained just one plant. This produced an estimate as follows:

$$N = \frac{22,500}{100} \times 1 = 255$$

Table 4.16 Numbers of plants counted in each of 20 quadrats and population estimates calculated from quadrats of a range of sizes

Quadrat number	Quadrat size											
	5		10		20		30		40		50	
	x	N	x	N	x	N	x	N	x	N	x	N
1	0	–	1	225	0	–	3	76	4	57	11	99
2	0	–	0	113	3	85	6	113	6	71	11	99
3	0	–	1	150	1	75	5	117	6	75	10	96
4	0	–	1	169	4	113	2	101	13	102	9	93
5	0	–	1	180	1	102	1	86	14	121	8	89
6	0	–	0	150	2	104	7	101	14	134	11	90
7	0	–	0	129	0	89	4	101	6	127	9	89
8	0	–	0	113	2	92	4	101	10	129	12	92
9	0	–	0	100	2	94	2	95	6	124	9	90
10	0	–	1	113	0	85	4	96	12	128	13	93
11	0	–	1	123	2	87	3	94	5	123	10	93
12	0	–	0	113	4	99	6	98	5	119	9	92
13	0	–	0	104	2	100	6	102	10	121	11	93
14	1	6	1	113	2	101	4	102	9	121	11	93
15	0	60	0	105	2	102	3	101	4	117	13	95
16	0	57	0	99	1	99	7	105	7	116	14	97
17	0	53	1	106	1	96	4	105	4	112	8	95
18	0	50	0	100	1	94	1	101	5	110	13	96
19	0	48	1	107	2	95	3	99	12	113	4	93
20	0	45	0	102	2	96	4	99	11	115	9	93

When the second quadrat was thrown no plants were recorded so the population estimate is:

$$N = \frac{22{,}500}{100} \times \frac{1+0}{2} = 113$$

The third quadrat recorded a single plant so the population estimate is:

$$N = \frac{22{,}500}{100} \times \frac{1+0+1}{3} = 150$$

and so on.

Q4.18.1 Examine the data in Table 4.16 and explain why there are so few population estimates calculated when the quadrat had sides of 5 units, and why they are all too low.

Q4.18.2 Why did the number of quadrats sampled have little effect on the population estimate when the quadrat had sides of 50 units?

Q4.18.3 Draw a graph of the effect of the number of quadrats sampled on the population estimate obtained when the quadrat has sides of:

a. 10 units;

b. 30 units.

Q4.18.4 What is the minimum number of quadrats necessary in order to obtain an accurate estimate of the population size (i.e., within plus or minus 1% of the actual size) when the quadrat has sides of:

a. 10 units;

b. 30 units?

Q4.18.5 What can be concluded about the relationship between quadrat size, the number of quadrats sampled and the accuracy of the estimates obtained?

References/Further Reading

Greig-Smith, P. (1952). The use of random and contiguous quadrats in the study of the structure of plant communities. *Annals of Botany New Series, 16*, 293–316.

Henderson, P., & Southwood, T. R. E. (2016). *Ecological methods* (4th ed). Chichester: Wiley.

CHAPTER 5

POPULATION GROWTH

This chapter contains exercises concerned with types of population growth, life tables and survivorship.

Part of a large plains zebra (*Equus quagga*) herd.

Examining Ecology. DOI: http://dx.doi.org/10.1016/B978-0-12-809354-2.00005-1

INTENDED LEARNING OUTCOMES

On completion of this chapter you should be able to:

- Recognise different types of population growth: exponential, logistic and boom-and-bust.

- Distinguish between density-dependent and density-independent factors by examining their effects on population size.

- Use Leslie matrices to study population growth.

- Construct a life table.

- Distinguish between static and dynamic life tables.

- Draw and interpret a survivorship curve.

- Distinguish between type I, II and III survivorship curves and identify organisms which exhibit each of these types.

- Demonstrate the use of log scales in appropriate situations when graphing population data.

- Construct a graph illustrating the age structure of a population.

- Distinguish between *r*- and *K*-strategists.

INTRODUCTION

DEFINING POPULATIONS

A population is a group of interbreeding organisms of the same species that live in a particular geographical area at a particular time. For example, we could study the population of lions in the Serengeti in 2016. A population may be further divided into demes: local groups of interbreeding organisms. The size of any population may increase due to births or immigration or decrease due to deaths or emigration. The population at any point in time is determined by these processes (Fig. 5.1).

Populations have characteristics that individuals cannot have. They have a density, a sex ratio, an age structure, a birth rate, a death

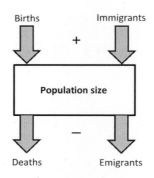

Figure 5.1 Factors affecting population size.

rate, an immigration rate, an emigration rate and a pattern of dispersion.

POPULATION GROWTH

Population growth has been studied in the field, in the laboratory using model systems such as microorganisms or insects, and modelled using mathematics (Rockwood, 2015).

In conditions where the environment is not limiting some populations are able to grow exponentially. Such growth occurs in laboratory studies of microorganisms and in the field when a species has been introduced to a new environment where it has no predators.

Where the environment is limiting a population that initially grew exponentially may level off so that its growth rate is zero. At this point it is said to have reached the carrying capacity of the environment. This type of growth is known as logistic growth.

A third type of growth is known as J-shaped or boom-and-bust growth. This may occur when a species invades a new environment and exhibits such rapid growth that the population suddenly crashes due to a sudden depletion of a resource, especially food, e.g., large mammalian herbivores introduced into an area where predators are absent may expand to the point where they overgraze the habitat and consequently die of starvation.

Stability in relation to population size may be more subtle than simply meaning a population whose numbers stay approximately the same. Some populations oscillate so that, although their numbers exhibit a cyclic increase and decrease, they never grow in an uncontrolled manner or become so low that they become extinct.

LIFE TABLES AND SURVIVORSHIP CURVES

A life table is a table of data showing the mortality rates of different age classes within a population of organisms (Pearl and Miner, 1935). Life tables are usually constructed separately for males and females. Static life tables are constructed by counting the number of animals in each age class present in a population at a single point in time. Dynamic life tables follow the survival of a cohort of animals born at the same time (e.g., in the same year). Most life tables are static because it is difficult to construct

dynamic life tables for long lived animals since they cannot be completed until all of the animals in the cohort have died. For some species this would take many decades. Static life tables suffer from the disadvantage that they assume environmental conditions for all of the age classes have remained the same throughout their life and that birth rates are stable from one year to the next.

The information in a life table may be used to produce a survivorship curve: a curve showing the pattern of mortality within a population with age, usually corrected so that the starting population is 1000. Graphs showing survivorship curves often use a logarithmic scale to indicate the number of survivors in each age class.

R- AND K-STRATEGISTS

Some species have evolved as r-strategists – they have a short generation time, are opportunistic, and produce large numbers of offspring – others as K-strategists – they have a long generation time, inhabit stable environments, and produce few young (MacArthur and Wilson, 1967). Many species fall somewhere between these two extremes.

The following exercises provide examples of different types of population growth and examine the methods used to study age structure, factors affecting population size, survivorship and population growth strategies.

References

MacArthur, R. H., & Wilson, E. O. (1967). *The theory of island biogeography*. Princeton, NJ: Princeton University Press.

Pearl, R., & Miner, J. (1935). Experimental studies on the duration of life. XIV. The comparative mortality of certain lower organisms. *The Quarterly Review of Biology, 10*, 60–79.

Rockwood, L. L. (2015). *Introduction to population ecology* (2nd ed). Chichester: Wiley-Blackwell.

EXPONENTIAL POPULATION GROWTH

Some species of bacteria are capable of reproducing every 20 minutes by dividing into two. Imagine a population which begins with a single organism and is allowed to grow in an unrestricted environment.

Q5.1.1 Complete Table 5.1 by doubling the number of bacteria every 20 minutes and by writing the $\log_{10}$ of this number in the last column.

Q5.1.2 Draw a graph showing the growth of the population (N) from time $= 0$ to time $= 140$ minutes.

Q5.1.3 Draw a graph of $\log_{10}$ of the number of bacteria over the same period of time.

Q5.1.4 What is the effect of plotting exponential growth data using a logarithmic scale on the y-axis on the shape of the graph?

Q5.1.5 What does this simple model of growth assume about death rates within the population?

Table 5.1 Growth of an imaginary population of bacteria

Time t (minutes)	Population size N (Number of bacteria)	$\log_{10}$ number of bacteria
0	1	0
20	2	0.301
40	4	0.602
60		
80		
100		
120		
140		

Q5.1.6 What would you expect to happen to the shape of this growth curve if some aspect of the environment started to have a limiting effect on growth, e.g., food became scarce?

Q5.1.7 Why do very few populations grow exponentially?

Reference/Further Reading

Monod, J. (1949). The growth of bacterial cultures. *Annual Review of Microbiology, 3*, 371–394.

EXERCISE 5.2

BOOM AND BUST POPULATION GROWTH

Figure 5.2 Reindeer (*Rangifer tarandus*).

Since 1891 reindeer (*Rangifer tarandus*) have been introduced into many parts of Alaska to replace the dwindling caribou herds which are essential to the economy of the Eskimo (Fig 5.2). In 1911, 25 reindeer were released on St. Paul Island, one of the Pribilof Islands in the Bering Sea. There were no predators on the island and little hunting pressure. The pattern of growth of this population is shown in Table 5.2.

Q5.2.1 Draw a graph showing changes in this population between 1911 and 1950.

Q5.2.2 Calculate the annual rate of increase in reindeer between 1926 and 1936 from the graph.

Q5.2.3 Explain why the population was able to grow so rapidly until 1938.

Table 5.2 Growth of an introduced reindeer population on St Paul Island

Year	Population size[a]
1911	25
1920	210
1925	250
1938	2000
1942	1160
1946	230
1950	8

Data from Scheffer (1951).
[a]Estimates only for some years.

Q5.2.4 Suggest a reason for the dramatic decline in the population between 1938 and 1950.

Q5.2.5 Why are such 'boom and bust' growth patterns relatively rare in nature?

Reference/Further Reading

Scheffer, V. B. (1951). The rise and fall of a reindeer herd. *The Scientific Monthly*, 73, 356–362.

LOGISTIC POPULATION GROWTH

Populations are unable to sustain exponential growth for very long because environments are not unlimited. Eventually growth rates are reduced by a shortage of food, space or some other factor. These factors constitute what is known as 'environmental resistance' and their effect upon growth may produce a logistic growth curve. The data in Table 5.3 illustrate this pattern of growth beginning with a population of 10 individuals.

Q5.3.1 Draw a graph showing the growth of this population with time. Extend the *y*-axis to 160.

Q5.3.2 Calculate the growth rate (individuals/ unit time) between time = 4 and time = 7.

Q5.3.3 What is the growth rate at time = 15?

Q5.3.4 The carrying capacity of the environment with respect to any population is the maximum number of individuals that can be supported. What is the carrying capacity for this population? Draw a horizontal line on the graph indicating this.

Q5.3.5 Assuming the original population of 10 organisms grew exponentially, doubling in size with every unit of time, calculate the population size from time = 0 to time = 4 and draw the position of this curve on your graph.

Q5.3.6 Shade the area between the two curves and label this area 'environmental resistance'.

Q5.3.7 Indicate with an arrow on the graph:

a. the region where there is least environmental resistance;

b. the region where there is most environmental resistance.

Reference/Further Reading

Tsoularis, A., & Wallace, J. (2002). Analysis of logistic growth models. *Mathematical Biosciences, 179*, 21–55.

Table 5.3 Logistic population growth in a hypothetical population

Time (arbitrary units)	0	1	2	3	4	5	6	7	8	9	10	11	12	13	14	15	16
Population size	10	15	21	29	39	51	64	75	84	91	95	97	99	99	100	100	100

DENSITY-DEPENDENT AND DENSITY-INDEPENDENT FACTORS AND POPULATION CONTROL

Figure 5.3 Cold weather: DD or DI?.

Population size may be affected by factors which act in a density-independent (DI) fashion on birth rate and death rate. Such factors may, for example, kill the same percentage of adult animals each year regardless of their density. Conversely, a density-dependent (DD) factor has a greater effect upon a population at high density than one at low density. For example, such a factor may be responsible for the death of 10% of adults in a low density population, but if the density doubled it might kill 30% of the population (Figs. 5.3 and 5.4).

Figure 5.4 European badger (*Meles meles*). Mortality due to road traffic accidents: DD or DI?

Table 5.4 The mortality pattern in two hypothetical species, A and B

Population density (m^{-2})	Species A		Species B	
	Population size	Mortality %	Population size	Mortality %
10	1	10	1	10
50	6	12	5	10
100	13		17	
200	25		45	
400	46		107	
800	83		263	

Table 5.5 Density-dependent and density-independent factors

Environmental variable	DD or DI?
Frost	
Predation	
Intraspecific competition	
Fire	
Parasitism	
Disease	
Road traffic	

Table 5.4 shows the mortality pattern with changes in density in two hypothetical animal species, A and B.

Q5.4.1 Complete Table 5.4 by calculating the percentage mortality at each density for both species.

Q5.4.2 Draw a graph of percentage mortality against density for both species.

Q5.4.3 Which species is exhibiting:

a. density-dependent mortality;

b. density-independent mortality?

Q5.4.4 Explain why environmental factors that affect mortality in a density-dependent fashion are more likely to act as mechanisms of population regulation than are those that act in a density-independent fashion.

Q5.4.5 Complete Table 5.5 indicating for each variable whether it is likely to act upon an animal population in a density-dependent (DD) or a density-independent (DI) fashion.

Reference/Further Reading

Coulson, T., Catchpole, E. A., Albon, S. D., Morgan, B. J. T., Pemberton, J. M., Clutton-Brock, T. H., … Grenfell, B. T. (2001). Age, sex, density, winter weather, and population crashes in soay sheep. *Science, 292,* 1528–1531.

EXERCISE 5.5

USING LESLIE MATRICES TO MODEL POPULATION GROWTH

Populations are not made up of identical individuals. Early studies of population growth took no account of the fact that populations generally contain individuals of two sexes and several age classes. These age classes exhibit differences in their survival (from one class to the next) and in their reproductive capacity (generally not being able to reproduce when either very young or very old). Although not perfect, the use of Leslie matrices overcomes some of these problems.

Traditionally, only the female age distribution and female life tables are used in population projections. Assuming a 1:1 sex ratio, the total population at any point in time is taken to be twice the female population.

The size of the population at time $= t + 1$ is calculated by multiplying the values in the Leslie matrix containing the fecundity (reproductive) rates and survival rates for each age class by the column vector (population vector) which contains the number of animals in each age class at time t (n_0, n_1, n_2 etc.):

$$\text{Leslie matrix} \times \text{population vector}$$

$$\begin{bmatrix} Number\ of \\ individuals \\ in\ each\ age\ class \\ at\ time\ t + 1 \end{bmatrix} =$$

$$\begin{bmatrix} Age\ specific \\ fecundity \\ and \\ survival\ rates \end{bmatrix} \times \begin{bmatrix} Number\ of \\ individuals \\ in\ each\ age\ class \\ at\ time\ t \end{bmatrix}$$

In mathematical notation, for a population with three age classes this is expressed as:

$$\begin{bmatrix} n_{0t+1} \\ n_{1t+1} \\ n_{2t+1} \end{bmatrix} = \begin{bmatrix} f_0 & f_1 & f_2 \\ p_0 & 0 & 0 \\ 0 & p_1 & 0 \end{bmatrix} \times \begin{bmatrix} n_{0t} \\ n_{1t} \\ n_{2t} \end{bmatrix}$$

$$= \begin{bmatrix} (f_0 \times n_{0t}) + (f_1 \times n_{1t}) + (f_2 \times n_{2t}) \\ (p_0 \times n_{0t}) + (0 \times n_{1t}) + (0 \times n_{2t}) \\ (0 \times n_{0t}) + (p_1 \times n_{1t}) + (0 \times n_{2t}) \end{bmatrix}$$

Imagine a population of animals in which there are three age classes; 1-, 2- and 3-years-old (females only). In this population the animals can reproduce as follows (Table 5.6):

Table 5.6 Age-specific reproductive rates

Age (years)	Offspring produced per year
1	0
2	8
3	10

The chance of an individual surviving from one age class to the next is (Table 5.7):

Table 5.7 Age-specific survival rates

From age class	To age class	Probability of survival
1	2	0.3
2	3	0.4
3	4	0

The relationship between the age classes in the population, the survival from one age class to another (p) and the number of offspring produced by each age class (f) are shown in Fig. 5.5.

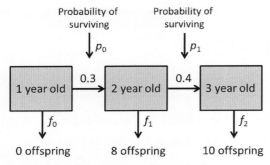

Figure 5.5 Growth parameters for a population consisting of three age classes.

The matrix below shows what will happen to the size and age structure of a population containing a single 3-year-old animal after one year.

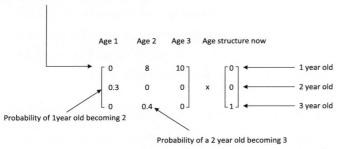

The age structure after one year is calculated as follows:

Number in age class x reproductive rate of age class (top row)

$$\begin{bmatrix} (0 \times 0) + (8 \times 0) + (10 \times 1) \\ (0.3 \times 0) + (0 \times 0) + (0 \times 1) \\ (0 \times 0) + (0.4 \times 0) + (0 \times 1) \end{bmatrix} = \begin{bmatrix} 10 \\ 0 \\ 0 \end{bmatrix} \begin{matrix} \text{Age 1} \\ \text{Age 2} \\ \text{Age 3} \end{matrix}$$

Number in age class x probability of survival to next age class (2nd and 3rd rows)

After one generation the population size has increase from one to 10 (females only) because the single 3-year-old individual has produced 10 offspring and then died (as 3 years is the maximum age attainable by this hypothetical species). The size of each age class in the next

generation may be calculated by multiplying the survival rates and reproductive rates for each age class by the new column vector:

$$\begin{bmatrix} 0 & 8 & 10 \\ 0.3 & 0 & 0 \\ 0 & 0.4 & 0 \end{bmatrix} \times \begin{bmatrix} 10 \\ 0 \\ 0 \end{bmatrix}$$

This process may be repeated indefinitely. In any generation the population size is the sum of all of the animals in each age class. This method may be extended to accommodate as many age classes as necessary.

The matrix below is for a population of 10 individuals: 10 1-year-old, no 2-year-old and no 3-year-old animals. The reproductive rates of 1-year-olds, 2-year-olds and 3-year-olds are 0, 6, and 10, respectively. The probability of a 1-year-old surviving to become a 2-year-old is 0.27 and the probability of a 2-year-old becoming a 3-year-old is 0.53.

$$\begin{bmatrix} 0 & 6 & 10 \\ 0.27 & 0 & 0 \\ 0 & 0.53 & 0 \end{bmatrix} \times \begin{bmatrix} 10 \\ 0 \\ 0 \end{bmatrix}$$

Q5.5.1 Calculate the size and age structure of this population for eight generations and enter the data into the table below (Table 5.8). Generation 1 has been completed for you.

Q5.5.2 Draw a line graph with four lines representing changes with time in:

a. age class 1;

b. age class 2;

c. age class 3;

d. the population size (all three age classes aged together).

Table 5.8 Growth of a population over eight generations

Generation	Number in each age class			Population size
	Age 1	Age 2	Age 3	
0	10	0	0	10
1				
2				
3				
4				
5				
6				
7				
8				

Q5.5.3 Comment on the pattern of growth shown by each line in your graph.

Note. You may find it useful to create a small spreadsheet to do these calculations.

Reference/Further Reading

Smith, G. C., & Trout, R. C. (1994). Using Leslie matrices to determine wild rabbit population growth and the potential for control. *Journal of Applied Ecology, 313,* 223–230.

LIFE TABLE FOR THE HONEY BEE

Population ecologists use life tables and survivorship curves to study the relationship between death rates and age in animals. Two Japanese biologists, Sakagami and Fukuda (1968) studied survivorship in the honey bee (*Apis* sp.). Some of their data are summarised in Table 5.9.

Table 5.9 A life table for the honey bee (*Apis* sp.)

Age (days) (x)	Survivors at start of age class x (l_x)	Deaths between age x and x + 1 (d_x)	Age specific death rate (q_x)
0–5	1000	1000 − 983 = 17	17/1000 = 0.017
5–10	983	983 − 960 = 23	23/983 = 0.023
10–15	960	960 − 943 = 17	17/960 = 0.018
15–20	943		
20–25	910		
25–30	727		
30–35	455		
35–40	139		
40–45	46		
45–50	6	6	1.00

Adapted from Sakagami & Fukuda (1968).

Q5.6.1 Complete the columns for d_x and q_x in Table 5.9 using the equations below:

a. $d_x = l_x - l_{x-1}$. For example, for age class 0–5 days, 17 = 1000 − 983.

b. $q_x = d_x/l_x$. For example, for age class 0–5 days, 0.017 = 17/1000.

Q5.6.2 Which age class has:

a. the lowest death rate;

b. the highest death rate?

Q5.6.3 Draw a survivorship curve for the honey bee by plotting l_x against age (x).

References/Further Reading

Pearl, R., & Miner, J. (1935). Experimental studies on the duration of life. XIV. The comparative mortality of certain lower organisms. *The Quarterly Review of Biology, 10*, 60–79.

Sakagami, S. F., & Fukuda, H. (1968). Life tables for worker honey bees. *Researches on Population Ecology, 10*, 127–139.

EXERCISE 5.7

SURVIVORSHIP OF DALL SHEEP

Muire (1944) made an intensive study of wolves and Dall mountain sheep (*Ovis d. dalli*) in Mount McKinley National Park in Alaska over a period of several years. Dall sheep possess horns which remain preserved for a long period after the animal dies. The age that a sheep dies can be determined from the horns. Muire collected data on 608 sheep thus providing an impressive set of data from which the age at which these sheep die may be determined.

Table 5.10 Life table for the Dall mountain sheep

Age (years) x	Survivors at beginning of age class l_x	Deaths d_x	Mortality rate % q_x
0–0.5	1000	$1000 - 946 = 54$	$54/1000 \times 100 = 5.4\%$
0.5–1	946		
1–2	801		
2–3	789		
3–4	776		
4–5	764		
5–6	734		
6–7	688		
7–8	640		
8–9	571		
9–10	439		
10–11	252		
11–12	96		
12–13	6		
13–14	3		

Based on data from Murie (1944).

Q5.7.1 Compete the life table provided (Table 5.10) by calculating the number of survivors in each age class (l_x), the number of deaths in each age class (d_x) and the mortality rate for each age class (q_x). The data have been corrected so that the youngest age class contains 1000 animals.

Q5.7.2 Draw a survivorship curve of these data by plotting l_x against age (x).

Q5.7.3 Which age class has:

a. the highest mortality rate;

b. the lowest mortality rate?

Q5.7.4 Is your completed table a static life table or a dynamic life table?

Reference/Further Reading

Murie, A. (1944). Dall sheep. In *Wolves of Mount McKinley*. Washington: National Parks Service Fauna No. 5.

TYPES OF SURVIVORSHIP CURVE

Living things show a variety of mortality patterns. Some species exhibit high mortality when very young; others tend to survive well when young and die mostly in old age (Fig. 5.6). These differences inevitably give rise to different shaped survivorship curves. Ecologists divide these survivorship curves into three basic types referred to as types I, II and III (Table 5.10).

The data in Table 5.11 represent the three types of survivorship curve. All data is scaled so that 10 represents the maximum lifespan and each population begins with 1000 individuals.

Q5.8.1 Draw a line graph showing the number of individuals surviving at each age using the data in Table 5.11. Draw three lines: one for each type of survivorship (I, II and III).

Q5.8.2 Describe the pattern of survivorship in each of the three types (I, II and III).

Table 5.11 Life tables for the three basic types of survivorship curve (I, II and III)

Age of organism	Number surviving Type			Log₁₀ number surviving Type		
	I	II	III	I	II	III
0	1000	1000	1000	3.000	3.000	3.000
1	1000	700	20	3.000	2.845	1.301
2	1000	350	4	3.000	2.544	0.602
3	1000	160	2	3.000	2.204	0.301
4	1000	79	1	3.000	1.898	0.000
6	1000	30	1	3.000	1.477	0.000
7	1000	16	1	3.000	1.204	0.000
8	1000	8	1	3.000	0.903	0.000
9	900	3	1	2.954	0.477	0.000
10	1	1	1	0.000	0.000	0.000

Q5.8.3 Draw a second graph (again with three lines) showing the log₁₀ of the number surviving at each age.

Q5.8.4 Describe the effect of plotting log₁₀ data on the shape of the curves.

Q5.8.5 Which type of survivorship curve does each of the following organisms exhibit?

  a. humans;

  b. parasites;

  c. songbirds.

Q5.8.6 In which type of survivorship curve do the mean life expectancy and maximum age at death more-or-less coincide?

Reference/Further Reading

Demetrius, L. (1978). Adaptive value, entropy and survivorship curves. *Nature, 275,* 213–214.

Figure 5.6 European robin (*Erithacus rubecula*). Song birds commonly exhibit type II survivorship.

THE AGE STRUCTURE OF CAPTIVE FEMALE AFRICAN ELEPHANTS

Zoos claim that they keep elephants as an insurance against their becoming extinct in the wild. However, this assumes that the zoo community can maintain a viable, self-sustaining population in captivity. Female African elephants (*Loxodonta africana*) may reproduce from the age of approximately 12 years (sometimes earlier in zoos) and may remain fertile until they are 60 years old.

Q5.9.1 Draw a graph showing the age structure of the elephant population represented by the data in Table 5.12.

Q5.9.2 What is the age range of this population?

Q5.9.3 Assuming that female elephants are capable of reproduction from the age of 12 years and none of the animals in this population is too old to ovulate, what percentage of the population was theoretically capable of reproducing at the time the data were collected?

Q5.9.4 Suggest reasons why there are so few very young elephants in this population.

Q5.9.5 What did this data suggest about the long-term prospects of the African elephant population in zoos in 1999?

Reference/Further Reading

Olson, D., & Wiese, R. J. (2000). State of the North American African elephant population and projections for the future. *Zoo Biology, 19,* 311–320.

Table 5.12 Age (years) of female African elephants (*Loxodonta africana*) in North American zoos at the beginning of January 1999

29	22	16	25	19	28	16	12	15	15	16	14
28	16	28	16	21	16	25	20	16	17	29	20
29	22	16	27	26	33	16	19	21	16	20	45
44	19	28	16	16	24	19	22	24	14	13	15
27	15	19	29	11	12	13	12	20	32	29	28
29	20	16	13	14	13	29	24	22	20	16	39
35	16	25	33	15	40	25	28	31	19	24	23
16	16	17	13	20	16	16	39	29	19	24	31
20	17	40	16	43	37	30	35	27	29	28	18
16	22	20	20	29	15	28	21	19	16	17	22
16	16	18	17	16	13	16	29	23	40	16	38
17	20	18	16	16	16	27	16	35	8	37	27
19	13	15	16	20	22	15	16	15	16	13	
16	23	19	28	16	26	16	15	13	24	13	
27	27	18	11	17	18	20	29	30	16	28	
16	17	28	28	30	16	41	48	15	12	17	
16	24	12	19	29	20	20	20	22	20	18	
21	28	12	15	29	29	20	28	20	28	21	

Data derived from Olson & Wiese (2000).

EXERCISE 5.10

OPPORTUNITY OR EQUILIBRIUM: *R*-STRATEGIST OR *K*-STRATEGIST?

MacArthur and Wilson (1967) coined the terms *r*-selection and *K*-selection for two different types of selection found in nature. This has given rise to the labelling of some species as *r*-strategists and others as *K*-strategists (Fig. 5.7). Most species fall somewhere between these two extremes. The terminology used here is derived from that used to describe population growth using the logistic equation.

A single species population growing in an unlimited environment has a rate of increase which may be described as:

$$\frac{dN}{dt} = rN$$

where,

N = population size;
t = time;
r = innate capacity for increase.

Figure 5.7 *Triceratops*. A dinosaur that lived in the United States in the late Cretaceous. *K*- strategist or *r*-strategist?

In a finite environment growth may be described by the logistic equation where:

$$\frac{dN}{dt} = rN\frac{(K - N)}{K}$$

where,

K = the asymptote (environmental carrying capacity).

Populations may be divided into two types based on their growth characteristics: *r*-strategists, in which the innate capacity for increase (*r*) is the most important; and *K*-strategists, in which the point at which the population levels off (*K*) is of overriding importance. Characteristically, *r*-strategists tend to be small in size and seize opportunities to colonise new habitats and expand their numbers rapidly. In contrast, *K*-strategists tend to be large with relatively stable, slow-growing populations.

Q5.10.1 In a population which grows logistically:

a. what happens to the growth rate (*dN/dt*) as the population size (*N*) approaches the carrying capacity of the environment (*K*)?

b. what is the growth rate when the population size reaches the carrying capacity (i.e., *N* = *K*)?

Q5.10.2 Complete Table 5.13 by stating whether each organism is an *r*-strategist or a *K*-strategist.

Table 5.13 Examples of *r*- and *K*-strategists.

Organism	Strategy (*r* or *K*)
African elephant	
Parasitic insects	
Dinosaurs	
Grasses	
Andean condor	
Albatross	
Bacteria	

Table 5.14 Characteristic of *r*- and *K*-strategists

Characteristic	*r*- or *K*-strategist
Long generation time	
Inhabits stable environments	
Exhibits high levels of migration	
Low mortality rate	
Colonises temporary habitats	
Vulnerable to extinction	

Q5.10.3 Complete Table 5.14 by stating whether each characteristic is typical of an *r*-strategist or a *K*-strategist.

Q5.10.4 What type of strategy is being exhibited by the organism whose growth is illustrated in Fig. 5.8? Explain your answer.

References and Further Reading

Gadgil, M., & Solbrig, O. T. (1972). The concept of r- and K-selection: Evidence from wild flowers and some theoretical considerations. *The American Naturalist, 106*, 14–31.

MacArthur, R. H., & Wilson, E. O. (1967). *The theory of island biogeography*. Princeton, NJ: Princeton University Press.

Pianka, E. R. (1970). On r- and K-selection. *The American Naturalist, 104*, 592–597.

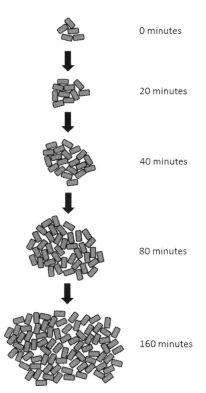

0 minutes

20 minutes

40 minutes

80 minutes

160 minutes

Figure 5.8 Growth of an imaginary organism. *K*-strategist or *r*-strategist?

CHAPTER 6

SPECIES INTERACTIONS

This chapter contains exercises concerned with interactions between and within species, including competition, predation, niche separation and biological control.

A lion (*Panthera leo*) moving a kill in the Serengeti National Park, Tanzania.

Examining Ecology. DOI: http://dx.doi.org/10.1016/B978-0-12-809354-2.00006-3

INTENDED LEARNING OUTCOMES

On completion of this chapter you should be able to:

- Distinguish between interspecific and intraspecific competition.

- Identify competitive exclusion in laboratory experiments.

- Describe the field methods that have been used to study competition and niche separation.

- Identify niche separation in closely related taxa from field data.

- Describe the historical development of the definition of the niche.

- Distinguish between the fundamental niche and the realised niche.

- Investigate the usefulness of the concept of n-dimensional hyperspace in defining the niche.

- Investigate the relationship between predator and prey numbers using data produced by a computer simulation.

- Identify factors which may affect the success of predators obtaining their prey.

- Explain the principles and benefits of biological control methods in protecting crops.

- Identify the role of disease in the regulation of wildlife populations.

- Determine the susceptibility of animal populations to future disease outbreaks from serology data.

INTRODUCTION

Animals and plants do not exist as single species populations but in biological communities within which they interact in complex ways. For example, individuals (and species) may compete with each other; some species feed on other species in interactions involving grazing, predation and parasitism; and some species cause diseases in others. Many of these interactions have been investigated using mathematical models and ecologists have subsequently attempted to test the predictions of these models in laboratory experiments and later in the field.

COMPETITION AND NICHE THEORY

Organisms may compete when they require similar resources and these resources are in short supply. There are two basic types of competition: that between different species (interspecific competition) and that between individuals of the same species (intraspecific competition). Early laboratory experiments on interspecific competition were conducted by the Russian biologist Georgii Gause using protozoa (Gause, 1934). These experiments were followed by field studies that examined the mechanisms used by species to coexist. Gause proposed the 'competitive exclusion principle' which states that complete competitors cannot coexist.

Each species in an ecosystem performs a specific role and has specific functions. It lives in a particular habitat, feeds on particular species and is food for other species. Ecologists call this role its niche. The definition of a niche, however, has changed with time. Early definitions described it in terms of the habitat occupied by an organism or its place in the food chain (Elton, 1927). However, in 1957 George Evelyn Hutchinson described the niche in terms of an *n*-dimensional hypervolume. He suggested that range of environmental variables that can be tolerated by a species can be described in *n* dimensions. The hypervolume created by such an approach represented the 'fundamental niche' of the species. The fundamental niche is the full range of environmental conditions an organism can occupy. However, limiting factors such as competition with other species, may prevent this. The 'realised niche' is that part of the fundamental niche that a species actually occupies as a result of the effects of various limiting factors in the environment (Fig. 6.1). Robert MacArthur and Richard Levins develop Hutchinson's ideas further describing resource utilisation in statistical terms (MacArthur and Levins, 1967).

Similar niches occur in different parts of the world. This has resulted in the evolution of organisms with similar characteristics and ecological roles in different habitats by the process of convergent evolution. For example, the antelopes found in African savannah and the kangaroos found in Australia are not closely related but they are both grazing mammals performing a similar ecological function.

KEYSTONE SPECIES

In many ecosystems, keystone species have been identified. These are animal and plant species that play a crucial role in the functioning of an ecosystem. For example, in North America, the bison (*Bison bison*) played an essentail role in maintaining the diversity of animals and plants on the plains due largely to its selective grazing of certain grass species. In parts of Africa, the elephant (*Loxodonta africana*) is a keystone species because it helps to maintain grasslands by destroying trees and it creates salt licks and digs waterholes used by other species. Once such keystone species are lost the ecosystems of which they are a part change dramatically.

PREDATION

Predators are animals that feed on other animals. When a population of predators feeds on a population of prey animals we would expect the predator numbers to increase and prey numbers to decrease. If the prey numbers become so low that the predators find them difficult to find we would expect the predators to decline due to lack of food, causing a consequent increase in prey numbers due to reduced predation pressure. This simple logic suggests that predator and prey species should interact in such a way as to produce

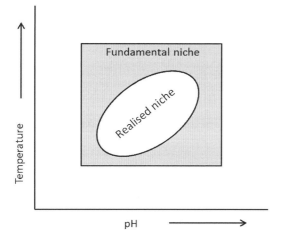

Figure 6.1 A two-dimensional representation of fundamental and realised niches.

oscillations in the numbers of each and this interaction can be modelled using mathematics.

In nature, of course, food webs are complex and most predators prey on more than a single species so such simple interactions seem unlikely to exist. Nevertheless, ecologists devised a number of simple laboratory experiments to investigate simple predator–prey relationships using model systems, notable among which were those of Huffaker (1958) who studied the dynamics of a species of predatory mite and a species of herbivorous mite that fed on oranges. Others began looking for oscillations in predator–prey systems in the wild, such as that apparently exhibited by the Canadian lynx (*Lynx canadensis*) and the snowshoe hare (*Lepus americanus*) (Elton and Nicholson, 1942).

BIOLOGICAL CONTROL

Increased awareness of the environmental damage done by many pesticides and a better understanding of the ecological relationships between predators and their prey has led to the development of biological control methods. These are often used in controlled conditions, for example greenhouses, where arthropod predators may be released to control the numbers of arthropod pests. These methods generally rely upon the use of natural predators or parasites to control agricultural or horticultural pests. However, biological control has also been used in natural ecosystems. For example, in Australia the cactus moth (*Cactoblastis cactorum*) was used to help with the control of the introduced prickly pear cactus, and genetically-modified mosquitoes that produce almost all male offspring are now being released in an attempt to prevent the spread of malaria by female *Anopheles* mosquitoes.

The following exercises examine competition between individuals of the same species and between individuals of different species, interactions between predators and their prey, niche theory and niche separation, the biological control of pests and the effects of disease on wildlife populations.

References

Elton, C. (1927). *Animal Ecology*. London: Sidgwick and Jackson.

Elton, C., & Nicholson, M. (1942). The ten-year cycle in numbers of the lynx in Canada. *Journal of Animal Ecology*, *11*, 215–244.

Gause, G. F. (1934). Experimental analysis of Vito Volterra's mathematical theory of the struggle for existence. *Science*, *79*, 16–17.

Huffaker, C. B. (1958). Experimental studies on predation: dispersion factors and predator-prey oscillations. *Hilgardia*, *27*, 343–383.

Hutchinson, G. E. (1957). Concluding remarks. *Cold Spring Harbor Symposia on Quantitative Biology*, *22*, 415–427.

MacArthur, R. H. (1958). Population ecology of some warblers of northeastern coniferous forests. *Ecology*, *39*, 599–619.

MacArthur, R., & Levins, R. (1967). The limiting similarity, convergence, and divergence of coexisting species. *The American Naturalist*, *101*, 377–385.

Park, T. (1954). Experimental studies of interspecies competition. II. Temperature, humidity, and competition in two species of *Tribolium*. *Physiological Zoology*, *27*, 177–238.

EXERCISE 6.1

COMPETITION IN TWO TREE SPECIES

Pine seedlings do not become established under forest stands where oak seedlings thrive. Pine seedlings are particularly susceptible to drought conditions. Kramer and Decker (1944) compared the photosynthetic rates of loblolly pine seedlings (*Pinus taeda*) and eastern red oak seedlings (*Quercus borealis maxima*) in an attempt to explain this phenomenon. They measured carbon dioxide uptake by the seedlings for one hour periods at 30°C (Table 6.1).

Q6.1.1 Draw a single graph of the data presented in Table 6.1.

Q6.1.2 The highest light intensity to which the seedlings were exposed represented full sunlight. At what proportion of full sunlight did each species achieve its maximum photosynthetic rate?

Q6.1.3 Explain why the oak seedlings survive under forest stands while the pine seedlings do not.

Q6.1.4 In 1949, Kozlowski repeated the observations of Kramer and Decker and found

Table 6.1 The effect of light intensity on photosynthesis in loblolly pine and eastern red oak

Light intensity (foot candles[a] $\times$ 10^3)	Rate of photosynthesis (mg CO_2 $100\,cm^{-2}$ of leaf surface)	
	Loblolly pine	Eastern red oak
0.3	0.4	1.7
1.8	2.0	5.3
3.3	2.5	6.1
4.8	2.7	5.6
6.3	3.1	5.5
7.8	3.2	5.6
9.3	3.6	5.3

Data derived from Kramer & Decker (1944).
[a]A foot candle = 10.764 lux.

that oak seedlings had a longer and more rapidly growing root system than that of pine seedlings. How would this give the oak an advantage in a forest environment?

References/Further Reading

Kozlowski, T. T. (1949). Light and water in relation to growth and competition of Piedmont forest tree species. *Ecological Monographs, 19*, 207–231.

Kramer, P. J., & Decker, J. P. (1944). Relation between light intensity and rate of photosynthesis of loblolly pine and certain hardwoods. *Plant Physioliology, 19*, 350–358.

EXERCISE 6.2

INTRASPECIFIC COMPETITION IN BARLEY

Intraspecific competition is competition between members of the same species and is commonly observed in crop plants. Table 6.2 shows the results of a greenhouse experiment which examined the effects of density on the germination and growth of barley. Plants were grown for two weeks in 10-cm diameter pots kept in identical environmental conditions and watered regularly. At the end of the experiment the plants were cropped by cutting the stems at the soil surface.

Q6.2.1 Calculate:

a. the percentage germination at each density;

b. the mean biomass of plants at each density two weeks after sowing.

The first row of the table has been completed for you.

Q6.2.2 Draw graphs to show the effect of density on:

a. germination rate (%);

b. mean wet biomass;

c. mean stem length.

(Note: Use the number of seeds germinated to indicate density not the number sown for graphs b. and c.).

Q6.2.3 Describe the effects of density shown by the graphs and suggest which environmental factor is most likely to be responsible.

Q6.2.4 Explain why knowledge of the effects of density is important to the crop farmer.

References/Further Reading

Donald, C. M. (1951). Competition among pasture plants. I. Intraspecific competition among annual pasture plants. *Crop and Pasture Science, 2,* 355–376.

Maddonni, G. A., & Otegui, M. E. (2004). Intra-specific competition in maize: early establishment of hierarchies among plants affects final kernel set. *Field Crops Research, 85,* 1–13.

Table 6.2 Effect of seed density on germination rate, biomass and stem length in barley

Number of seeds sown per pot	Number of plants that germinated	Germination rate (%)	Total wet biomass of plants (g)	Mean wet biomass of plants (g)	Mean stem length (cm)
50	43	86	17.0	0.395	25.6
100	88		21.4		19.2
200	168		27.7		14.3
400	332		40.9		13.6

EXERCISE 6.3

COMPETITIVE EXCLUSION

In 1934, Georgii Gause, a Russian microbiologist, examined competition between two species of protozoa from the genus *Paramecium*. He grew pure cultures of each species (i.e., the species was grown alone) and a mixed culture of the two species grown together. Both species were fed on the same bacteria.

Q6.3.1 Draw scatter diagrams to show the following from the data in Table 6.3:

    a. Graph 1 The growth of *P. aurelia* in pure and mixed cultures.

    b. Graph 2 The growth of *P. caudatum* in pure and mixed cultures.

Do not join the points in these graphs. Draw smooth curves through the points to represent the general trend of growth in each.

Q6.3.2 State two important differences between the growth curves of the two species grown in pure cultures.

Q6.3.3 How did the pattern of growth change in each species in a mixed culture?

Q6.3.4 Explain why there were differences in the growth curves of both species when grown in a mixed culture compared with their growth in pure cultures.

Q6.3.5 This experiment demonstrates the 'competitive exclusion principle.' What does this principle say?

Table 6.3 Growth of *Paramecium aurelia* and *P. caudatum* in pure and mixed cultures

Time (day)	Number of organisms 0.5 cm^{-3}			
	P. aurelia		*P. caudatum*	
	Pure	Mixed	Pure	Mixed
2	27	18	31	13
4	177	142	75	51
6	319	168	119	31
8	398	239	125	40
10	451	283	158	20
12	442	301	132	11
14	469	-[a]	110	20
16	471	292	128	9

Based on Gause (1934).
[a]No recording made.

Reference/Further Reading

Gause, G. F. (1934). Experimental analysis of Vito Volterra's mathematical theory of the struggle for existence. *Science, 79,* 16–17.

EXERCISE 6.4

COMPETITION IN FLOUR BEETLES

In the 1950s, Thomas Park and his students at the University of Chicago conducted a series of laboratory experiments to investigate competition between two species of flour beetle: *Tribolium confusum* and *T. casteneum* (Park, 1954). Park established populations of each species living alone and compared them with cultures containing both species living together.

Competition between these two species of beetle is complex. It is not competition for food, but a special form of mutual predation as adult and larval forms of both species cannibalise their own eggs and pupae.

One of Park's experiments considered the effect of different artificially created climates on the outcome of competition. These climates consisted of combinations of three different temperatures (24°C, 29°C and 34°C) and two relative humidity levels (30% and 70%). This produced six different 'climates'. Several cultures were established for each climate and Park recorded the percentage of cultures in which each species 'won'. The results of this experiment are shown in Fig. 6.2

Q6.4.1 Identify the two climates in which the results of competition were completely consistent, i.e., the same species always won.

Table 6.4 shows the results of growing each *Tribolium* species alone. For example, *confusum > casteneum* means that *confusum* produced more individuals than *casteneum* when each species was kept separate from the other (i.e., when they were not competing).

Q6.4.2 Identify the two climates in which the outcome of competition could not be predicted from the results obtained from growing single species cultures in the six climates.

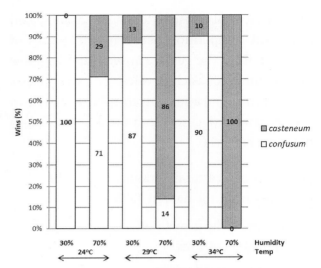

Figure 6.2 The effect of temperature and humidity on competition between *Tribolium confusum* and *T. casteneum*. Based on Park (1954).

Table 6.4 Relative numbers of each species when kept alone in each climate

Climate	Single species numbers
Hot–moist	*confusum = casteneum*
Hot–dry	*confusum > casteneum*
Temperate–moist	*confusum < casteneum*
Temperate–dry	*confusum > casteneum*
Cold–moist	*confusum < casteneum*
Cold–dry	*confusum > casteneum*

Q6.4.3 If you were asked to repeat part of Park's experiment and grow a single culture of both *Tribolium* species in temperate–dry conditions, explain why you would not be able to predict which species would win.

Reference/Further Reading

Park, T. (1954). Experimental studies of interspecies competition. II. Temperature, humidity, and competition in two species of *Tribolium*. *Physiological Zoology*, 27, 177–238.

EXERCISE 6.5

DO CORMORANTS AND SHAGS COMPETE FOR FOOD?

The cormorant (*Phalacrocorax carbo*) (Fig. 6.3) and the shag (*P. aristotelis*) are closely related seabirds found around the coast of the British Isles. Their geographical distributions overlap and they appear to feed in the same areas of the sea but are they competing for the same food organisms or have they evolved to specialise in particular food types? Lack (1945) claims that Collinge (1924-7) suggested that the two species eat largely the same types of foods.

The data in Table 6.5 are based on an analysis of the stomach contents of 188 shags and 27 cormorants taken in Cornwall, England by Steven (1933).

Q6.5.1 Draw a bar chart showing the percentage of stomachs of each bird species that contained each food type.

Figure 6.3 Juvenile cormorant (*Phalacrocorax carbo*).

Q6.5.2 Does the information in Table 6.5 support Collinge's belief that shags and cormorants eat the same types of foods?

Q6.5.3 Does the evidence from Steven's study suggest that cormorants and shags compete for food?

Q6.5.4 Suggest behavioural differences in the two bird species that allow them to have overlapping geographical distributions.

References/Further Reading

Collinge, W. E. (1924-7). *The food of some British wild birds* (2nd ed). Published by the author. 217–221.

Lack, D. (1945). The ecology of closely related species with special reference to the cormorant (*Phalacrocorax carbo*) and shag (*P. aristotelis*). *Journal of Animal Ecology*, *14*, 12–16.

Steven, G. A. (1933). The food consumed by shags and cormorants around the shores of Cornwall (England). *Journal of the Marine Biological Association of the United Kingdom* (New Series*), *19*, 277–292.

Table 6.5 Foods taken by cormorants and shags around the coast of Cornwall

Food	Percentage of stomachs in which found	
	Shag	Cormorant
Ammodytes spp. (sand eel)	51	0
Clupea sprattus (sprat)	11	4
Pleuronectids (flatfish)	3	52
Crangonidae (shrimps)	3	33
Palaemonidae (prawns)	5	15
Ctenolabrus rupestris (gold-sinny wrasse)	13	17
Labrus bergylta (Ballan wrasse)	6	15
Callionymus spp. (dragonet)	10	11

After Steven (1933).

COMPETITION BETWEEN ANTS AND RODENTS

Brown and Davidson (1977) studied interspecific completion between seed-eating ants and rodents in the deserts of the southwestern United States. They performed field experiments in 36-m diameter plots of homogeneous desert scrub. In two plots, rodents were completely removed by trapping; in another two plots, ants were killed with insecticide; in a further two plots, both ants and rodents were removed. Two additional plots were kept as controls with neither ants nor rodents removed. The results of the experiment are presented in Fig. 6.4, including the modal seed sizes taken by rodents and ants.

Q6.6.1 What was the effect of removing the ant colonies on the size of the rodent population?

Q6.6.2 What was the effect of removing the rodents on the number of ant colonies?

Q6.6.3 The seed density was defined as 1.0 in the control plots where both ants and rodents were present.

a. Why did this value remain the same when either the ants were removed or the rodents were removed?

b. Why did it increase when both were removed?

Q6.6.4 What difference in food preference helps the ants and the rodents coexist in the desert?

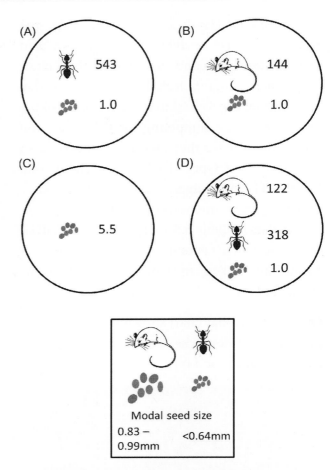

Figure 6.4 A representation of the results of Brown and Davidson (1977) showing the numbers of ant colonies, numbers of rodents and seed density in plots containing (A) ants only, (B) rodents only, (C) ants and rodents removed and (D) ants and rodents present.

Reference/Further Reading

Brown, J. H., & Davidson, D. W. (1977). Competition between seed-eating rodents and ants in desert ecosystems. *Science, 196,* 880–882.

EXERCISE 6.7

DEFINING THE NICHE

The definition of the niche has changed considerably with time. The term was first coined by Joseph Grinnell in 1917. He used it to refer to the habitat of a species (Grinnell, 1917). Ten years later, Charles Elton took a functional approach to defining the niche and saw it as the equivalent to the position a species occupies in a food web (Elton, 1927).

In 1957, Dr G. Evelyn Hutchinson of Yale University made the concluding remarks at the end of the Cold Spring Harbor Symposium on Quantitative Biology in the United States. These words changed the way that ecologists think about niches. Hutchinson's idea essentially proposed this (Fig. 6.5):

1. Consider a variable x that can be measured along an ordinary linear scale. The environmental states that permit an organism to survive may be indicated on this scale. So, if we consider temperature, our organism can survive between two temperature limits: one high and one low. This scale now represents the range of temperatures that this organism can survive as a straight line drawn between these two limits.

2. Now add a second, independent, variable (y), for example, the food sizes it eats. These too can be expressed in terms of an upper and lower limit on a linear axis. If this axis is drawn at right angles to the x-axis (in the traditional manner of a graph), we can express the limits between which our organism can survive in terms of both temperature and the food sizes available as an area, i.e., in two dimensions.

3. Now add a third dimension and a third axis (z) representing a third set of limits. This might be, for example, the height the organism feeds in the forest. This too would have an upper and lower limit. Using these values plotted against the z-axis we can now turn our area into a volume representing the limits for survival of our organism in terms of three dimensions: temperature, food size and feeding height.

4. Now add a fourth dimension and a fifth and so on…

> *In this way an n-dimensional hypervolume is defined, every point in which corresponds to a state of the environment which would permit the species…to exist indefinitely.*
> Hutchinson (1957).

This idea came to be known as the Hutchinsonian niche. It is the fundamental niche of the species, i.e., the niche that it could potentially occupy. In reality, a species occupies a smaller niche than this: its realised niche (Fig. 6.1). Hutchinson's idea revolutionised the

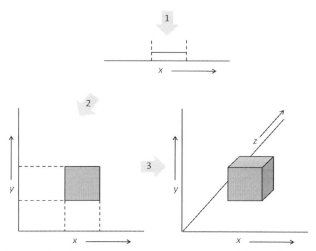

Figure 6.5 Diagrammatic representations of the niche using one, two and three dimensions.

Table 6.6 Environmental requirements of three imaginary species

Species	Prey size (x)	Height in vegetation (y)	Altitude above sea level (z)
K	1–4	10–40	15–200
L	3–7	20–60	300–450
M	1–2	60–80	250–500

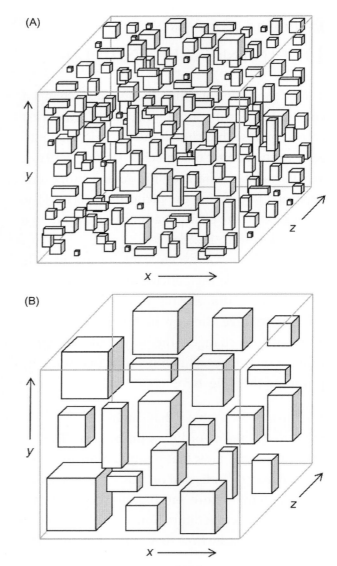

(A)

(B)

Figure 6.6 Representations of niches in two ecosystems. Which is the tropical forest and which is the tundra?

way ecologists think about the niche although, clearly, it is a theoretical concept that cannot be expressed graphically because we cannot draw in n dimensions, and even if we could, we cannot ever hope to measure all of the factors in the environment that affect a species' survival.

Table 6.6 contains data which define the niches of three hypothetical species K, L and M in terms of the prey size each takes, the height in vegetation where it is found and the altitude at which it occurs.

Q6.7.1 Draw a single axis (x) representing prey size.

   a. Mark on this axis the ranges of the prey sizes taken by species K, L and M.

   b. Which species appear to have overlapping niches?

Q6.7.2 Draw two axes representing prey size (x) and the height in the vegetation where the species occurs (y).

   a. Draw three areas representing the niches of species K, L and M with respect to these two variables and shade any area where they overlap.

   b. Which species appear to have overlapping niches now?

Q6.7.3 Add a third axis (z) to this graph.

   a. Turn your three areas into volumes using the data in Table 6.6. The volumes so created represented the niches of species K, L and M in three dimensions.

   b. Do the niches of any of the three species overlap now a third dimension has been added?

Q6.7.4 Explain what has happened to the extent to which the niches of species K, L and M overlap as you have increased the number of parameters used to measure this from one to two and from two to three.

Q6.7.5 Fig. 6.6 shows the hypothetical niches of organisms in two ecosystems measured in three dimensions, each representing a different environmental variable. State which represents a tropical rainforest ecosystem and which represents tundra, and explain your decision.

References/Further Reading

Elton, C. (1927). *Animal ecology*. London: Sidgwick and Jackson.

Grinnell, J. (1917). The niche-relationships of the California Thrasher. *The Auk, 34,* 427–433.

Hutchinson, G. E. (1957). Concluding remarks. *Cold Spring Harbor Symposia on Quantitative Biology, 22,* 415–427.

EXERCISE 6.8

NICHE SEPARATION IN WARBLERS

This study concerns the ecology of four closely related warblers living in coniferous forests in North America: the alpine warbler, the blue warbler, the carmine warbler and the diamond warbler. These birds are imaginary but the exercise is based on a study conducted by MacArthur (1958).

Observations were made in six sample areas of woodland. Trees were divided into five vertical zones (zone 1 at the apex and zone 5 at the base). Trees in each sample area were observed for the same amount of time. On each occasion when a warbler was seen a record was made of its species

and the height zone in which it was present. A total of 78 recordings were made in the sample areas (Fig. 6.7).

Q6.8.1 Construct a table showing the percentage of recordings of each species within each of the zones using the format shown in Table 6.7.

Q6.8.2 Draw a stacked bar chart showing, for each species, the percentage of observations made in each height zone. Note that the height of each bar will be the same as the total for each species (100%).

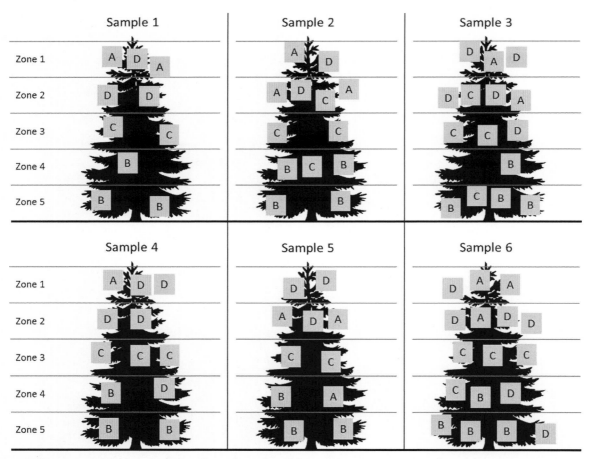

Figure 6.7 The vertical distribution of four hypothetical warbler species in a coniferous forest showing the locations where individuals belonging to each species were recorded. Key: A = alpine warbler; B = blue warbler; C = carmine warbler; D = diamond warbler.

Table 6.7 Summary table for the data on warbler distribution collected from Fig. 6.7

Height zone	Warbler species							
	Alpine		Blue		Carmine		Diamond	
	Number of recordings	Percentage of recordings	Number of recordings	Percentage of recordings	Number of recordings	Percentage of recordings	Number of recordings	Percentage of recordings
1								
2								
3								
4								
5								
Total		100		100		100		100

Q6.8.3 Suggest reasons why these four species are able to coexist in the same coniferous forests.

Q6.8.4 How could recording the time of each sighting, as well as the height, shed more light on the ecology of these species?

Reference/Further Reading

MacArthur, R. H. (1958). Population ecology of some warblers of northeastern coniferous forests. *Ecology, 39,* 599–619.

EXERCISE 6.9

NICHE SEPARATION IN TROPICAL MONKEYS

Tropical forests are complex three-dimensional ecosystems so it would not be surprising to find that some forest species separate themselves out vertically rather than horizontally. Monkeys belonging to the genera *Cercopithecus* and *Miopithecus* inhabit the tropical forests of West and Central Africa, including the blue monkey (*C. mitis*) (Fig. 6.8). They are active, gregarious animals that normally live in troops of 30 to 40 individuals.

Q6.9.1 Use the information in Table 6.8 to complete Table 6.9 by placing an 'X' in the forest layers inhabited by each monkey species.

Q6.9.2 Which two layers of the forest are used by the largest number of species?

Q6.9.3 How does the vertical separation of monkey species help them to coexist?

Figure 6.8 A blue monkey (*Cercopithecus mitis*).

Table 6.8 Preferred habitats of *Cercopithecus* species

Species		Preferred habitat
A	Hamlyn's monkey (*Cercopithecus hamlyni*)	Shrubs
B	Diana monkey (*C. diana*)	Tall and medium trees
C	Black-cheeked white-nosed monkey (*C. ascanius*)	Medium and small trees, shrubs and low herbage
D	Great white-nosed monkey (*C. nictitans*)	Medium trees
E	Moustached monkey (*C. cephus*)	Small trees
F	Talapoin (*Miopithecus talapoin*)	Shrubs
G	De Brazza's monkey (*C. neglectus*)	Shrubs and low herbage
H	Blue monkey (*C. mitis*)	Tall, medium and small trees

Table 6.9 Summary table for the information in Table 6.8 on preferred monkey habitats

Forest layer	Species							
	A	B	C	D	E	F	G	H
Tall trees								
Medium-sized trees								
Small trees								
Shrubs								
Low herbage								

Q6.9.4 Tropical forests support vast numbers of flowering plants. Why is this important to the monkeys?

Q6.9.5 It is difficult for animals to see long distances in thick forest. How would you expect members of a troop of monkeys to communicate?

References/Further Reading

Ganzhorn, J. U. (1989). Niche separation of seven lemur species in the eastern rainforest of Madagascar. *Oecologia*, 79, 279–286.

Gartlan, J. S., & Struhsaker, T. T. (1972). Polyspecific associations and niche separation of rain-forest anthropoids in Cameroon, West Africa. *Journal of Zoology*, 168, 221–265.

EXERCISE 6.10

PREDATOR–PREY SIMULATION

The relationship between predator and prey populations has been studied by computer simulation using equations which form part of a mathematical model devised by Lotka and Volterra. The data in Table 6.10 show the results of such a simulation when the prey population begins with 20 individuals and the predator population begins with six individuals.

Q6.10.1 Draw a graph showing the changes in the numbers of predators and prey with time. Both lines should be drawn on the same graph.

Q6.10.2 Explain why the relationship exhibited here may be described as stable.

Q6.10.3 How far apart are the peaks in:

a. the prey population;

b. the predator population?

Q6.10.4 Explain why the peaks in the predator population occur after those in the prey population.

Q6.10.5 This simulation assumes one prey species and one predator species in an imaginary ecosystem, and is based on mathematical equations. Why is it likely to be too simplistic to describe accurately what happens in nature?

Table 6.10 The results of a simulation of the relationship between predator and prey numbers in hypothetical populations

Time 1 (arbitrary units)	Prey numbers	Predator numbers
0	20	6
1	28	6
2	38	8
3	46	11
4	39	15
5	21	14
6	14	10
7	13	8
8	16	6
9	21	6
10	30	7
11	40	9
12	46	12
13	36	15
14	19	13
15	13	10
16	13	8
17	17	6
18	23	6
19	32	7
20	42	9
21	46	13
22	33	15
23	17	12
24	13	9

References/Further Reading

Berryman, A. A. (1992). The orgins and evolution of predator-prey theory. *Ecology, 73*, 1530–1535.

Curds, C. R. (1971). A computer-simulation study of predator-prey relationships in a single-stage continuous-culture system. *Water Research, 5*, 793–812.

Smith, J. M., & Slatkin, M. (1973). The stability of predator-prey systems. *Ecology, 54*, 384–391.

EXERCISE 6.11

PREDATION OF FOXES ON RABBITS

Figure 6.9 A red fox (*Vulpes vulpes*).

Foxes (*Vulpes vulpes*) feed upon rabbits (*Oryctolagus cuniculus*) (Figs. 6.9 and 6.10). Foxes are not very successful at catching adult rabbits on the surface. Instead, they remove nestlings by digging into rabbit warrens. Myers and Parker (1965) studied attacks made on rabbit warrens by foxes in different regions of eastern Australia (Table 6.11).

Q6.11.1 Complete Table 6.11 by calculating the percentage of all rabbit warrens attacked by foxes in each soil type.

Q6.11.2 Rank the soil types in order from that where the percentage of warrens attacked by foxes is lowest to that where it is highest and tabulate your results.

Figure 6.10 A rabbit (*Oryctolagus cuniculus*).

Table 6.11 Predation by foxes on rabbits living in burrows made in different soils in eastern Australia

Soil	Number of warrens		Percentage of total attacked	Number of nests destroyed by foxes
	Total	Attacked by foxes		
Stony banks	829	0	0	0
Loam	871	66		78
Stony hills	806	11		12
Stony pediments	371	46		58
Desert loam	324	148		255
Sand dunes	3309	2339		2550

Adapted from Myers & Parker (1965).

Q6.11.3 How many times more likely is a warren in a sand dune area to be attacked by foxes than one in a loam?

Q6.11.4 Explain why, in all soil types where warrens were attacked, the number of nests destroyed by foxes was higher than the number of warrens attacked.

Q6.11.5 Use evidence from this data to determine which soil types are the safest for rabbits to construct their warrens. Explain your answer.

Reference/Further Reading

Myers, K., & Parker, B. S. (1965). A study of the wild rabbit in climatically different regions in eastern Australia: I. Patterns of distribution. *CSIRO Wildlife Research*, 10, 33–72.

THE EFFECTS OF DISEASE, CLIMATE AND VEGETATION CHANGE ON LARGE MAMMAL POPULATIONS

The following text is an extract from a paper on the ecology of Ngorongoro Crater by Oates and Rees (2013).

Historically, wild herbivore numbers in Tanzania were regulated by rinderpest (an infectious viral disease of ungulates; Dobson 1995). Numbers of wildebeest and buffalo in the Serengeti National Park increased rapidly during the 1960s and 1970s, while numbers of other grazers changed only slightly or not at all (Sinclair, 1979). The increase coincided with the eradication of rinderpest in the wild ungulate populations in the early 1960s. Rinderpest had been suppressing wildebeest numbers in particular because of their high susceptibility (Sinclair, 1979). Dobson (1995) believed that the disease may have affected the density of buffalo and wildebeest in the Ngorongoro area for most of the 1900s. There has been a significant increase in the number of buffalo in the crater since the eradication of rinderpest (Estes & Small, 1981, Homewood & Rodgers, 1991) and regression analysis suggested that wildebeest numbers increased by 38% between 1961 and 1973 (Runyoro et al., 1995). In 1958, a rinderpest outbreak caused declines in eland and buffalo populations (Dobson, 1995).

Kissui and Packer (2004) concluded that disease, possibly exacerbated by an increased susceptibility due to inbreeding and an increasing human population, was the causative agent holding the lion population below carrying capacity. The 1962 decline was caused by a plague of biting flies Stomoxys calcitrans (Fosbrooke, 1963) resulting from heavy rains (Kissui & Packer, 2004). The bites weakened the lions, resulting in starvation. Some weakened individuals preyed on Masai livestock and were speared (Fosbrooke, 1963). The 1994 decline coincided with an outbreak of canine distemper virus (CDV) in the Serengeti and was associated with a severe drought. Drought can bring more animals into contact at watering holes and at kills, while flood can improve conditions for the persistence of pathogens. The 1997 outbreak coincided with the El Niño rains but the exact cause of the decline is unknown (Kissui & Packer, 2004). In 2001, a die-off in the lion population was attributed to a combination of CDV (Kissui & Packer, 2004), the tick-borne protozoan disease babesiosis (Nijhof et al., 2003, Estes et al., 2006) and the weakening effects of a severe drought in 1999–2000 (Brett, 2001, Mills et al., 2003, Kissui & Packer, 2004). One thousand buffalo, 250 wildebeest, 100 zebras, 22 lions and five rhinoceros were recorded as having died (Brett, 2001, Mills et al., 2003, 2006), with babesiosis confirmed in two of the dead rhinoceros (Nijhof et al., 2003, Fyumagwa et al., 2004). The loss of rhinoceros was regarded as serious enough to warrant the unprecedented action of treating the remaining population with the antiprotozoal drug Diminazine aceturate (Berenil Hoechst; Fyumagwa et al.,

2004). *The large mammals in the crater were observed to be carrying high tick and biting fly loads in 2001 (Fyumagwa et al., 2004). As buffalo numbered over 5000 individuals prior to the drought, it is possible that the species was at the upper limit of the carrying capacity of the crater and therefore more likely to succumb to the drought and associated disease (Estes et al., 2006). Additionally, an increase in human settlements in the area may have led to an increase in the number of domestic dogs (Kissui & Packer, 2004), major carriers of CDV in Tanzania (Cleaveland et al., 2001).*

Early explorers described luxuriant vegetation in the crater (e.g., Barns, 1921). The removal of the Masai in 1974 may have been the cause of changes in the vegetation due to the associated removal of traditional grassland management, in which controlled burning and livestock grazing kept the grass sward short and palatable. Short grass suits small to medium grazers such as wildebeest (Demment & van Soest, 1985) and browsers such as rhinoceros (Brett, 2001, Mills et al., 2003, 2006). The grass sward subsequently became longer and less palatable, and therefore more suited to less selective feeders such as buffalo. Runyoro et al. (1995) found that buffalo and hartebeest numbers increased after the removal of the Masai, while wildebeest, zebras, Thomson's gazelles, Grant's gazelles, rhinoceros and elands declined. Both the presence of buffalo and the longer grass favour the growth and spread of disease-carrying ticks (Brett, 2001, Mills et al., 2003). Fyumagwa et al. (2007) investigated the babesiosis outbreak and concluded that the eviction of the Masai, combined with the increased rainfall associated with the 1997–98 El Niño event,

improved conditions for tick vectors. A decrease in the annual amount of rainfall from approximately 950 mm in 1963 to 800 mm in 2000 may have contributed to a reduction in available grazing and browsing for herbivores (Mills et al., 2003).

An increase in invasive weeds (e.g., Gutenbergia cordifolia and Bidens shimperi), possibly caused by the change in grassland management, may have reduced available food for many herbivore species (Henderson, 2002, Trollope et al., 2002, Mills et al., 2003, Estes et al., 2006); this has been cited as a particular problem for rhinoceros (Brett, 2001, Mills et al., 2003, 2006). Since the drought of 2000, efforts have been made to improve the grass sward for grazers and control invasive species by burning and mowing (Anonymous, 2007). It is unclear whether these activities have been successful.

The Lerai Forest, previously an important calving refuge and home range for many rhinoceros (Goddard, 1967b), has decreased in size and become fragmented (Fosbrooke, 1972, Kabigumila, 1993, Trollope et al., 2002) to such an extent that rhinoceros stopped using it (Mills et al., 2006). The lack of suitable calving areas is believed to have increased rhinoceros calf mortality (Mills et al., 2003). The forest's decline has been attributed to an increase in the salinity of the soil caused by the flooding of Lake Magadi (Fosbrooke, 1972, Trollope et al., 2002, Mills, 2006), a fungal infection in the trees and the action of elephants that remove tree bark (Trollope et al., 2002). Mills et al. (2003) recorded that forest regeneration towards the crater rim was occurring but was not producing suitable calving habitat.

[Note that organisms are only given their scientific names in this extract when they have not appeared earlier in the full paper. The convention in *Mammal Review* is to omit the parentheses within which scientific names often appear. Readers interested in the references in the extract should refer to Oates and Rees (2013).]

Q6.12.1 Draw a simple flow chart explaining the effects that the removal of the Masai appeared to have on the large mammal populations of the crater.

Q6.12.2 Explain how vegetation change in the crater adversely affected the black rhinoceros population.

Q6.12.3 What factors were thought to be holding the lion population below its carrying capacity?

Q6.12.4 Why are invasive plants species a problem in the crater and what steps have been taken to control them?

Q6.12.5 What effect did rinderpest have on the wildebeest population of the crater?

Q6.12.6 What is the significance of an increase in the number of domestic dogs in the area?

Q6.12.7 Suggest reasons for the decline of the Lerai Forest.

Reference/Further Reading

Oates, L., & Rees, P. A. (2013). The historical ecology of the large mammal populations of Ngorongoro Crater, Tanzania, east Africa. *Mammal Review*, *43*, 124–141.

EXERCISE 6.13

BIOLOGICAL CONTROL

Ecologists are increasingly using natural predators to control pests instead of using pesticides. Mites live on cucumbers grown in glasshouses, causing damage to the crop. Table 6.12 compares the effect of using a pesticide to control these mites with the effect of using a predatory mite species.

Q6.13.1 Draw a graph showing the effect of both methods of control on the number of pest mites. Both lines should appear on the same graph.

Table 6.12 Numbers of mites found on cucumbers when controlled using pesticides and biological control

Time (days)	Number of pest mites 5 cm^{-2}	
	Control method 1	Control method 2
15	2.1	2.9
30	14.3	2.6
45	1.9	2.7
60	20.3	2.8
75	1.2	2.4
90	15.1	2.2
105	1.7	2.5
120	24.5	1.9
135	1.2	2.3
150	4.0	2.1
165	0.9	2.4
190	9.7	2.6
205	1.1	2.2

Q6.13.2 Which control method (1 or 2) involved pesticides and which involved biological control? Explain why you have decided this.

Q6.13.3 Label the two lines on your graph to indicate the control method.

Q6.13.4 Draw arrows on your graph to indicate the times when the pesticides were applied.

Q6.13.5 If the population of the predatory mite had also been drawn on the graph what form would you have expected the line to take and why?

Q6.13.6. List some advantages of using biological control methods.

Q6.13.7. List some reasons why biological control of pests is not widely used.

Reference/Further Reading

Hussey, N. W., Parr, W. J., & Gould, H. J. (1965). Observations on the control of *Tetranychus urticae* Koch on cucumbers by the predatory mite *Phytoseiulus riegeli* Dosse. *Entomologia Experimentalis et Applicata, 8,* 271–281.

SEALS AND PHOCINE DISTEMPER VIRUS

Figure 6.11 A bull grey seal (*Halichoerus grypus*).

Two major phocine distemper virus (PDV) outbreaks occurred in harbour seals (*Phoca vitulina*) in northwestern European coastal waters in 1988 and 2002, causing the death of tens of thousands seals. Bodewes et al. (2013) tested serum samples collected in Dutch coastal waters from 2002 to 2012, from harbour seals (*Phoca vitulina*) and grey seals (*Halichoerus grypus*) (Fig. 6.11) for the presence of PDV-neutralising antibodies.

Q6.14.1 Using the data in Table 6.13, determine if there is any evidence that PDV-neutralising antibody is more likely to be found in one of

Table 6.13 The number of individual harbour and grey seals in which PDV-neutralising antibody was detected

Species	PDV-neutralising antibody	
	Present	Absent
Harbour seal	70	353
Grey seal	4	33

Based on Bodewes et al. (2013).

these species than the other? Use a χ^2 test (2 × 2 contingency table) to determine this:

a. state your null hypothesis;

b. perform the test showing each step clearly;

c. determine whether or not the results are statistically significant (state the level).

Table 6.14 Serology data for harbour seals (*Phoca vitulina*)

Year	Pups No.	Pups Positive	Juveniles No.	Juveniles Positive	(Sub)adults No.	(Sub)adults Positive
2002	9	7	35	18	0	N/A
2003	38	21	12	6	0	N/A
2004	16	1	13	0	3	1
2005	19	5	16	0	0	N/A
2006	22	4	12	0	2	2
2007	7	0	36	0	1	N/A
2008	4	0	41	0	1	1
2009	27	1	23	0	1	1
2010	3	0	8	0	1	1
2011	16	1	33	0	1	1
2012	0	0	21	0	0	N/A

Adapted from Bodewes et al. (2013).

Statistical tests are explained in Chapter 10, Statistics.

Q6.14.2 Draw a bar chart showing the percentage of harbour seal pups that tested positive for PDV-neutralising antibodies in each year between 2002 and 2011 (Table 6.14).

Q6.14.3 Mark the end of the most recent serious outbreak of PDV on your graph.

Q6.14.4 Approximately what percentage of juvenile harbour seals had antibodies to PDV prior to the end of the year in which the last outbreak of PDV occurred?

Q6.14.5 Explain, with reference to your graph and the data in Table 6.14, why Bodewes et al. were concerned about the possibility of a future outbreak of PDV.

Q6.14.6 The seals used in this study were admitted to the Seal Research and Rehabilitation Centre (SRRC) in Pieterburen, the Netherlands for rehabilitation. How might this method of sampling the seal population have affected the results of the study?

Reference/Further Reading

Bodewes, R., Morick, D., van de Bildt, M. W. G., Osinga, N., García, A. R., Sánchez Contreras, G. J., Osterhaus, A. D. M. E. (2013). Prevalence of phocine distemper virus specific antibodies: Bracing for the next seal epizootic in north-western Europe. *Emerging Microbes & Infections*, 2, e3. http://dx.doi.org/10.1038/emi.2013.2.

CHAPTER 7

BEHAVIOURAL ECOLOGY AND ECOLOGICAL GENETICS

This chapter contains exercises concerned with the behavioural adaptations that animal species have evolved to survive and the genetic mechanisms that ensure that organisms can adapt to an ever changing environment.

Giant day gecko (*Phelsuma grandis*)

Examining Ecology. DOI: http://dx.doi.org/10.1016/B978-0-12-809354-2.00007-5

INTENDED LEARNING OUTCOMES

On completion of this chapter you should be able to:

- Suggest morphological characteristics of animals that could be used to distinguish between and identify individuals within the same species.

- Identify an individual animal from an identification record.

- Estimate the location of an animal from compass bearings obtained from radio-tracking equipment.

- Identify the direction of movement of an animal from information obtained from radio-tracking equipment.

- Calculate the time spent by an animal on a variety of behaviours: a time budget.

- Compare the merits of different feeding strategies.

- Determine the outcome of contests between animals using game theory methods.

- Explain how the degree of association between individual animals in a group may be measured.

- Use graphical methods to determine the winner of a contest.

- Explain how studies of industrial melanism have provided evidence to support the theory of evolution.

- Calculate effective population size.

- Use graphical methods to demonstrate the relationship between sex ratio and effective population size.

- Demonstrate the effect of inbreeding on the level of homozygosity in a population.

- List examples of inbreeding in isolated populations of wild animals.

- Discuss the contribution of Charles Darwin and others to the study of evolution.

- Identify the existence of heavy metal tolerance in plants from field data.

- Explain the evolutionary significance of heavy metal tolerance in plants.

- Explain the effect of genetic drift on genetic variation within a population.

- Recognise a normal distribution.

INTRODUCTION

The ecology of an organism cannot be properly understood without an understanding of its genetics and, if it is an animal, its behaviour. Ecological genetics and behavioural ecology developed into major disciplines in their own right in the second half of the 20th century.

Behavioural ecology is the scientific study of the ecological and evolutionary basis of animal behaviour, and its role in adapting an organism to its environment, particularly the way in which behaviour contributes to reproduction and survival. Hypotheses in behavioural ecology assume that behaviour is optimised.

The science of ecological genetics developed largely out of the work of E.B. Ford, who was interested in the role of natural selection in the functioning of ecosystems and had a particular interest in butterflies and moths (Ford, 1964). Ford ended his career as Professor of Ecological Genetics at the University of Oxford. His work, and that of many others who followed him, has led to a better understanding of evolution and been of great importance in developing methods to help conserve populations and breed organisms that can survive in heavily polluted environments.

EVOLUTIONARILY STABLE STRATEGIES AND GAME THEORY

An evolutionarily stable strategy (ESS) is an inherited strategy (usually behavioural) which, if practised by most members of a population, cannot be supplanted during evolution by a different strategy. The term was first used by John Maynard Smith who was interested in the use of mathematics in studying ecology, evolution and behaviour, including the use of game theory.

Figure 7.1 Behavioural ecologists calculate the outcome of contests between animals based on probability theory.

According to Smith (1984) evolutionary game theory is:

> … *a method of analysing the evolution of phenotypes (including types of behaviour) when the fitness of a particular phenotype depends on its frequency in the population.*

The method was first applied to pairwise contests between animals, where there is usually some associated asymmetry. This may be, for example, in relation to their size, age, sex or status. Game theory predicts that the contestants will use the asymmetry to settle the contest and studies have found that this is indeed the case (Fig. 7.1). Game theory has also been used to study feeding strategies in animals and has led to the development of optimal foraging theory (Pyke, 1984).

INDUSTRIAL MELANISM

The Industrial Revolution provided a unique opportunity for scientists to study evolution. Deposits of soot from badly regulated factories and the chimneys of domestic properties coated the surfaces of buildings and trees so that they turned black. After London's Great Smog of 1952 – caused by fog mixing with smoke – the

UK Parliament passed the Clean Air Act 1956 which introduced smoke control areas and other measures to reduce air pollution. As a result air pollution decreased and in the following decades the many dirty buildings were cleaned.

Under normal circumstances, evolution proceeds at a very slow rate. However, the rapid change in the background colour of much of the environment caused by heavy industry and coal fires caused a change in the proportion of dark and light colour morphs of a common moth species in the United Kingdom, *Biston betularia*. The dark form was dominant in polluted areas and the light form was dominant in clean areas.

METAL TOLERANCE IN PLANTS

Some plants exposed to high levels of heavy metals – because they grow on mine spoil heaps – have evolved a tolerance to particular elements such as zinc, copper and lead. This has allowed botanists to breed metal-tolerant strains of grasses that have been used to reclaim land contaminated with metals. This work was pioneered by Prof. Tony Bradshaw and his colleagues at the University of Liverpool in the 1970s (Smith and Bradshaw, 1979).

GENETIC CONSERVATION

An understanding of genetics is essential if we are to conserve wildlife effectively. Healthy populations of species can only be maintained if they contain sufficient genetic diversity to protect against threats from changes to the environment, such as the emergence of disease, new predators or changes to the climate. Small populations are likely to suffer from the effects of inbreeding due to a reduction in the choice of mates which results in breeding with close relatives. This can cause an increase in the frequency of congenital diseases and deformities similar to that observed in some heavily inbred breeds of domestic dogs.

Genetic diversity in some endangered species is extremely low. The Florida panther is a subspecies of cougar (*Puma concolor*). By the mid-1980s only about 40 animals remained in the wild, creating a genetic bottleneck. Individuals suffered from a number of physical and physiological abnormalities, including kinked tails, sperm defects, heart defects (atrial septal defects), undescended testicles (cryptorchidism) and vaginal papillomas.

When rare species are bred in captivity studbook keepers determine which individuals are used for breeding and studbooks are kept recording each mating to prevent inbreeding.

The following exercises examine methods used to study animal behaviour, including the identification and tracking of individual animals in the wild, territoriality, hunting and feeding strategies, and contests for mates. They also consider interactions between genetics and ecology by examining the process of evolution, inbreeding, adaptations to pollution and the genetic problems that affect the conservation of species.

References

Ford, E. B. (1964). *Ecological genetics*. London: Chapman and Hall.

Pyke, G. H. (1984). Optimal foraging theory: A critical review. *Annual Review of Ecology and Systematics, 15,* 523–575.

Smith, J. M. (1984). Game theory and the evolution of behaviour. *Behavioral and Brain Sciences, 7,* 95–101.

Smith, R. A. H., & Bradshaw, A. D. (1979). The use of metal tolerant plant populations for the reclamation of metalliferous wastes. *Journal of Applied Ecology, 16,* 595–612.

IDENTIFYING INDIVIDUAL ANIMALS

Most studies of behavioural ecology require that the researcher knows the identity of all of the individuals in the study population. It is more useful to know that individual A attacked individual B 21 times and individual C 5 times than it is to know that 26 attacks took place within this group. Individual elephants may be distinguished by differences in their ears and tusks. Often, at least initially, researchers rely upon photographic records of animals to aid identification (Fig. 7.2).

Q7.1.1 List four different characteristics of elephant ears that would assist in the identification of individual animals.

Q7.1.2 Why would photographic records of ears need to be updated from time to time?

Q7.1.3 Why would it be important to photograph both ears for each elephant?

Q7.1.4 What other morphological characteristics of elephants would be useful in the identification of individuals?

Q7.1.5 Why are tusk shape and size not very useful in identifying female Asian elephants (*Elephas maximus*)?

Q7.1.6 Cynthia Moss studied the social organisation of hundreds of elephants in Amboseli National Park in Kenya (Moss, 1988). She assigned a name to each elephant because she found identification numbers difficult to remember. Table 7.1 lists examples of some of the elephants Moss studied. Divide the list into two family groups and explain how you have done this.

Q7.1.7 Moss used the first three letters of each name as a code. Why was Alyce spelt with a 'y'?

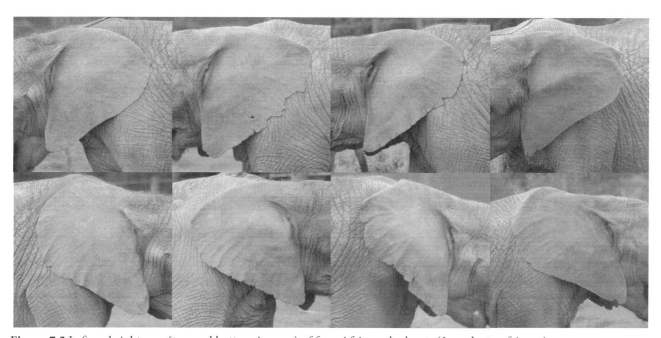

Figure 7.2 Left and right ears (top and bottom images) of four African elephants (*Loxodonta africana*).

Table 7.1 The names of elephants from two family groups studied by Moss (1988) in Amboseli

Annabel	Tio
Tess	Agatha
Amy	Tabatha
Tasmin	Turner
Tara	Alison
Alyce	Tia
Tamar	Amelia
Taddeus	Abigail

Q7.1.8 One of these two families contained a female called 'Wart Ear'. Suggest why her name was not changed to fit in with the naming pattern you have identified from the names given in Table 7.1.

Reference/Further Reading

Moss, C. (1988). *Elephant Memories. Thirteen Years in the Life of an Elephant Family.* Glasgow: William Collins & Co. Ltd.

EXERCISE 7.2

RADIO-TRACKING ANIMALS

Many studies of animal ecology track the movements of individual animals to determine their patterns of daily movement, home range size, migration pattern, or some other aspect of their biology. This may be done using a radio collar, the signals from which may be detected by a portable directional antenna. If compass bearings to the collar can be determined from two different locations at more or less the same time it is possible to estimate the position of the collar by a process known as triangulation (Fig. 7.3).

Q7.2.1 On a piece of A4 paper, draw a line 15 cm (6 inches) long. Mark the left-hand

end 'Location 1' and the right hand end 'Location 2'

Q7.2.2 Table 7.2 indicates the bearings taken from locations 1 and 2 to the collar at various times during a single day. On the diagram you have already produced in answer to Q7.2.1, add four points representing the location of the animal at these times, labelling each point with the appropriate time.

Q7.2.3 In which general compass direction is the animal moving?

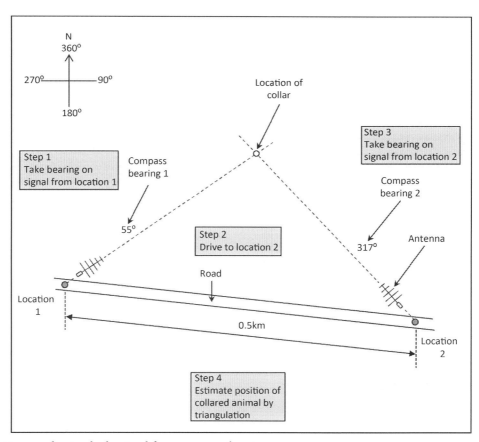

Figure 7.3 Locating a radio-tracked animal from compass bearings.

Table 7.2 Compass bearings obtained for four position fixes during radio-tracking

Time	Compass bearing from location	
	1	2
10.07	51	306
11.30	59	298
12.14	65	287
13.25	71	281

Q7.2.4 What assumption does this method make about the location of the collared animal while the researcher travels from location 1 to location 2?

Q7.2.5 What factor is most likely to limit the amount of time over which a radio-tracking study may be conducted?

Q7.2.6 Under what circumstances should a transmitter be attached to a harness rather than a collar?

Q7.2.7 Discuss the ethical issues that need to be considered before using radio-tracking equipment to study animal movements.

Reference/Further Reading

The American Society of Mammologists (1987). Acceptable field methods of mammalogy, preliminary guidelines prepared by the american society of mammalogists. *Journal of Mammalogy*, 68(Supp 1–18), 4. November p.13.

EXERCISE 7.3

ACTIVITY BUDGETS IN ASIAN ELEPHANTS

How does an elephant spend its day? One of the most basic behavioural studies that may be undertaken on an animal is the construction of an activity budget. This determines how much of its time it spends on basic activities such as feeding, sleeping, walking, etc. and expresses the time spent on these behaviours as a percentage of the duration of the study. Often this is achieved by sampling behaviour at fixed points in time (e.g., every 5 minutes) and then estimating the time spent on each activity as a percentage of all recordings. So, if 100 recordings were made and on 23 of these occasions the animal was feeding, then this activity occupied 23% of the animal's time.

The data in Table 7.3 shows the activity of two adult female Asian elephants (*Elephas maximus*), A and B, in a zoo (Rees, 2009). The meaning of the codes used for each behaviour is provided in Table 7.4

Q7.3.1 Complete Table 7.5 by counting the number of times each behaviour was recorded and then calculating the percentage of all recordings when animal A was engaged in each behaviour on each day. Repeat for animal B. Percentages should be calculated using the total number of actual recordings not the total number of samples. The grey cells in Table 7.3 indicate times when no recording was made because the animal was out of sight.

Q7.3.2 For each elephant, produce a stacked bar chart in which each day is represented by a bar divided into sections indicating the amount of time spent on each behaviour. To facilitate easy comparison, the behaviours should occur in the same vertical sequence in each bar. The total height of each bar represents 100%.

Table 7.3 Behaviours recorded in two Asian elephants (*Elephas maximus*) living in a zoo using 5-minute instantaneous scan sampling for three days

Time	Elephant A			Elephant B		
	Day 1	Day 2	Day 3	Day 1	Day 2	Day 3
10:00	S	S	F	F	L	F
10:05	S	ST	F	F	DT	F
10:10	ST	ST	L	F	S	F
10:15	ST	ST	L	F	S	F
10:20	ST	ST	F	F	S	F
10:25	ST	S	F	F	S	F
10:30	ST	ST	F	F	L	DG
10:35	ST	ST	F	ST	S	DT
10:40	ST	L	DT	S	L	DT
10:45	ST	L	DT	S	S	S
10:50	ST	L	S	S	ST	S
10:55	ST	S	ST	L	S	S
11:00	L	ST	S	S	ST	S
11:05	S	F	ST	ST	S	S
11:10	S	S	ST	S	S	S
11:15	ST	ST	ST	S	L	ST
11:20	S	S	ST	L	ST	S
11:25	S	ST	ST	S	ST	L
11:30	F	S	ST	F	ST	S
11:35	F	S	S	F	S	S
11:40	F	ST	ST	F	ST	AGG
11:45	ST	ST	ST	S	ST	S
11:50	ST	S	ST	L	ST	L
11:55	ST	ST	ST	L	ST	S
12:00	ST	ST	ST	L	ST	S
12:05	ST	DT	ST	L	ST	S
12:10	F	ST	ST	F	ST	L
12:15	F	ST	ST	F	ST	S
12:20	F	ST	ST	F	ST	L
12:25	F	F	L	L	F	S
12:30	ST	F	S	L	F	S
12:35	ST	S	S	S	ST	S
12:40	ST	ST	S	F	ST	L
12:45	ST	ST	S	ST	ST	L
12:50	L	ST	ST	L	ST	S
12:55	ST	ST	ST	ST	ST	L
13:00	ST	ST	ST	S	ST	S
13:05	ST	ST	ST	S	ST	S
13:10	ST	ST	ST	S	ST	S
13:15	ST	ST	ST	S	ST	ST
13:20	ST	ST	ST	O	ST	ST
13:25	O	ST	ST	O	DT	S
13:30	ST	ST	ST	L	ST	ST
13:35	S	ST	ST	S	ST	ST
13:40	S	ST	ST	L	ST	S
13:45	S	ST	ST	S	ST	S
13:50	ST	S	ST	S	ST	S
13:55	ST	ST	ST	L	S	S

Table 7.4 Behaviour codes for the data in Table 7.3

Symbol	Behaviour
AGG	Aggression
DG	Digging
DT	Dusting
F	Feeding
L	Walking (locomotion)
S	Standing
SEX	Sexual behaviour
ST	Stereotyping
0	No recording made

Table 7.5 Analysis of the data in Table 7.3

Symbol	Behaviour	Day 1		Day 2		Day 3	
		No.	%	No.	%	No.	%
AGG	Aggression						
DG	Digging						
DT	Dusting						
F	Feeding						
L	Walking (locomotion)						
S	Standing						
SEX	Sexual behaviour						
ST	Stereotyping						
0	No recording made						
Total recordings made			–		–		–

Q7.3.3 Comment on the differences in the activity budgets of each elephant on each of the three days and on the differences in the patterns of behaviour of the two animals.

Q7.3.4 Which of the two elephants spent the most time exhibiting stereotypic behaviour over the 3-day period?

Reference/Further Reading

Rees, P. A. (2009). Activity budgets and the relationship between feeding and stereotypic behaviors in Asian elephants (*Elephas maximus*) in a zoo. *Zoo Biology, 28,* 79–97.

FEEDING STRATEGIES IN LIZARDS

Desert lizard species fall into two natural groups with respect to the way in which they obtain food: those that 'sit-and-wait', ambushing prey when it passes by, and those that are 'widely-foraging' active hunters. The two groups may be distinguished by their behaviour, physiology and morphology (Table 7.6).

Q7.4.1 Match each of the characteristics listed in Table 7.6 with the appropriate type of lizard in Table 7.7.

Q7.4.2 Copy the axes shown in Fig. 7.4 and then draw straight lines representing the movements of lizards that:

    a. 'sit-and-wait';

    b. 'forage widely'.

Q7.4.3 Which of the two types of lizard would you expect to:

    a. encounter predators least often?

    b. be most vulnerable to predators that 'sit-and-wait'?

Explain why.

Q7.4.4 When attacked, many lizard species shed their tails which continue to wriggle while

Table 7.6 Characteristics of predatory lizards

Characteristic	Alternatives	
Prey type	Active	Sedentary
Morphology	Streamlined	Stocky
Brain size	Large	Small
Volume of prey captured/day	Low	High
Endurance capacity	High	Limited
Learning ability	Limited	Enhanced learning and memory
Daily metabolic cost	Low	High

Table 7.7 Characteristics of 'sit-and-wait' and 'widely-foraging' lizards

Characteristic	Sit-and-wait	Widely-foraging
Prey type		
Morphology		
Brain size		
Volume of prey captured/day		
Endurance capacity		
Learning ability		
Daily metabolic cost		

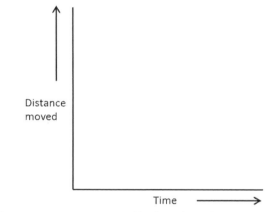

Figure 7.4 Distance moved by lizards with time.

the lizard escapes. This phenomenon is called caudal autotomy.

    a. What function does the wriggling tail perform?

    b. Suggest how the phenomenon of caudal autonomy could be used to study predation frequency in lizards.

Q7.4.5 Some lizards shed their tail if it is bitten by a venomous snake. Why is this useful behaviour in these circumstances?

Reference/Further Reading

Huey, R. B., & Pianka, E. R. (1981). Ecological consequences of foraging mode. *Ecology*, *62*, 991–999.

ROARING CONTESTS IN RED DEER

Many species of animals have evolved a means of assessing the strength of conspecifics without fighting. This often involves signalling systems that one individual may use to assess the strength of another such as the size of horns or antlers, vocalisations or visual displays, such as claw waving in male fiddler crabs (*Uca mjoebergi*) (Morrell, Backwell, & Metcalfe, 2005).

Roaring is used as a means of assessment in red deer (*Cervus elaphus*) (Fig. 7.5). Table 7.8 contains data on the frequency of roars heard in a contest between two red deer stags over a period of 40 minutes.

Q7.5.1 Calculate the total number of roars made by each stag over the 40-minute period (Table 7.8).

Q7.5.2 Calculate the mean number of roars per minute made by each stag during the 40-minute period.

Q7.5.3 Draw a graph showing the number of roars per 4-minute period during the 40-minute study for stags A and B.

Q7.5.4 Decide which stag won this contest and explain your reasoning.

Table 7.8 A roaring contest over a harem by two adult male red deer stags

Time (minutes)	Roars per 4 minutes	
	Stag A	Stag B
4	2	1
8	1	2
12	2	1
16	15	1
20	13	3
24	24	8
28	23	11
32	27	20
36	20	13
40	33	11
Total roars		

Based on a graph of data collected by Clutton-Brock and reproduced in Dawkins & Krebs (1978).

Q7.5.5 Was it possible to predict which stag would win the contest after:

a. 8 minutes?

b. 16 minutes?

Q7.5.6 How do these stags benefit from using roaring to assess each other's strength?

References/Further Reading

Dawkins, R., & Krebs, J. R. (1978). Animal signals: Information or manipulation? In J. R. Krebs & N. B. Davies (Eds.), *Behavioural Ecology. An Evolutionary Approach*. Oxford: Blackwell Scientific Publications.

Morrell, L. J., Backwell, P. R., & Metcalfe, N. B. (2005). Fighting in fiddler crabs *Uca mjoebergi*: What determines duration? *Animal Behaviour, 70*, 653–662.

Figure 7.5 Red deer (*Cervus elaphus*) stags.

EXERCISE 7.6

TERRITORIALITY IN FERAL CATS

Feral cats exist all over the world. They originate from domestic cats (*Felis catus*) but have effectively returned to the wild and, when present in sufficient numbers, have a social organisation similar to that of lions (*Panthera leo*). Social groups generally consist of females with their young, sometimes with resident adult males (Fig. 7.6). Other wandering males (nomads) move between these groups.

Table 7.9 shows the association indices between pairs of male cats living in the grounds of a large hospital in Cheshire, United Kingdom, before and after they were neutered (sterilised) as part of a control programme. A value of 1.0 indicates that the individuals were always seen together; a value of zero indicates they were

Figure 7.6 Feral cats: (A) male 63; (B) female 10 (left), male 44 (right).

Table 7.9 Association indices between 12 adult male feral cats (*Felis catus*) living within a single colony, before (upper value) and after (lower value) being neutered

Cat	27	28	34	44	50	52	53	49	68	23	29
28	0.41 / 0.06										
34	0.12 / 0.05	0.26 / 0.12									
44	0.09 / 0.02	0.21 / **0.66**	0.17 / 0.18								
50	0 / 0	0 / 0	0 / 0	0 / 0							
52	0 / 0.04	0 / 0	0 / 0	0.10 / 0	0.11 / 0						
53	0.08 / 0.04	0.12 / 0	0 / 0.07	0.15 / 0	0 / 0	0.09 / 0					
49	0.32 / 0.08	0.34 / **0.60**	0.16 / 0.14	0.22 / **0.61**	0 / 0	0 / 0	0.13 / 0				
68	0 / 0	0 / 0	0 / 0	0 / 0	0 / **0.59**	0 / 0	0 / 0	0 / 0			
23	0.05 / 0	0.07 / 0.09	0.06 / 0.03	0.11 / 0.09	0 / 0	0 / 0	0.03 / 0.16	0.08 / 0.06	0 / 0		
29	0.31 / 0.07	0.22 / 0.11	0.05 / 0.16	0.12 / 0.06	0 / 0	0 / 0	0.14 / 0	0.34 / 0.10	0 / 0	0.08 / 0.04	
65	0 / 0	0 / 0	0 / 0	0 / 0	0.08 / 0.08	0 / 0	0 / 0	0 / 0	0.15 / 0.04	0 / 0	0 / 0
Cat	27	28	34	44	50	52	53	49	68	23	29

Adapted from Rees (1982). Values >0.5 are indicated in bold.

never seen together. Each animal was assigned a number for the purpose of identification.

Q7.6.1 What evidence exists in the data in Table 7.9 that male feral cats exhibit a territorial system similar to that of cheetahs (*Acinonyx jubatus*) in which individuals use the same areas but at different times?

Q7.6.2 Complete Table 7.10 showing the number of associates for each male cat before and after neutering.

Q7.6.3 Suggest a reason for the changes in associations observed after neutering.

Q7.6.4 Which pair of male cats was not seen associating when entire (before being neutered) but were closely associated when neutered?

Table 7.10 The number of associates of each male cat before and after neutering

Cat	No. of associates	
	Before neutering	**After neutering**
27		
28		
34		
44		
50		
52		
53		
49		
68		
23		
29		
65		
Mean		

Q7.6.5 Which male cat had the smallest number of male associates before he was neutered?

Reference/Further Reading

Rees, P.A. (1982). *The ecology and management of feral cat colonies*. Unpublished PhD Thesis, University of Bradford UK. Available from <http://www.bl.uk/reshelp/findhelprestype/theses/ethos/>.

SAFETY IN NUMBERS: FLOCKING AS AN ANTIPREDATOR DEVICE

Figure 7.7 A woodpigeon (*Columba palumbus*).

Predators that hunt prey species that live in groups are faced with a recurring dilemma: which individual to choose as a target. There is a long-established belief that, for herd or flock animals, there is safety in numbers. Living in a group clearly reduces each individual's chance of being taken by a predator, but does the size of a group reduce the likelihood of a predator being able to take any prey individual at all?

Kenward (1978) tested the hypothesis that flock size in woodpigeons (*Columba palumbus*) (Fig. 7.7) affected the attack success of goshawks (*Accipiter gentilis*) (Table 7.11).

Q7.7.1 Draw a bar chart indicating the relationship between flock size in woodpigeons and the hunting success of goshawks.

Q7.7.2 What conclusion can be drawn from these data about the effect of flock size on hunting success?

Table 7.11 The effect of woodpigeon flock size on the hunting success of goshawks

Flock size (individuals)	Attack success (%)
1	77
2–10	57
11–50	16
50+	8

Q7.7.3 How may this conclusion be explained?

Q7.7.4 Suggest how natural selection could have caused flocking behaviour to evolve.

Q7.7.5 How could natural selection cause the hunting ability of goshawks to improve with time?

Reference/Further Reading

Kenward, R. E. (1978). Hawks and doves: Factors affecting success and selection in goshawk attacks on wood-pigeons. *Journal of Animal Ecology*, 47, 449–460.

THE EFFECT OF PREY DENSITY ON TERRITORY SIZE IN AN AVIAN PREDATOR

Many species exhibit plasticity in their behaviour when environmental conditions change.

Temeles (1987) studied the behaviour of a large bird of prey, the northern harrier (*Circus cyaneus*), in California, United States. Table 7.12 shows the response of this species to changes in the availability of food.

Q7.8.1 Draw a graph of the relationship between prey availability and territory size in the northern harrier.

Q7.8.2 Suggest why 'mouse availability' is defined as the number of prey that were active rather than the number that were present.

Table 7.12 The effect of prey availability on feeding territory size in the northern harrier (*Circus cyaneus*)

Mouse availability[a]	Territory size (ha)
0.5	123
1.6	83
6.0	50
6.1	9
6.8	17
8.5	7
11.5	4

Data derived from Temeles (1987).

[a]Defined as number of prey per 0.25 hectares that were active when harriers were hunting.

Q7.8.3 What behavioural response does the harrier exhibit when the density of active prey is low?

Q7.8.4 In theory, how many harrier territories could an area of 1000 ha support if mouse availability throughout the area was 1.6 individuals per 0.25 ha?

Q7.8.5 How would this change if mouse availability rose to 8.5 individuals per 0.25 ha?

Reference/Further Reading

Temeles, E. J. (1987). The relative importance of prey availability and intruder pressure in feeding territory size regulation by harriers. *Circus cyaneus. Oecologia, 74,* 286–297.

FLEXIBILITY IN THE SOCIAL BEHAVIOUR OF PIED WAGTAILS

Davies (1976) studied pied wagtails (*Motacilla alba*) in an open grassland area near Oxford in England. He followed a pied wagtail from dawn to dusk during 11 days in winter. This individual spent time feeding in its own territory but when food here was scare it left the territory to feed with flock birds nearby (Table 7.13; Fig 7.8).

Q7.9.1 Draw a scatter graph showing the relationship between the percentage of the day spent feeding on the territory and the number of prey items taken per minute while on the territory.

Q7.9.2 When there was little food available on the territory the wagtail still spent some of its time there. Explain why?

Table 7.13 The relationship between feeding rate and time spent in territory in a pied wagtail

Feeding rate in territory (items/min)	Percentage of day spent in territory
3.9	11
5.5	12
7.0	23
7.5	35
8.2	47
9.5	43
11.0	32
14.5	51
13.8	96
19.6	24
19.8	100

Data derived from Davies (1976).

Figure 7.8 A pied wagtail (*Motacilla alba*).

Q7.9.3 Davies was not able to ring most of the pied wagtails in his study. Suggest how he could tell them apart.

Q7.9.4 Davies established that wagtails spent 90% of the day feeding and took one prey item every 4 seconds throughout the day. He found that up to 500 wagtails used two reed bed roosts near his study area. How many food items (insects) would 500 wagtails take if feeding for 90% of a 12-hour feeding period when taking food items at a rate of one every 4 seconds?

Reference/Further Reading

Davies, N. B. (1976). Food, flocking and territorial behaviour of the pied wagtail (*Motacilla alba yarrellii* Gould) in winter. *Journal of Animal Ecology*, 45, 235–253.

OPTIMAL FORAGING THEORY: USING GAME THEORY TO STUDY ECOLOGICAL STRATEGIES

Organisms need to develop ecological strategies to ensure their survival. Ecologists sometimes use game theory in order to study these strategies.

Imagine a predator that feeds on two different types of prey: one that can run quickly and one that is much slower. Each time the predator encounters a prey organism it must decide whether to run quickly to catch the prey or to save energy and run slowly. It cannot distinguish between the two prey types (fast and slow) until the chase begins and then it is too late to change strategy.

The predator must feed in order to obtain energy. Both types of prey provide the same amount of energy but running quickly uses more energy than running slowly. Running slowly after a fast prey will use energy without resulting in a kill; running quickly after a slow prey is likely to result in a kill but will waste energy. So, what strategy should the predator adopt to maximise its energy intake? Game theory attempts to answer this question (Tables 7.14–7.18).

Q7.10.1 Complete Tables 7.16 – 7.18 and calculate the energy pay-off to the predator when it attacks four prey organisms (two fast and two slow) using the following strategies:

a. All fast attacks;

b. All slow attacks;

c. Fifty percent fast attacks and fifty percent slow attacks.

Table 7.14 Predator's energy costs and gains

Energy cost/gain	Energy (arbitrary units)
Energy gained by catching a slow prey (gross)	10.0
Energy gained by catching a fast prey (gross)	10.0
Cost to predator of running slowly	−2.5
Cost to predator of running quickly	−5.0

Table 7.15 A game matrix indicating the outcome of the two strategies

Predator strategy	Fast prey (gain = 10.0)	Slow prey (gain = 10.0)
Slow attack (cost = −2.5)	0−2.5 = −2.5[a]	10.0−2.5 = 7.5
Fast attack (cost = −5.0)	10.0−5.0 = 5.0	10.0−5.0 = 5.0

[a]The predator does not catch the prey so loses 2.5 units of energy.

Table 7.16 All fast attacks

Predator strategy	Prey type	Energy gain
Fast	Fast	
Fast	Fast	
Fast	Slow	
Fast	Slow	
Total		

Table 7.17 All slow attacks

Predator strategy	Prey type	Energy gain
Slow	Fast	
Slow	Fast	
Slow	Slow	
Slow	Slow	
Total		

Table 7.18 Fifty percent fast attacks and fifty percent slow attacks

Predator strategy	Prey type	Energy gain
Fast	Fast	
Slow	Fast	
Fast	Slow	
Slow	Slow	
Total		

Q7.10.2 Assuming that half of the prey are of the slow type and half are of the fast type, which of the three strategies examined here should the predator adopt in order to maximise its energy gain?

Q7.10.3 Would your answer to this question change if the predator could predict the speed of its prey before the chase started, and why?

References/Further Reading

Dugatkin, L. A., & Reeve, H. K. (Eds.), (1998). *Game theory and animal behaviour*. Oxford: Oxford University Press.

Pyke, G. H. (1984). Optimal foraging theory: A critical review. *Annual Review of Ecology and Systematics, 15,* 523–575.

Smith, J. M. (1984). Game theory and the evolution of behaviour. *Behavioral and Brain Sciences, 7,* 95–101.

EVOLUTIONARILY STABLE STRATEGIES: REPRODUCTION IN DUNG FLIES

Figure 7.9 Dung flies on a cow pat (*Scathophaga* sp.).

An ESS is an inherited strategy (usually behavioural) which, if practised by most members of a population, cannot be supplanted during evolution by a different strategy.

Parker and Stuart (1976) studied reproductive strategies in dung flies (*Scathophaga*). Female dung flies visit fresh cow pats looking for a place to lay eggs (Fig. 7.9). Males congregate there seeking females to mate. It takes about 100 minutes for a male to fertilise all of one female's eggs. The male guards the female while she lays eggs then flies off to find another female. If he leaves before the end of 100 minutes there is a chance that another male will mate with her and fertilise some of her eggs. Guarding and searching took an average of 2.5 hours

(156 minutes). The male is faced with a dilemma. Should he copulate for the full 100 minutes to fertilise all of the eggs of each female or stop copulating sooner so that he can make a start on searching for his next mate? Which strategy would produce the most offspring? Somewhere between 0 and 100 minutes there is an optimum copulation duration which will result in the male leaving the maximum number of offspring when moving between females.

Behavioural ecology theory predicts that the actual duration of copulation will be near the optimum. Parker and Stuart measured the proportion of eggs fertilised for various periods of copulation in the laboratory. Their results are presented in Table 7.19.

Table 7.19 The relationship between time spent copulating and the proportion of eggs fertilised

Time spent copulating with female (mins)	Proportion of eggs fertilised
9.0	0.51
22.0	0.72
35.5	0.86
68.0	0.98
100.0	0.99

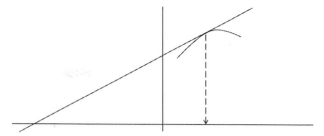

Figure 7.10 Estimating optimum copulation time from a graph.

Q7.11.1 Draw a scatter graph of the proportion of eggs fertilised against time spent copulating. The *x*-axis of your graph should be from −160 to +100 minutes.

Q7.11.2 Draw a smooth curve through the points.

Q7.11.3 Draw a line from −156 on the *x*-axis to the point where it touches your curve. Draw a vertical line from this point and determine where it crosses the *x*-axis (Fig. 7.10). This value is the optimum time a male fly should spend copulating before leaving to search for another female.

Q7.11.4 The optimum copulation time recorded by Parker and Stuart in the field was 35.5 minutes. How close is your predicted value to this?

Reference/Further Reading

Parker, G. A., & Stuart, R. A. (1976). Animal behaviour as a strategy optimizer: Evolution of resource assessment strategies and optimal emigration thresholds. *American Naturalist, 110*, 1055–1076.

INDUSTRIAL MELANISM: HIDING IN PLAIN SIGHT

Figure 7.11 Two stone terraced houses in a village in Lancashire. The walls on the right have been cleaned but those on the left still retain the black colouration caused by pollution. These houses were built for the employees of the large cotton mill that existed in the village until around 1970.

Trees, walls and building were blackened by pollution (especially soot) during the Industrial Revolution in Britain and many species of lichens died out (Fig. 7.11). In 1956, Parliament passed the Clean Air Act. This improved the air quality in urban areas by introducing smoke control areas in some towns and cities. Thereafter air quality continued to improve and the general environment in urban areas became cleaner as soot-blackened stone buildings were cleaned.

This change in air pollution levels provided an important opportunity for biologists to study the process of evolution. The peppered moth (*Biston betularia*) is polymorphic. It has two basic colour morphs. The typical form (*typica*) is white but a melanic form exists (*carbonaria*) which is black, and there is an intermediate form known as *insularia*. This polymorphism is genetically controlled.

The first melanic form was found in Manchester (the heart of the Industrial Revolution) around 1850 (Sheppard, 1975). It thereafter increased in frequency, reaching more than 95% of the population of the species in some urban areas.

Kettlewell (1973) showed that in urban areas *carbonaria* was at a selective advantage because it could hide from predatory birds on dark surfaces. In clean rural areas *typica* was at an advantage because it could hide amongst grey-coloured lichens on trees and walls.

Q7.12.1 Draw a graph showing the change in the percentage of the *carbonaria* variety at different distances west of the city of Liverpool using the data in Table 7.20.

Q7.12.2 Describe the pattern shown by your graph.

Q7.12.3 Explain the differences in the percentage of the *carbonaria* found at the locations listed in Table 7.20.

Q7.12.4 You are a biologist working in Britain in the 1960s. In order to study the effects of predation on *Biston* colour morphs, you pin 50 dead *typica* moths and 50 dead *carbonaria* moths on trees in a park in an industrial area where the bark has been blackened by soot and the same number of each morph on trees in a park in a

Table 7.20 Percentage of *carbonaria* in random samples of *Biston betularia* collected from sites in Merseyside and across North Wales

Location	carbonaria %	Approximate distance from Liverpool (km)
Bangor	3.1	160
Birkenhead	93.3	15
Liverpool	97.3	0
Mostyn	60.8	45
Old Colwyn	11.8	100
Prestatyn	46.4	60
Rhyl	46.2	70

Based on data in Bishop (1972).

rural area well away from any sources of industrial pollution. You return to these sites a week later. What would you expect to find and why?

References/Further Reading

Bishop, J. A. (1972). An experimental study of the cline of industrial melanism in *Biston betularia* (L.) (Lepidoptera) between urban liverpool and rural North Wales. *Journal of Animal Ecology*, 41, 209–243.

Kettlewell, B. (1973). *The evolution of melanism.* Oxford: Clarenden Press.

Sheppard, P. M. (1975). *Natural selection and heredity* (4th ed.). London: Hutchinson University Library.

ZINC TOLERANCE IN *AGROSTIS CAPILLARIS*

Toxic metal tolerance has evolved in a number of species of plants, particularly grasses found in association with mine waste. The spoil heaps associated with such mines contain concentrations of metals which are normally toxic to most plants but many are colonised by a small number of species. The data in Table 7.21 show the occurrence of a tolerant ecotype of the grass *Agrostis capillaris* along a transect across a mine and an adjacent pasture in North Wales.

Q7.13.1 Draw a line graph indicating the change in zinc tolerance along the transect.

Q7.13.2 Draw a vertical line on the graph where you think the boundary occurs between the mine waste (contaminated with zinc) and the uncontaminated pasture. Indicate on the graph which side of the line represents the contaminated soil.

Table 7.21 Zinc tolerance of *Agrostis capillaris* plants along a transect in North Wales

Distance along transect (m)	Index of zinc tolerance[a]
0	78
25	85
43	96
47	77
52	29
64	12
81	17
100	22

[a]A high value for the index indicates a high level of tolerance.

Q7.13.3 Would you describe the loss of tolerance at the boundary between the mine and the pasture as abrupt or gradual?

Q7.13.4 The plants growing near and either side of the boundary flower at different times. How does this help to maintain genetic differences in the population?

Q7.13.5 Would you expect to find an increase or a decrease in self-fertility in the tolerant populations? Explain your answer.

Q7.13.6 Why would you expect tolerance to be rare in plant species whose reproduction is predominantly vegetative?

Q7.13.7 What practical use may be made of the phenomenon of metal tolerance in grasses in industrial areas?

References/Further Reading

Bradshaw, A. D. (1993). Restoration ecology as a science. *Restoration Ecology*, 1, 71–73.

Karatagliss, S. S., McNeilly, T., & Bradshaw, A. D. (1986). Lead and zinc tolerance of *Agrostis capillaris* L. and *Festuca rubra* L. across a mine-pasture boundary at Minera, North Wales. *Phyton*, 26, 65–72.

INBREEDING AND HOMOZYGOSITY

Inbreeding is the mating of closely related individuals. It causes an increase in homozygosity in a population. In other words, the proportion of a population which is homozygous dominant or homozygous recessive for a particular gene will increase and the proportion of heterozygotes will decrease.

The effect of inbreeding on homozygosity is best understood by considering a hypothetical example. Imagine a theoretical self-fertilising animal that is heterozygous for the allele A and therefore possesses the genotype 'Aa.' If this animal were to self-fertilise and produce four offspring, one half of them would be heterozygotes (Aa) and therefore identical to the parents, a quarter would be homozygous dominant (AA), and a quarter would be homozygous recessive (aa) (Fig. 7.12 (inset)). The homozygous individuals produced will only create offspring with the same genotypes as themselves, so AA individuals will only produce more AA individuals and aa individuals will only produce more aa individuals. If we assume that each animal produces four offspring in each generation, the result is that the proportion of heterozygotes (Aa) in the population as a whole halves with each generation (Fig. 7.12).

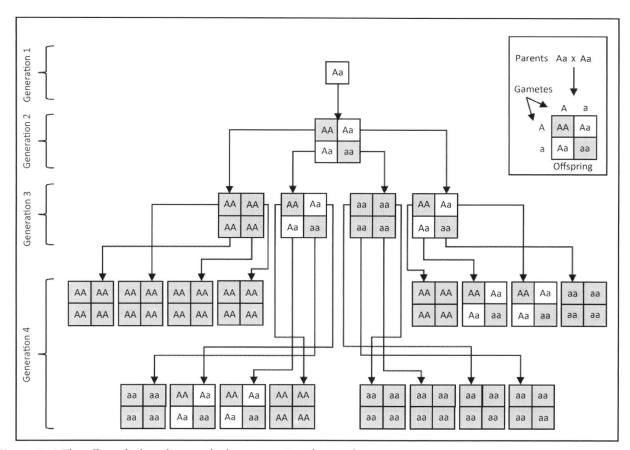

Figure 7.12 The effect of inbreeding on the homozygosity of a population.

Q7.14.1 Draw a graph showing the change in the percentage of homozygotes in the population between generation 1 and generation 8.

Q7.14.2 What proportion of the population in this hypothetical example will be homozygous recessive (aa) after eight generations?

Q7.14.3 What proportion of the population will be heterozygous (Aa) after eight generations?

Q7.14.4 If the recessive allele (a) caused a disease, what proportion of the population would have the disease in generation 7?

Q7.14.5 Explain why inbreeding is a potential problem for animals kept in zoos and how this can be reduced.

References/Further Reading

Frankham, R., & Ralls, K. (1998). Conservation biology: Inbreeding leads to extinction. *Nature, 392*, 441–442.

Hedrick, P. W., & Kalinowski, S. T. (2000). Inbreeding depression in conservation biology. *Annual Review of Ecology and Systematics*, 139–162.

EXERCISE 7.15

THE PROBLEM OF GENETICALLY ISOLATED POPULATIONS: INBREEDING IN LION AND BLACK RHINOCEROS

The following text is an extract from a paper published in 2013 by Louise Oates and Paul Rees entitled *The historical ecology of the large mammal populations of Ngorongoro Crater, Tanzania, east Africa* (Fig. 7.13).

Inbreeding depression may affect the crater's rhinoceros population (Moehlman et al., 1996; Mills et al., 2006) and Mills et al. (2006) reported that the dominant male had sired the majority (12 of 19 individuals) of the rhinoceros. However, neither Moehlman et al. (1996) nor Mills et al. (2006) found evidence of any deleterious effects resulting from inbreeding. Nevertheless, in order to address the possible future threat of

inbreeding, five rhinoceros were moved from Addo Elephant National Park, South Africa to the crater in 1998. It is believed that these individuals were carrying babesiosis (Hilsberg et al., 2003, Mills et al., 2006). However, it is unclear whether two individuals that subsequently died of babesiosis during the drought of 2000 were residents as claimed by Fyumagwa et al. (2004) or translocated animals as claimed by Nijhof et al. (2003).

The Ngorongoro lions are an isolated population and the 1962 reduction in numbers, together with limited immigration of new individuals, resulted in a much reduced genetic diversity, confirmed by

Figure 7.13 The Lerai forest in Ngorongoro Crater is an important breeding area for the black rhinoceros population. A permanent ranger station is located on the edge of the forest.

DNA analyses (O'Brien et al., 1987; Yuhki & O'Brien, 1990; Packer et al., 1991b). Computer simulations have shown that the population may have passed through similar genetic bottlenecks prior to 1962. Inbreeding may increase the population's susceptibility to disease outbreaks (Packer et al., 1991b) and may reduce fertility through an increase in sperm and testicular abnormalities (Wildt et al., 1987, Munson et al., 1996). Strong coalitions of lions may have prevented immigration of other males and resulted in the genes of members of the coalitions being overrepresented in the population (Packer et al., 1991a).

Note that scientific names of species appeared earlier in the full paper so are not given in this extract. Readers interested in the papers mentioned in the extract should refer to the original work by Oates and Rees (2013).

Q7.15.1 What long-term problems could result from what is known about the parentage of the black rhinoceros population prior to 2006?

Q7.15.2 Why were black rhinoceros from South Africa translocated to the crater in 1998?

Q7.15.3 What evidence supports the suggestion that the rhinoceros translocated into the crater were responsible for deaths from babesiosis?

Q7.15.4 Is there any evidence of deleterious effects from inbreeding in the black rhinoceros population?

Q7.15.5 What aspect of lion behaviour in the crater may have damaging genetic effects in the long-term?

Q7.15.6 What evidence exists that the lion population has passed through a genetic bottleneck in the past?

Reference/Further Reading

Oates, L., & Rees, P. A. (2013). The historical ecology of the large mammal populations of Ngorongoro Crater, Tanzania, east Africa. *Mammal Review, 43*, 124–141.

EXERCISE 7.16

CALCULATING EFFECTIVE POPULATION SIZE

If the total population size of a species is 100 animals, all males, and the species is only capable of reproducing sexually, the population is doomed to become extinct when the last animal dies. Without any females, this population has an effective size of zero. Effective population size is an important concept in conservation genetics.

The effective population size is a measure of how well a particular population maintains genetic diversity from one generation to the next. It is usually lower than the actual population size. This may be because the sex ratio is unequal, some animals are beyond their reproductive age, or for some other reason.

Small populations lose genes as a result of chance events. For example, the last two individuals possessing the gene for a particular character might be killed by severe weather, thereby removing this gene from the population: a process known as genetic drift.

The effective population size is defined as the size of an ideal population that would lose genetic variation by genetic drift at the same rate. In other words, a population of 300 individuals may have an effective population size of, say, 250 because there are too few females in the population. So this population of 300 actually loses genetic variation at the same rate as an ideal population of 250.

The effective population size is essentially a measure of the number of individuals that are effectively contributing genes to the next generation. Effective population size (N_e) may be calculated as:

$$N_e = \frac{4(N_f \times N_m)}{N_f + N_m}$$

where,

N_f = number of breeding females; and

N_m = number of breeding males.

Strictly speaking, this formula only applies to stable populations with nonoverlapping generations where individuals mate at random.

Example

In a population of 423 tigers (*Panthera tigris*) there are 190 males and 233 females. Assuming all individuals are capable of breeding, the effective population size is:

$$N_e = \frac{4\left(N_f \times N_m\right)}{N_f + N_m}$$

$$= \frac{4 \times (233 \times 190)}{233 + 190}$$

$$= 418.6$$

Q7.16.1 Complete Table 7.22 by calculating the effective population size for each population.

Table 7.22 The effective population size of populations with a variety of sex ratios

Breeding males	Breeding females	Actual population size	Effective population size
0	100	100	
10	90	100	
20	80	100	
30	70	100	
40	60	100	
50	50	100	
60	40	100	
70	30	100	
80	20	100	
90	10	100	
100	0	100	

Q7.16.2 Draw a line graph showing the effect of changing the number of breeding males on the effective population size.

Q7.16.3 Which ratio of breeding males to breeding females produces the highest effective population size?

Q7.16.4 Explain how artificial insemination may be used to increase the effective population size of a captive population of animals held in zoos:

a. where individuals of the species are difficult to move and using only captive specimens;

b. using individuals from wild populations of the species without taking them into captivity as part of a breeding programme.

Reference/Further Reading

Charlesworth, B. (2002). Effective population size. *Current Biology*, *12*, R716–R717. http://dx.doi.org/10.1016/S0960-9822(02)01244-7.

EXERCISE 7.17

CHARLES DARWIN: CAREFUL SCIENTIST OR CLUMSY AMATEUR NATURALIST?

Charles Darwin has been credited with having had 'the single best idea that anyone has ever had' (Dennett, 1996): evolution by natural selection. Students' and teachers' understanding of Darwin's contribution to biology inevitably comes largely from school and college textbooks. But, many of these textbooks present an idealised view of Darwin's development of his theory of evolution. He is often depicted as an elderly gentleman in busts, statues and photographs (Fig. 7.14), but these were created after he became famous; he was young and inexperienced when he collected specimens while travelling around the world on HMS Beagle. When he left England on his five-year voyage Darwin was just 22 years old. The following extract is part of a paper that discusses textbooks' misconceptions about Darwin (Rees, 2007).

… texts imply that Darwin made careful collections of specimens from around the world and that he understood their importance at the time of collection.

What impressed him most were the distinct variations between the species which inhabited [the Galapagos Islands]…
Toole and Toole (1987).

In particular he was intrigued by the characteristic distribution of species of tortoises and finches.
Green, Stout, and Taylor (1984).

Darwin remained in England after returning from his voyage on HMS Beagle, so most of his collecting was undertaken when he was very young. When the Beagle set sail he was aged just 22. Most textbook photographs show him much later

in life – commercial photography only came into existence after Darwin returned from his voyage – but when he began collecting specimens he was an inexperienced and rather disorganised graduate in divinity. He was appointed to the position of naturalist on HMS Beagle more because his social status made him a suitable companion for the captain, Robert FitzRoy, than for his abilities as a naturalist. Darwin was his second choice.

During his visit to the Galapagos Islands, Darwin rarely bothered to label any of the specimens he collected by island because he did not think it important. Although he was told during the final days of his visit that many trees and tortoises (Fig. 7.15) were unique to each island, by then it was too late and his collecting was finished. As the Beagle crossed the Pacific he

Figure 7.14 Charles Darwin. Library of Congress Reproduction Number: LC-USZ62-52389 (b&w film copy neg.).

Figure 7.15 Giant Galapagos tortoise (*Geochelone nigra*).

ate the tortoises, and the carapaces – the most obvious clue to the adaptive radiation of the species – were thrown into the sea (Desmond and Moore, 1991).

On returning from his voyage, Darwin presented the Zoological Society of London with 80 mammals and 450 birds on condition that they were all to be mounted and described. He had to rely upon experts to identify and catalogue his collection because he lacked the expertise to do this work himself. During a visit to the Linnaean Society, Darwin was forced to admit that "[he] knew no more about the plants, which [he] had collected, than the Man in the Moon" (Desmond and Moore, 1991).

…textbooks frequently contain diagrams and descriptions of Galapagos (Darwin's) finches, and the reader is generally left to infer that the phenomenon of adaptive radiation which they illustrate was discovered by Darwin.

Darwin was particularly interested in the Galapagos finches for he could see in them the key to understanding the evolutionary process.

Roberts (1976).

One of Darwin's great achievements was the idea that the different types of finch on the Galapagos Islands could have evolved because populations of one species had been isolated on the separate islands

Bailey and Hurst (2001).

Darwin had great difficulty telling the finches in the Galapagos Islands apart and mixed up the samples of birds collected from different islands. At the end of his stay he admitted to an "inexplicable confusion" (Desmond and Moore, 1991). He could not separate them into species and did not appreciate the significance of the shapes of their bills. Darwin believed that the group contained finches, wrens, 'Gross-beaks' and 'Icteruses' (relatives of blackbirds). The specimens were badly labelled and not considered to be particularly important by Darwin. He had no sense that they were members of a closely related group with bills adapted to the exploitation of particular niches (Desmond and Moore, 1991). It was John Gould, Superintendent of stuffed birds at the Zoological Society, who recognised that the specimens represented 12 species of closely related finches (Gould, 1837). It was only later that Darwin appreciated the evolutionary significance of this. The more detailed work on resource utilisation by Galapagos finches was undertaken much later, principally by David Lack (Lack, 1947). One modern … text (Roberts, Reiss and Monger, 2000) acknowledges that Darwin did not initially understand the significance of the Galapagos finches:

At the time these did not interest him as much as some of the other organisms. But later he saw in these finches the key to understanding the evolutionary process.

A second text (Boyle and Senior, 2002) goes further and acknowledges the role of Gould, but without naming him:

Interestingly, Darwin did not immediately recognise the significance of all of these different species; he simply collected a large number of specimens in a bag and presented them to a bird specialist on his return home.

Acknowledgement of Lack's role in studying adaptive radiation in Galapagos finches, if it appears at all, often consists of a reference in the title of a figure showing differences between their bills. Some texts, however, do more fully describe his contribution (e.g., Green, Stout, and Taylor, 1984).

Readers interested in the texts mentioned in this extract should refer to the original work by Rees (2007).

Q7.17.1 What evidence exists that Darwin did not appreciate the true significance of the specimens that he collected at the time they were acquired?

Q7.17.2 To what extent did Darwin's work depend upon the assistance of more experienced scientists?

Q7.17.3 Is there any evidence that the young Darwin was a rather careless field naturalist?

Q7.17.4 Galapagos finches are sometimes called 'Darwin's finches'. Would it be more appropriate to call them 'Lack's finches'?

Q7.17.5 Is it scientifically legitimate for authors of college textbooks to oversimplify Darwin's appreciation of the process of evolution in the early days of his career?

References/Further Reading

Dennett, D. C. (1996). *Darwin's dangerous idea: Evolution and the meaning of life.* London, UK: Penguin.

Desmond, A., & Moore, J. (1991). *Darwin.* London, UK: Michael Joseph.

Rees, P. A. (2007). The evolution of textbook misconceptions about Darwin. *Journal of Biological Education, 41,* 53–55.

WEIGHT DISTRIBUTION OF SEEDS AND NATURAL SELECTION IN THE HORSE CHESTNUT TREE

The best known distribution found in biology is the 'normal' or Gaussian distribution. This occurs in variables such as height, weight, wingspan and length of pregnancy. It takes the form of a smooth, symmetrical, bell-shaped curve, where the mean of the population corresponds to the highest point on the curve (see Chapter 10: Statistics). Often such distributions are the result of the interactions of several genes and the effects of the environment.

The seed of the horse chestnut tree (*Aesculus hippocastanum*) is commonly known as a 'conker' (Figs. 7.16–7.18). These seeds fall from the trees in the autumn (fall) inside spiky green seed cases. Table 7.23 contains the weights of a sample of 200 conkers.

Q7.18.1 For this sample of conker weights calculate:

    a. the mean;

    b. the standard deviation; and

    c. the range.

Q7.18.2 Draw histograms showing the distribution of conker weights using the following class intervals:

    a. 1.01–3.00 g, 3.01–6 g, 6.01–9.00 g…etc.;

Figure 7.17 Horse chestnut seed cases.

Figure 7.16 A horse chestnut tree (*Aesculus hippocastanum*).

Figure 7.18 Some of the conkers measured in the study.

Table 7.23 The weights of 200 conkers measured in grams to the nearest 0.01 g

6.50	7.58	4.15	3.41	8.12	5.74	3.62	3.98
6.23	6.56	3.94	3.44	10.81	8.39	8.21	8.49
4.11	6.52	6.39	6.04	9.12	5.73	4.89	8.92
4.87	3.72	5.25	3.85	9.61	3.94	2.97	7.66
7.30	6.08	2.18	4.96	6.79	5.25	7.10	7.68
10.11	6.73	5.87	3.58	11.25	4.79	8.16	7.85
5.99	8.77	5.46	2.64	6.85	1.98	10.24	6.54
3.66	4.16	4.14	4.84	4.12	1.91	6.83	6.56
5.37	7.05	6.29	3.98	4.84	5.77	3.73	2.48
3.75	2.81	5.09	2.74	4.74	8.33	4.50	7.06
9.13	5.58	4.67	3.42	9.31	11.46	5.96	9.04
5.27	6.53	4.46	4.92	7.13	5.95	5.15	3.80
6.21	2.02	7.79	3.96	13.64	10.40	7.00	6.46
8.33	4.53	4.61	5.77	7.07	5.91	5.43	5.17
7.11	4.06	6.92	3.40	6.18	8.30	9.66	9.77
5.90	4.43	6.33	3.70	6.30	6.91	5.22	4.59
11.45	5.27	5.74	1.79	8.83	6.88	10.07	5.84
6.31	4.31	5.20	2.61	5.40	11.52	6.32	14.66
9.18	4.03	3.68	2.17	9.39	4.81	9.80	9.82
1.12	6.94	5.70	3.14	10.84	6.39	5.50	6.05
2.60	6.68	4.36	2.46	7.59	12.64	5.79	13.34
9.53	8.99	3.87	2.87	6.61	4.23	8.62	5.29
5.99	2.13	4.65	3.22	5.05	7.78	7.81	4.64
3.23	4.40	2.87	3.50	3.94	6.81	6.54	8.97
8.02	4.18	5.42	7.32	5.05	7.66	7.17	6.93

b. 1.01–2.00 g, 2.01–3.00 g, 3.01–4.00 g…etc.;

c. 1.01–1.50 g, 1.51–2.00 g, 2.01–2.50 g…etc.

Q7.18.3 Describe how the class interval used to graph the distribution of weights affects the appearance of the graphs.

Q7.18.4 Discuss whether or not the weights of conkers in this study exhibit a normal distribution?

Q7.18.5 Children gather and remove conkers when they fall from the trees. They may use them to play a tradition game of 'conkers' in which a conker is suspended from a piece of string and smashed against that of an opponent with the intention of breaking it. How could this help to explain the distribution of weights in this sample?

Reference/Further Reading

Conran, R. J. (1980). Nature study: Its rehabilitation. *Education 3–13, 8*, 12–18.

CHAPTER 8

ENVIRONMENTAL POLLUTION AND PERTURBATIONS

This chapter contains exercises concerned with environmental pollution and changes in the normal state of ecosystems, or perturbations, caused by human activities.

A wind turbine

Examining Ecology. DOI: http://dx.doi.org/10.1016/B978-0-12-809354-2.00008-7

INTENDED LEARNING OUTCOMES

On completion of this chapter you should be able to:

- Calculate the degree of concentration of a pesticide as it passes along a food chain.

- Draw a graph of the effect of insecticide concentration on insect mortality and estimate the LD_{50}.

- Recognise taxon-specific responses to toxic chemicals.

- Distinguish between the causes of natural and cultural eutrophication.

- Predict the effects of eutrophication on the chemistry and biology of a water body.

- Describe the effects of organic pollution on the ecology of a river.

- Draw a graph illustrating the decay of a radioactive isotope and explain the persistence of radioactive isotopes in the environment.

- Analyse and explain the effects of acid rain on biodiversity in lakes.

- Calculate diversity indices and use them to distinguish between polluted and unpolluted ecosystems.

- Describe the effect of thermal pollution on poikilotherms.

- Use graphical methods to explain the effects of a temperature inversion on air pollution.

- Explain the relationships between urbanisation, forestry and hydrology.

- Discuss the effect of climate change on polar ecosystems.

INTRODUCTION

Pollution is the release into the environment (air, water or land) of any substance that is harmful to ecosystems or organisms, including people. The amount of harm done by any particular pollutant will depend upon various factors, including its toxicity, its extent and its persistence in the environment. Clearly, a pollution incident involving a harmful substance that has low toxicity, is spread over a small geographical area and only persists for a short time in the environment before being broken down is less serious than an incident involving a highly toxic and persistent substance over a very large area.

An environmental perturbation is a change in, or disturbance of, the normal state of the environment, e.g., deforestation, a drought, climate change. The cutting down of a forest may increase runoff, interfere with nutrient cycles and reduce biodiversity. Climate change may result in an increase in global temperature causing changes in the distribution of wild animals and plants and affecting geographical patterns of agriculture.

PESTICIDES IN FOOD CHAINS

There has been great concern about the presence of pesticides in food chains since the early 1960s. The American biologist Rachel Carson published her book *Silent Spring* in 1962, warning of the dangers of continuing to use DDT as an insecticide and predicting a world in which much of nature would be destroyed; where the wild birds would fall silent (Carson, 1962). DDT accumulates and is concentrated in organisms. If it is applied to crops in a low concentration, each step in the food chain causes further concentration until it is present at toxic levels, especially in top carnivores.

WATER POLLUTION

Water is essential to all life but humans have used it to dispose of their waste for hundreds of years. Sewage and chemical pollutants have been poured into rivers and oceans without any thought for the ecological consequences. Organic wastes, such as sewage, animal manure, and dairy waste, can cause fertilisation of water resulting in an overgrowth of plants (eutrophication). This can have serious consequences for wildlife. Heavy metals such as mercury may enter marine food chains and can have devastating effects on human health when it passes along these food chains and into marine organisms we use as food.

RADIOACTIVITY

The presence of radioactive isotopes in the environment is of great concern because many have very long half-lives resulting in their persisting in ecosystems for extremely long periods. Radioactivity can cause genetic mutations which may not be apparent in the individuals that are directly exposed to it but which are transmitted to their offspring. Thus radioactivity poses a hidden threat which may take a generation or more to manifest itself.

TOXICITY

The toxicity of a substance is the extent to which it produces harmful effects on an organism. Some substances are beneficial in low concentrations but toxic in high concentrations. For example, vitamin A is important to human health but toxic to humans in high concentrations. However, polar bears (*Ursus maritimus*) and some other Arctic top predators are tolerant of vitamin A and their livers contain high levels.

POLLUTION MONITORING, BIOINDICATORS AND DIVERSITY INDICES

Some species act as pollution monitors because of their tolerance or intolerance of particular substances. They may be thought of as indicator species or 'bioindicators'. A lichen is an association of a green alga and a fungus living together symbiotically for their mutual benefit. Some species are tolerant to air pollution but others are highly intolerant. Consequently, the number and types of lichens growing in an area may be used to assess the quality of the air.

The biological diversity of a particular area may be assessed using a biodiversity index. A number of indices exist, all of which are dimensionless, so have no units. Each index makes different assumptions so it is not possible to compare values unless they are obtained using the same index. These indices are often used to measure changes in ecosystems, for example, those caused by pollution.

CLIMATE CHANGE

Decades of industrial activity have altered the chemistry of the atmosphere to such an extent that the world is now experiencing changes to the global climate that will be extremely difficult to limit. The Industrial Revolution led to the uncontrolled construction of factories which burned coal to create steam to power their machinery (Fig. 8.1). Burning coal releases carbon dioxide, sulphur dioxide, particles and mercury into the atmosphere posing a risk to human health, causing acid rain and changing the heat balance

Figure 8.1 The brick chimney that is all that remains of the former Halliwell Bleachworks in Bolton, England, founded in 1739.

of the planet. Reducing the pollution from fossil fuels is one of the greatest challenges humans face.

The following exercises provide examples of the ways in which pollutants move and accumulate in ecosystems, the effect of pollutants on biological systems including their toxicity, the detection of pollution using diversity indices and bioindicators, the effects of urbanisation and deforestation on ecosystems and changes to the global ecosystem caused by climate change.

Reference

Carson, R. (1962). *Silent spring*. Boston, MA: Houghton Mifflin.

THE BIOLOGICAL CONCENTRATION OF DDT RESIDUES IN FOOD CHAINS

Dichlorodiphenyltrichloroethane (DDT) is a chlorinated hydrocarbon insecticide. Once released into the environment it persists for long periods of time. Fig. 8.2 shows the concentration of DDT residues (DDT, DDD* and DDE**) measured in organisms from an aquatic ecosystem in the United States. All of these residues are toxic.

Q8.1.1 Rank the organisms from those with the lowest to those with the highest concentration of DDT residues in their bodies.

Q8.1.2 Draw a bar chart to show the variation in residue concentrations in these organisms (using the ranked values, excluding the value for water and beginning with plankton).

Q8.1.3 Examine one of the food chains in this ecosystem by:

a. drawing a simple food chain showing the relationships between the needlefish, silverside minnow, cormorant, plankton and water. Note that predators are coloured grey; the herring gull is a scavenger.

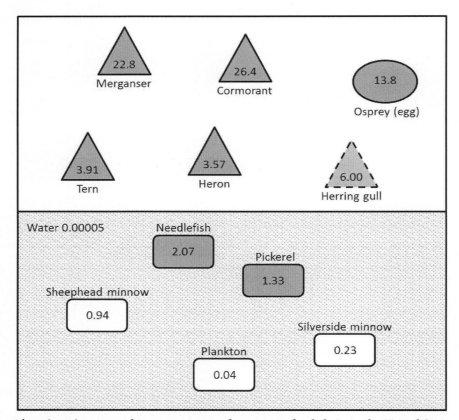

Figure 8.2 DDT residues (ppm) measured in components of an aquatic food chain in the United States.

Based on Woodwell, Wurster & Isaacson (1967).

b. showing the residue concentration in each component of this food chain and indicating the number of times the concentration has increased at each stage. For example, if the concentration is 10 ppm in a herbivore and 100 ppm in the carnivore that eats it, the concentration has increased by 100/10 = 10 times (i.e., ×10).

Q8.1.4 How many times greater is the concentration of DDT residues in the cormorant than in the water?

Q8.1.5 Suggest two different ways in which fish may absorb DDT.

Q8.1.6 Explain why top carnivores are extremely susceptible to poisoning by DDT residues.

Q8.1.7 In birds, there is evidence that DDE interferes with the action of an enzyme that assists in the transport of calcium carbonate from the blood to the egg gland. Suggest how exposure to sublethal doses of DDE may affect the population growth of birds of prey.

*dichlorodiphenyltrichloroethane.
**dichlorodiphenylethane.

Reference/Further Reading

Woodwell, G. M., Wurster, C. F., & Isaacson, P. A. (1967). DDT residues in an east coast estuary: A case of biological concentration of a persistent insecticide. *Science, 156*, 821–824.

PESTICIDES AND EGGSHELLS

You have been asked to examine a number of eggs of sparrowhawks (*Accipiter nisus*) held in the collections of three museums in Britain. The year in which each egg was collected was recorded and you have been asked to measure the thickness of each eggshell to determine if there has been any change in thickness with time. The higher the eggshell index, the thicker the shell.

Dichlorodiphenyltrichloroethane (DDT) is an organochlorine insecticide which was used during World War II to control the insects that transmit malaria and typhus and thereafter as an agricultural insecticide. Its use led to the decline in a number of nontarget species, especially birds of prey, as it accumulated in their bodies and caused a reduction in eggshell thickness. Consequently, chicks were lost when the eggshells cracked prematurely.

Q8.2.1 Draw a graph using the data in Table 8.1 showing the pattern of change in eggshell thickness between 1902 and 1964.

Q8.2.2 Is there any evidence of a sudden change in eggshell thickness? If so when?

Q8.2.3 Suggest reasons for any pattern observed.

Q8.2.4 DDT use was banned in 1970. What effect would you have expected this ban to have on birds of prey populations?

Table 8.1 Eggshell index of sparrowhawk eggs from hypothetical museum collections

	Specimen	Year	Eggshell index
Museum A	A	1964	1.10
	B	1938	1.46
	C	1930	1.36
	D	1912	1.40
	E	1935	1.35
Museum B	F	1915	1.42
	G	1962	1.11
	H	1952	1.20
	I	1926	1.33
	J	1932	1.43
	K	1909	1.35
	L	1955	1.09
	M	1943	1.40
	N	1950	1.10
	O	1905	1.40
	P	1960	1.14
Museum C	Q	1958	1.12
	R	1918	1.37
	S	1921	1.45
	T	1940	1.42
	U	1902	1.32
	V	1946	1.25

Reference/Further Reading

Ratcliffe, D. A. (1970). Changes attributable to pesticides in egg breakage frequency and eggshell thickness in some British Birds. *Journal of Applied Ecology, 7,* 67–107.

Exercise 8.3

The Toxicity of Pesticides and Herbicides

The LD_{50} test determines the dose of a chemical which is lethal to 50% of a test population of organisms. It is used widely to assess the efficacy of pesticides and to assess the ecological risks from potential pollutants. Fig. 8.3 shows the results of exposing individuals of a species of fly to different concentrations of an insecticide for a period of two hours.

Q8.3.1 Draw a graph of the effect of insecticide dose on percentage mortality.

Q8.3.2 Estimate the LD_{50} of the insecticide from the graph by drawing a horizontal line from the 50% mortality point (on the y-axis) to

the graph line and then a vertical line from the point of intersection down to the x-axis.

Q8.3.3 Table 8.2 shows the toxicity of a number of herbicides to mammals. Identify the herbicide which is:

a. least toxic to mammals;

b. most toxic to mammals.

Laboratory tests examined the effect of the oral administration of 25 mg of the insecticide carbophenothion kg^{-1} body weight on a variety of bird species (Fig. 8.4). The results are presented in Table 8.3.

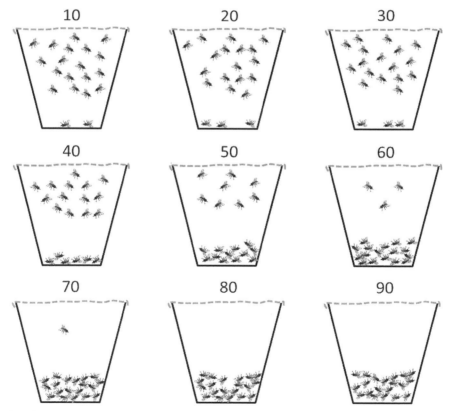

Figure 8.3 The results of an experiment to determine the LD_{50} of a hypothetical experiment in which flies have been given doses of insecticide ranging from 10 to 90 (mg m^{-2}). Each receptacle contains 20 flies. Those at the bottom of the receptacle are dead.

Table 8.2 Mammalian toxicity of selected herbicides

Chemical	LD$_{50}$ (mg kg^{-1} body weight)
Sodium arsenite	10–50
Paraquat	150
2, 4, 5 - T	100–300
2, 4 - D	300–1000
Cacodylic acid	830
Asprin	1775
Picloram	8200

Table 8.3 The effect of 25 mg of the insecticide carbophenothion kg^{-1} body weight on a variety of bird species

Species	Effect
Canada goose (*Branta canadensis*)	None
Pink-footed goose (*Anser brachyrhynchus*)	Lethal
Pigeon (*Columba livia*)	None
Chicken (*Gallus domesticus*)	None
Greylag goose (*Anser anser*)	Lethal

Figure 8.4 A dose of carbophenothion that will kill a greylag goose (*Anser anser*) has no effect on (A) a chicken (*Gallus domesticus*) or (B) a Canada goose (*Branta canadensis*).

Q8.3.4 With reference to the information in Table 8.3, comment on the usefulness of an LD$_{50}$ for a whole class of animals such as mammals or birds.

Reference/Further Reading

Westlake, G. E., Bunyan, P. J., & Stanley, P. I. (1978). Variation in the response of plasma enzyme activities in avian species dosed with carbophenothion. *Ecotoxicology and Environmental Safety, 2*, 151–159.

EXERCISE 8.4

CULTURAL EUTROPHICATION

When a new lake is formed it begins in an oligotrophic (nutrient-poor) state and slowly becomes enriched with nutrients as it is colonised by organisms which eventually die and decay. This natural process is known as eutrophication and normally occurs very slowly. When organic matter of human origin enters a lake it may quickly become enriched and suffer serious damage (Fig. 8.5). This process is known as cultural eutrophication. Table 8.4 shows changes in the concentration of chlorophyll in a lake associated with a discharge of untreated sewage.

Q8.4.1 Why is the concentration of chlorophyll a useful measure of the effect of the sewage on the lake?

Q8.4.2 Draw a graph showing the change in chlorophyll concentration with time between 1971 and 1979.

Q8.4.3 Draw and label an arrow on your graph indicating when sewage discharges began and another indicating approximately when they ceased.

Table 8.5 The effect of eutrophication on a lake

Variable	Normal condition (unpolluted)	Eutrophic condition
Biological oxygen demand		
Dissolved oxygen		
Fish numbers		
Water transparency		
Submerged plant numbers		
Phosphates		
Nitrates		
Sedimentation rate		
Primary productivity		
Bottom fauna species diversity		

Table 8.6 Which pollutants cause eutrophication?

Detergent	Dairy waste	Lead
Paper waste	Zinc	Farm slurry
Fertiliser	Herbicide	Mercury

Q8.4.4 Complete Table 8.5 by adding the words 'high' or 'low' in the empty cells to indicate the most likely effects of eutrophication on a lake.

Q8.4.5 Indicate in Table 8.6 those pollutants which could cause cultural eutrophication.

Q8.4.6 Why is eutrophication likely to be a greater problem in lakes than in rivers?

Reference/Further Reading

Smith, V. H., Joye, S. B., & Howarth, R. W. (2006). Eutrophication of freshwater and marine ecosystems. *Limnology and Oceanography, 51*, 351–355.

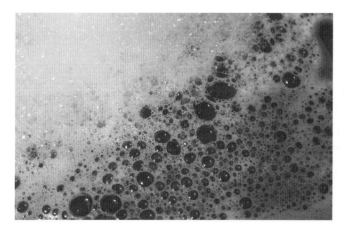

Figure 8.5 Does detergent cause eutrophication?

Table 8.4 Chlorophyll concentration in a lake during July–August between 1971 and 1979

Year	1971	1972	1973	1974	1975	1976	1977	1978	1979
Chlorophyll µg L^{-1}	3.0	2.0	3.5	12.0	15.3	22.6	12.7	9.5	3.9

ORGANIC POLLUTION OF FRESHWATER ECOSYSTEMS

Organic pollution has a very damaging effect on freshwater ecosystems because of its effect on water chemistry. The number of organisms found at two locations (A and B) on the same river is shown in Table 8.7. The two locations were similar except for the presence of organic pollution at one of them.

The extent of organic pollution in water may be determined by calculating the biological oxygen demand (BOD). This is a method which determines the amount of oxygen used by a water sample kept in the dark for a period of five days at 20°C. The oxygen is used up by the aerobic organisms in the water during the process of respiration. The more organic matter present, the more aerobic organisms will be able to grow in the sample and the more oxygen will be used up. So, a high BOD indicates a high level of organic pollution and a low BOD indicates a low level of organic pollution.

Organic pollution also affects the diversity of organisms within a water body. High diversity indicates clean water and low diversity indicates polluted water.

Q8.5.1 Examine the data in Table 8.7. Using information on the ecology of some of the river organisms in Table 8.8 determine which site is contaminated with organic pollution?

Q8.5.2 Account for the presence of *Tubifex* at site B but not at site A.

Q8.5.3 Account for the presence of *Baetis* at site A but not at site B.

Q8.5.4 Calculate the percentage decrease in numbers of *Gammarus* from site A to site B.

Table 8.7 Organisms present at two locations on the same river

Organism	Number per m² of river bed	
	A	B
Baetis (mayfly)	60	0
Asellus (hoglouse)	15	512
Chironomus (midge)	0	229
Ecdyonurus	19	0
Gammarus (shrimp)	40	4
Tubifex	0	207
Polycelis (flatworm)	15	0
Limnaea (snail)	8	0
Ancylus	6	0
Hydracarina	37	0
Elmis	9	0
Perla	14	0

Table 8.8 Notes on the ecology of selected river invertebrates

Taxa	Ecology
Chironomus	A genus of midges that contain haemoglobin-like proteins that allow them to tolerate low oxygen levels.
Tubifex	A small annelid worm whose blood contains haemoglobin allowing it to survive in low oxygen levels.
Asellus	Water louse. A freshwater crustacean that resembles a woodlouse. Associated with standing water. Lives among rotting leaves and other debris feeding on this and dead animals. Can tolerate low oxygen levels and polluted water.
Baetis	A genus of mayflies. Tolerate moderate levels of nutrient enrichment.
Gammarus	Freshwater shrimp. A crustacean that is intolerant of polluted waters and low oxygen levels. Feeds on detritus.

Table 8.9 Selected metrics for sites A and B

Variable	Site A	Site B
BOD		
Dissolved oxygen		
Diversity index		
Fish numbers		

Q8.5.5 Complete Table 8.9 with the words 'high' or 'low'.

Q8.5.6 Why is BOD measured in the dark?

References/Further Reading

Mason, C. F. (2002). *Biology of freshwater pollution*. London: Pearson Education, Prentice Hall.

Whitehurst, I. T. (1991). The Gammarus: Asellus ratio as an index of organic pollution. *Water Research*, *25*, 333–339.

THE EFFECTS OF ACID RAIN ON INVERTEBRATE DIVERSITY IN NORWEGIAN LAKES

The term 'acid rain' was first used to describe the acidity of the polluted rain falling around Manchester, England, in 1872 by Robert Angus Smith, the Chief Inspector of the Alkali Inspectorate United Kingdom at the time (Alma, 1993). Acid rain is any wet precipitation with a pH lower than 5.6. Acidification is caused by air pollution containing oxides of nitrogen and sulphur which produce nitric and sulphuric acids, respectively. In some areas hydrochloric acid is also important. Acid breaks down calcium carbonate in soil water and aquatic ecosystems. Earthworms and molluscs need calcium to survive and are absent from soils where calcium levels are low.

Q8.6.1 Draw a scatter diagram illustrating the effect of pH on the number of species of molluscs and crustaceans in the Norwegian lakes studied by Økland (Table 8.10).

Q8.6.2 Identify the highest pH at which the number of species is zero.

Q8.6.3 Identify the lowest pH at which the number of species is 17.

Q8.6.4 Between which two consecutive pH values on the scale is the greatest decline in species observed?

Q8.6.5 When these data were collected, what was the most likely source of the acidity in Norwegian lakes?

Q8.6.6 Why does acidic water cause a particular problem for molluscs?

Q8.6.7 Which common sedimentary rock may be added to lakes to reduce acidity?

References/Further Reading

Alma, P. (1993). *Environmental concerns*. Cambridge: Cambridge University Press.

Økland, K.A. (1980). Mussels and crustaceans: Studies of 1000 lakes in Norway. Ecological Impact of Acidification, In: Drablos, D., & Tollan, A. (Eds.) *Ecological impact of acid precipitation*. Proceedings of an International Conference, Sandefjord, Norway, March 11–14, 1980. pp. 324–325.

Table 8.10 The number of common species of molluscs and crustaceans found in near-neutral waters in Norway

pH	Species[a]	pH	Species[a]	pH	Species[a]	pH	Species[a]	pH	Species[a]
5.1	5	6.5	17	4.5	0	6.4	17	4.7	2
6.1	16	5.8	9	6.7	17	5.0	3	6.8	17
6.6	17	5.3	6	7.0	17	6.9	17	5.5	6
4.4	0	6.2	16	6.3	17	5.4	6	6.0	15
5.6	7	4.8	3	4.9	3	4.6	0	5.2	6
5.7	7	5.9	11						

Adapted from Økland (1980).
[a]Common molluscs and crustacean species only.

EXERCISE 8.7

DIPPERS AND ACID RAIN

Figure 8.6 A dipper (*Cinclus cinclus*).

Dippers (*Cinclus cinclus*) are small birds that are frequently found associated with streams in the United Kingdom (Fig. 8.6). They feed on insect larvae, especially those of caddis flies (Trichoptera). Unpolluted rainwater is acidic because carbonic acid is formed when carbon dioxide reacts with water in the air producing a pH which does not usually fall below 5.7. Acid rain is produced when sulphur dioxide and nitrogen oxide from pollution react with water molecules in the atmosphere. When this falls on a body of water the pH of the water will drop. This may have a deleterious effect on aquatic organisms and the food chains they support.

Q8.7.1 Draw graphs showing the relationship between:

    a. stream pH and dipper density (using the data in Table 8.11);

Table 8.11 The relationship between stream pH and dipper density

Stream water pH	Breeding pairs of dipper km^{-1}
5.5	0.40
6.0	0.60
6.5	0.61
7.0	0.62
7.5	0.85
8.0	0.70

Table 8.12 The relationship between stream pH and territory length in dippers

Stream water pH	Territory length of breeding pairs of dipper (km)
5.5	2.10
5.8	1.65
6.0	1.90
6.5	1.80
7.0	1.00
7.5	0.90

Table 8.13 The relationship between dipper abundance and caddis fly abundance

Relative abundance of caddis fly larvae	Dipper pairs 10 km^{-1}
10	1
100	5
800	8

b. stream pH and dipper territory length (using the data in Table 8.12); and

c. the relationship between caddis fly larvae abundance and dipper abundance (using the data in Table 8.13).

Q8.7.2 In Table 8.13, why is caddis fly larva abundance:

a. expressed as 'relative abundance'?

b. expressed without any units?

Q8.7.3 Using the graphs produced in answer to Q8.7.1, explain the relationship between acid rain and dipper abundance.

Reference/Further Reading

Vickery, J. (1991). Breeding density of Dippers *Cinclus cinclus*, Grey Wagtails *Motacilla cinerea* and common sandpipers *Actitis hypoleucos* in relation to the acidity of streams in south-west Scotland. *Ibis, 133,* 178–185.

EXERCISE 8.8

MONITORING RIVER POLLUTION USING DIVERSITY INDICES

Diversity indices are sometimes used to monitor freshwater pollution. Table 8.14 shows the numbers of organisms found at two sampling sites on a river. Conditions were similar at both sites but site A was upstream of the point of entry of a polluting discharge and site B was downstream of it.

Q8.8.1 Complete Table 8.14 and then, using the formulae below, calculate for each site:

a. Simpson's Index of diversity (D);

b. Menhinick's Index of diversity (d).

Simpson's Index is calculated as:

$$D = 1 - \sum p_i^2$$

Table 8.14 The mean number of individuals per m^3 of river bed. (Each organism (genus) listed was represented by a single species only)

Organism	Site A	p_i	p_i^2	Site B	p_i	p_i^2
Ancylus	8			–		
Asellus	17			9		
Baetis	52			3		
Chironomus	2			5		
Ecdyonurus	19			2		
Elmis	11			2		
Gammarus	31			1		
Hydropsyche	7			4		
Limnaea	11			–		
Perla	12			1		
Polycelis	13			–		
Tubifex	–			–		
Total		1.0			1.0	

where,

p_i = the proportion of all individuals in species i.

Menhinick's Index is calculated as:

$$d = S / \sqrt{N}$$

where,

S = the total number of species present at the site;

N = the total number of individuals (from all species).

Q8.8.2 What is the effect of pollution on the two diversity indices?

Q8.8.3 What would be the effect on Simpson's Index of altering the data for site A as follows: *Ancylus* = 17; *Asellus* = 8?

Q8.8.4 Why would it not be possible to compare the diversity indices for sites A and B in any meaningful way if Simpson's Index was used for site A and Menhinick's Index was used for site B?

Reference/Further Reading

Peet, R. K. (1974). The measurement of species diversity. *Annual Review of Ecology and Systematics*, 5, 285–307.

EXERCISE 8.9

THERMAL POLLUTION AND WATER FLEAS

Many industrial processes use water for cooling. Warm water is then discharged into a nearby river or lake and may cause problems for various species of animals. The water flea (*Daphnia*) is a small freshwater crustacean. Its heart rate is affected by the environmental temperature (Table 8.15), and the solubility of oxygen varies with temperature as indicated in Table 8.16.

Q8.9.1 Draw a scatter diagram showing the relationship between oxygen solubility in water and temperature (Table 8.15). Draw a line-of-best fit through the points.

Table 8.15 The solubility of oxygen in water of different temperatures

Temperature (°C)	Oxygen concentration (ppm)
0	14.6
4	13.1
8	11.9
12	10.8
16	10.0
20	9.2
24	8.5
28	7.9
32	7.4

Table 8.16 The effect of temperature on heart rate in *Daphnia*

Temperature (°C)	Heart rate (min^{-1})
5	112
10	197
15	240
20	298
25	381
30	400

Q8.9.2 Draw a scatter diagram showing the relationship between heart rate in *Daphnia* and temperature (Table 8.16). Draw a line-of-best fit through the points.

Q8.9.3 What type of correlation is shown by each of the two graphs?

Q8.9.4 Homiothermic animals maintain a more or less constant body temperature, above that of their surroundings. The body temperature of poikilothermic animals cannot be regulated physiologically and tends to vary with the environmental temperature. Is *Daphnia* a homiotherm or a poikilotherm?

Q8.9.5 What would be the effect of a discharge of warm water into a lake on the *Daphnia* present if the lake water was originally at 14°C and the warm water raised it to 26°C? (Use the graph line you have constructed in answer to question Q8.9.1 to predict these effects.)

Q8.9.6 Is the increased temperature caused by the discharge likely to be beneficial or damaging to the *Daphnia*?

Reference/Further Reading

Goss, L. B., & Bunting, D. L. (1976). Thermal tolerance of zooplankton. *Water Research*, 10, 387–398.

EXERCISE 8.10

TEMPERATURE INVERSIONS AND AIR POLLUTION

Energy from the sun heats the ground which in turn warms the air in the atmosphere. The temperature of the atmosphere normally decreases with height, i.e., distance from the surface of the Earth. However, a temperature inversion may occur in the atmosphere when cold air becomes trapped under warm air, for example, on a hot day following a cold night. This can affect the dispersion of atmospheric pollutants (Fig. 8.7).

The data in Table 8.17 show the temperature at various heights above ground in an urban area on a hot day in summer. Changes in atmospheric carbon monoxide concentration in an urban street in a 24-hour period are shown in Table 8.18.

Q8.10.1 Draw a graph of the data in Table 8.17, plotting the temperature data against the x-axis.

Figure 8.7 Smog in New York City, 1953. Library of Congress Reproduction Number: LC-USZ62-114346 (b&w film copy neg.)

Table 8.17 Temperature at various heights above an urban area on a hot day in summer

Temperature (°C)	Height above ground (m)
− 5	0
0	250
5	500
0	1500
− 3	2000
− 8	3000
− 13	4000
− 20	5000

Table 8.18 Levels of carbon monoxide in a busy urban street throughout a 24-hour period

Time of day	Carbon monoxide (ppm)
04:00	1.0
08:00	2.5
12:00	16.5
16:00	20.0
20:00	10.0
24:00	3.5
04:00	1.0

Q8.10.2 Why is it appropriate to plot the data for the dependent variable on the *x*-axis, against normal convention?

Q8.10.3 Draw a graph showing the change in carbon monoxide levels in an urban street using the data in Table 8.18.

Q8.10.4 What is likely to be the most important source of the carbon monoxide in this street?

Q8.10.5 If a temperature inversion was present above this street how would it affect the dispersion of the carbon monoxide and other gaseous pollutants produced at or near ground level?

Q8.10.6 Most of the oxygen carried in the blood combines temporarily with haemoglobin in the red blood cells, forming oxyhaemoglobin. This oxygen is collected from the air in the lungs and delivered to the tissues in the body. When carbon monoxide (CO) is present in the atmosphere it is inhaled and combines more or less irreversibly with haemoglobin forming a stable compound called carboxyhaemoglobin and thereby reducing the blood's capacity to carry oxygen. At high concentrations of CO this may cause difficulty in breathing. What would be the worst time of the day for someone with breathing difficulties (e.g., an asthmatic) to be walking along the street referred to above?

References/Further Reading

Greenburg, L., Jacobs, M. B., Drolette, B. M., Field, F., & Braverman, M. M. (1962). Report of an air pollution incident in New York City, November 1953. *Public Health Reports*, *77*, 7.

Lu, R., & Turco, R. P. (1995). Air pollutant transport in a coastal environment—II. Three-dimensional simulations over Los Angeles basin. *Atmospheric Environment*, *29*, 1499–1518.

THE EFFECT OF SULPHUR DIOXIDE POLLUTION ON GROWTH IN RYEGRASS

A number of studies have examined the effects of sulphur dioxide on plant growth. Bell and Clough (1973) measured the yield of 144 plants of S23 ryegrass (*Lolium perenne*) after exposure to two different levels of sulphur dioxide in the laboratory. There were no obvious injuries to the plants but a number of sublethal effects were recoded (Table 8.19).

Q8.11.1 For each of the measures of yield, calculate the percentage decrease in productivity in the plants exposed to the higher level of sulphur dioxide compared with that achieved at the lower level of exposure.

Q8.11.2 Briefly describe the effects of the higher level of sulphur dioxide on the plants.

Q8.11.3 What is the likely to be the principal route of entry of sulphur dioxide into the plants?

Table 8.19 Yield of ryegrass (*Lolium perenne*) after 26 weeks' growth from seed in air at two concentrations of sulphur dioxide

Yield (means of 144 plants)	SO_2 exposure ($\mu g\ m^{-3}$)		Decrease in productivity (%)
	9	191	
Number of tillers	25.18	14.84	41.1
Number of living leaves	85.61	47.31	
Dry weight of leaves (g)	0.791	0.388	
Leaf area (cm^2)	417.2	203.6	

Q8.11.4 Suggest two ways in which wind speed may reduce the effect of sulphur dioxide on plants in the field.

Q8.11.5 Why are sublethal effects of pollution like those illustrated here difficult to detect?

Q8.11.6 Cowling and Koziol (1978) exposed ryegrass to sulphur dioxide at a concentration of 400 $\mu g\ m^{-3}$ and recorded no effect on yield but some leaf damage. Their plants were packed much more densely than those studied by Bell and Clough. Why might this have had an effect?

Q8.11.7 What difficulties might you encounter in attempting field studies of the effects of sulphur dioxide on plant growth?

References/Further Reading

Bell, J. N. B., & Clough, W. S. (1973). Depression of yield in ryegrass exposed to sulphur dioxide. *Nature, 241,* 47–49.

Cowling, D. W., & Koziol, M. J. (1978). Growth of Ryegrass (*Lolium perenne* L.) Exposed to SO$_2$. I. Effects on photosynthesis and respiration. *Journal of Experimental Botany, 29,* 1029–1036.

TRENDS IN GREENHOUSE GAS EMISSIONS IN THE UNITED KINGDOM

A greenhouse gas is one which traps heat inside the atmosphere in a similar manner to the way in which the heat from sunlight is trapped behind glass. These gases include carbon dioxide (CO_2), methane (CH_4), nitrous oxide (N_2O), and fluorinated gases such as hydrofluorocarbons and perfluorocarbons. As the concentration of these gases in the atmosphere increases the global climate warms and weather patterns are disrupted. A relatively small increase in global temperature is causing glaciers and ice caps to melt. These changes are altering the distribution and abundance of organisms and have the potential to disrupt agricultural systems and patterns of disease as the distributions of insects and other vectors change.

The sources of greenhouse gases are many and varied, including agricultural animals (Fig. 8.8) and waste incineration (Fig. 8.9). The sources of these gases in the United Kingdom between 1990 and 2013 are shown in Table 8.20. Table 8.21 explains the sources from each sector in more detail.

Q8.12.1 Calculate the percentage drop in carbon emissions between 1900 and 2013 for each sector.

Q8.12.2 Which sector has shown:

a. the greatest percentage drop in emissions;

b. the smallest percentage drop in emissions?

Q8.12.3 Positive values in Table 8.20 represent emissions of carbon to the atmosphere. What do negative values represent?

Q8.12.4 Suggest two reasons why LULUCF (land use, land use change and forestry) data show negative values since 2005?

Q8.12.5 Explain why electric vehicles are a good thing in city centres but will themselves make little or no difference to greenhouse gas emissions (Fig. 8.10).

Q8.12.6 Renewable sources of energy such as wind (Fig. 8.11), solar and tidal power have the

Figure 8.8 Cattle produce the greenhouse gas methane.

Figure 8.9 A household waste incinerator.

Table 8.20 Sources of greenhouse gas emissions (United Kingdom and Crown Dependencies, 1990–2013)

Sector	Emissions (MtCO$_2$e[a])						
	1990	1995	2000	2005	2010	2012	2013
Energy supply	278.8	238.8	221.5	231.4	206.7	203.5	189.7
Transport	121.7	122.2	126.8	130.7	120.3	118.0	116.8
Business	115.4	113.7	117.2	109.5	94.4	88.4	90.9
Residential	80.6	81.9	89.0	86.0	87.8	77.3	77.6
Agriculture	66.0	65.1	61.4	57.3	54.6	54.0	53.7
Waste management	69.3	71.5	66.8	53.0	31.5	26.3	22.6
Industrial processes	60.0	50.9	27.2	20.4	12.5	10.5	12.8
Public	13.5	13.3	12.1	11.2	9.8	9.3	9.5
LULUCF	4.0	3.3	0.8	−2.9	−4.3	−5.0	−5.3
Total	809.4	760.6	722.8	696.6	613.3	582.2	568.3

Adapted from Anonymous (2015).

[a]MtCO$_2$e = million metric tons of carbon dioxide equivalent. This measure can aggregate different greenhouse gases into a single measure, using global warming potentials. One unit of carbon is equivalent to 3.664 units of carbon dioxide.

Table 8.21 Key to sources of greenhouse gases for Table 8.22

Sector	Source of emissions
Energy supply	Fuel combustion for electricity and other energy production sources.
Business	Combustion in industrial/commercial sectors, industrial off-road machinery and refrigeration and air conditioning.
Transport	Aviation, road transport, railways, shipping, fishing and aircraft support vehicles.
Public	Combustion of fuel in public sector buildings.
Residential	Fuel combustion for heating/cooking, garden machinery and fluorinated gases released from aerosols/metered dose inhalers.
Agriculture	Livestock, agricultural soils, stationary combustion sources and off-road machinery.
Industrial processes	Industry, except for those associated with fuel combustion.
Land use, land use change and forestry (LULUCF)	Forestland, cropland, grassland, settlements and harvested wood products.
Waste management	Waste disposed of to landfill sites, waste incineration, and the treatment of wastewater.

Figure 8.10 (A) Vehicles powered by fossil fuels; (B) an electric vehicle.

Figure 8.11 An ecobuilding powered by solar panels and a wind turbine.

potential to make significant contributions to the world's energy needs in the future. Write a short paragraph explaining why they are unlikely to have much effect on greenhouse gas emissions in the foreseeable future.

Reference/Further Reading

Anonymous (2015). *2013 UK greenhouse gas emissions, final figures*. London: Department of Energy and Climate Change.

DECAY AND CONCENTRATION OF RADIOISOTOPES

In the United Kingdom, Heysham Nuclear Power Station in Lancashire is an advanced gas-cooled reactor which uses uranium dioxide as fuel, enriched to 3% U-235 (^{235}U) (Fig. 8.12). Environmentalists are opposed to the use of nuclear power stations to produce electricity because of the potential leaks of nuclear material into the environment which remain radioactive for extremely long periods of time. The accident at the nuclear power station at Chernobyl in

Figure 8.12 Heysham Nuclear Power Station, Lancashire, England. This facility uses pellets made of uranium-238 as fuel.

Ukraine in 1986 released strontium-90 (^{90}Sr) that contaminated an area of thousands of square kilometres and resulted in the resettlement of over 200,000 people.

The time taken for half of the atoms in a given amount of radioactive material to decay is called its half-life. This varies from one radioisotope to another.

Q8.13.1 Examine the decay pattern of radioactive nuclei by calculating the missing values in Table 8.22.

Q8.13.2 The half-life of strontium-90 is 28 years. Draw a graph of the decay of radioactive nuclei over a period of six half-lives for strontium-90. Add two scales to the x-axis by labelling it in half-lives (i.e., 0, 1, 2, etc.) and also in years (i.e., 0, 28, 56, etc.).

Table 8.22 Decay pattern of radioactive nuclei

Number of half-lives	Percentage of original sample remaining
0	100.00
1	50.00
2	25.00
3	
4	
5	
6	

Table 8.23 Half-lives of selected radioisotopes

Radioisotope	Half-life
Uranium-238	4.5×10^9 years
Iodine-131	8.6 days
Uranium-235	7.1×10^8 years
Plutonium-339	24,400 years
Strontium-90	28 years

Q8.13.3 Which of the elements listed in Table 8.23 would disappear from the environment at:

a. the slowest rate;

b. the fastest rate?

Q8.13.4 If a quantity of plutonium-339 was released into the environment what percentage of the original amount would still be present after 73,200 years?

Q8.13.5 How many days would it take for a quantity of spilled iodine-131 to reduce to 1/64th of the original amount?

Table 8.24 The amount of strontium-90 in various components of a hill pasture ecosystem in Britain

Component	^{90}Sr (pCi g^{-1} dry matter)
Soil	0.112
Grass	2.500
Sheep bone	80.000

Q8.13.6 Table 8.24 indicates the amount of strontium-90 in various components of a hill pasture ecosystem in Britain. Calculate the degree of concentration that has occurred between:

a. the soil and the grass; and

b. the grass and the sheep bone.

Q8.13.7 What are the implications of this concentration process for humans?

Reference/Further Reading

Grynpas, M. D., & Marie, P. J. (1990). Effects of low doses of strontium on bone quality and quantity in rats. *Bone*, *11*, 313–319.

THE HUBBARD BROOK STUDY: DEFORESTATION AND NUTRIENT LOSS

Figure 8.13 Deforestation. This forest is in mid-Wales.

One of the largest and best-known studies of deforestation (Fig. 8.13) was carried out on the Hubbard Brook in New Hampshire, United States. As part of this study, one forested watershed of 15.6 ha (A) – the treatment watershed – had all of its vegetation cut down and left in place in winter. Herbicide was sprayed on the area for the next three summers to prevent regrowth. A second watershed (B) of a similar size was left untouched as a control (Fig. 8.14). Over the three years subsequent to cutting the total water flow from the deforested watershed increased by about 31%.

Q8.14.1 What would have happened to the additional water lost in the deforested watershed if the trees had not been cut down?

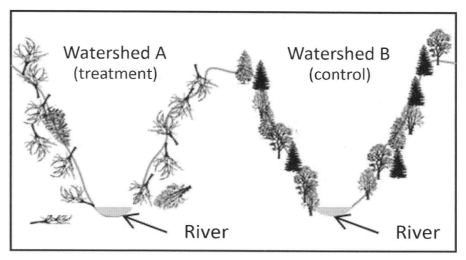

Figure 8.14 Diagrammatic representation of the study at Hubbard Brook.

Q8.14.2 What is the source of the nutrients that naturally occur in streams and rivers?

Q8.14.3 Why are the levels of many of the nutrients in the deforested watershed higher than those in the control watershed?

Q8.14.4 Positive values in Table 8.25 indicate a net loss of nutrients. What does a negative value indicate?

Q8.14.5 In which two forms has nitrogen been either lost or gained in these watersheds?

Q8.14.6 In what form is nitrogen lost from watershed A after deforestation?

Q8.14.7 Suggest reasons why the effects of deforestation recorded here are unlikely to be observed as a result of normal forestry operations.

Table 8.25 Net losses of nutrients for two watersheds of Hubbard Brook (summer 1966 – summer 1969)

Element	Net losses (kg ha^{-1})	
	Deforested watershed	**Control watershed**
Ca^{2+}	77.7	9.0
Mg^{2+}	15.6	2.6
K^+	30.3	1.5
Na^+	15.4	6.1
Al^{3+}	21.1	3.0
NH^{4+}	−1.6	−2.2
NO^{3-}	114.1	−2.3
SO_4^{2-}	2.8	4.1
Cl^-	1.7	−1.2

Adapted from Bormann & Likens (1979).

Net loss = streamwater outputs – atmospheric inputs.

Reference/Further Reading

Bormann, F. H., & Likens, G. E. (1979). *Pattern and process in a forested ecosystem.* New York: Springer-Verlag.

EXERCISE 8.15

INTERACTIONS BETWEEN URBANISATION, FORESTRY AND HYDROLOGY

Changes of land use profoundly affect the hydrology of an area. Urbanisation results in the covering of land with impervious materials and the channelling of runoff into sewers. The removal of vegetation destabilises soil and disrupts the hydrological and nutrient cycles. Table 8.26 compares sediment losses (yields) from an urban and a rural basin in the United States.

Q8.15.1 Calculate the sediment yield in the rural and urban basins in terms of $t\ m^{-1}\ y^{-1}$ using the data in Table 8.26.

Q8.15.2 Account for the difference between the two values obtained.

Q8.15.3 Calculate the percentage increase in average annual sediment discharge when forest cover was reduced from 80% to 20% (Table 8.27).

Table 8.26 A comparison of sediment yields in urban and rural drainage areas in the United States

Basin	Drainage area (mi^{-2})	Total sediment yield $(t\ y^{-1})$
Watts Branch (rural)	3.7	1910
Little Falls Branch (urban)	4.1	9530

Source: From data published in Leopold (1968).

Table 8.27 Changes in sediment discharge in the Potomac River basin following deforestation

	Percentage of forest cover	
	80%	**20%**
Average annual sediment discharge $(t\ mi^{-2}\ y^{-1})$	50	400

Based on data published in Walk & Keller (1963).

Q8.15.4 Account for this increase in sediment yield.

Q8.15.5 What would you expect to happen to the nutrient status of the soil after deforestation?

Q8.15.6 In the 1970s during storm conditions the levels of lead in the wastewater flowing in sewers increased. What was the most likely source of this lead?

Q8.15.7 A study of the temperature of water in a brook which drained a highly urbanised area in the United States found that during storm flows in August temperatures were 10–12°F above prestorm levels. What is the source of this additional heat?

References/Further Reading

Leopold, L. B. (1968). Hydrology for urban land plannning – A guidebook on the hydrologic effects or urban land use. In *Geological Survey Circular*. USA: US Geological Survey. Department of the Interior.554

Wark, J. W., & Keller, F. J. (1963). Preliminary study of sediment sources and transport in the Potomac River basin: Interstate Comm. on the Potomac River basin, Washington DC. *Tech. Bull.*, *1963-11*, 28.

EXERCISE 8.16

LICHENS AS MONITORS OF AIR POLLUTION

Atmospheric pollution may be studied by examining the distribution of lichens. Lichens consist of an association of a green alga and a fungus living together for their mutual benefit as symbionts. Some are very susceptible to air pollution while others are highly tolerant (e.g., *Xanthoria parietina*; Fig. 8.15). The number of lichen species recorded at different distances from the centre of the industrial city of Newcastle-upon-Tyne in England is shown in Table 8.28.

Q8.16.1 Draw a graph showing the relationship between the number of species of lichens recorded and distance from the city centre.

Q8.16.2 How may the observed distribution of species be explained?

Q8.16.3 Lichens may be used as monitors of air pollution rather than using chemical methods to analyse the air itself. Suggest one advantage of using lichens in air pollution studies and one advantage of using chemical analysis.

Table 8.28 Number of lichen species recorded at different distances from Newcastle-upon-Tyne, England

Miles from city centre	Number of lichen species
1	4
2	8
3	14
4	30
5	29
6	38
7	47
8	51
9	67
10	65

Figure 8.15 *Xanthoria parietina* is a bright orange lichen that is tolerant of air pollution.

Q8.16.4 In the north of England the prevailing winds blow from the west. What difference would you expect to find in the number of lichen species recorded at different distances from a city in this region if you compared those along a transect running east from the city centre with those along a transect running west from the city centre?

Reference/Further Reading

Conti, M. E., & Cecchetti, G. (2001). Biological monitoring: Lichens as bioindicators of air pollution assessment—a review. *Environmental Pollution, 114,* 471–492.

FERAL PIGEONS AS BIOINDICATORS OF LEAD LEVELS IN THE ENVIRONMENT

Lead is a heavy metal that is extremely toxic to biological systems. It is known to affect the functioning of the nervous system and has been widely removed from various products such as paint and petrol in recent years. Once lead enters the body it is distributed to major organs such as the brain, kidneys and liver and is stored in the bones and teeth. Lead is a cumulative toxicant and no known level of lead exposure is considered safe. It is particularly harmful to children.

Cai and Calisi (2016) measured lead (Pb) levels in the tissues of 825 feral pigeons in eight neighbourhoods of New York City – mostly from Manhattan – over a period of five years (Fig. 8.16). They also determined the number of children under 18 years old whose blood lead levels were $\geq 10\,\mu g\,dL^{-1}$ in each neighbourhood.

Figure 8.16 A feral pigeon (*Columba livia*).

Table 8.29 Lead levels in pigeons in New York City

Season	Mean blood Pb level in pigeons µg dL^{-1}
Spring	18.6
Summer	22.3
Fall (Autumn)	16.2
Winter	17.8

Q8.17.1 Draw a bar chart of the data in Table 8.29.

Q8.17.2 Suggest possible reasons why there is a seasonal difference in blood lead levels in pigeons.

Q8.17.3 Draw a scatter diagram of the data in Table 8.30 with the mean pigeon blood level on the *x*-axis.

Q8.17.4 Draw a line of best fit through the points. (See Chapter 10: Statistics).

Q8.17.5 Calculate a correlation coefficient for the data (see Chapter 10: Statistics).

Q8.17.6 What does this data tell us about the relationship between lead levels in the blood of pigeons and those in the blood of children?

Q8.17.7 Is monitoring lead levels in pigeons likely to be a useful indicator of the exposure of children to lead pollution in cites?

Q8.17.8 The pigeons used in this study were taken to a wildlife rehabilitation centre in New York City because they were either visibly ill or exhibiting abnormal behaviour. Why would abnormal behaviour suggest lead poisoning?

Table 8.30 The relationship between pigeon blood lead levels and blood levels in children in Manhattan

Mean pigeon blood Pb µg dL^{-1} per neighbourhood	Rate of children with elevated blood Pb levels per neighbourhood (per 1000 tested) younger than 18 years of age with blood Pb ≥ 10 µg dL^{-1}
16.7	1.2
18.9	1.4
19.6	1.7
19.0	2.3
19.6	3.0
20.1	3.1
22.6	2.6
23.3	2.7

Data derived from Cai & Calisi (2016).

Q8.17.9 Suggest a reason why the rates of elevated lead levels were calculated based on blood Pb ≥ 10 µg dL^{-1}.

Q8.17.10 The use of lead in gasoline in the United States was banned in 1995. However, this past use still contributes to lead levels in urban environments today. Why?

Q8.17.11 Demolition of old buildings can increase particulate lead levels in the air by several hundred times (Farfel et al., 2003). What is the source of this lead?

References/Further Reading

Cai, F., & Calisi, R. (2016). Seasons and neighbourhoods of high lead toxicity in New York City: The feral pigeon as a bioindicator. *Chemosphere, 161*, 274–279.

Farfel, M. R., Orlova, A. O., Lees, P. S. J., Rohde, C., Ashley, P. J., & Chisolm, J. J. (2003). A study of urban housing demolitions as sources of lead in ambient dust: Demolition practices and exterior dust fall. *Environmental Health Perspectives, 111*, 1228–1234.

WHERE HAS ALL THE ICE GONE? POLAR BEARS AND CLIMATE CHANGE

One characteristic of global climate change is a gradual warming of the oceans. This has affected the extent of sea ice at the poles. Polar bears (*Ursus maritimus*) travel and hunt on the ice and find it harder to find and stalk prey when the ice packs break up in summer (Fig. 8.17). Ice provides polar bears with camouflage and when the ice retreats the polar bears are left stranded on land.

Q8.18.1 Draw a graph comparing the extent of sea ice in 2105 with the average extent between 1981 and 2010 using the data in Table 8.31.

Q8.18.2 Calculate the sea ice extent on 15 September 2015 as a percentage of the average value between 1981 and 2010 on the same date.

Q8.18.3 Calculate the rate of loss of sea ice between 1st March and 1st June 2015 in millions km^2 day^{-1}.

Q8.18.4 What percentage of the sea ice present on 1st March 2015 had been lost by 15th September 2015?

Q8.18.5 Calculate the total area of ice lost (millons km^2) in 2015 between the point in time when sea ice was at its maximum and when it was at its minimum area.

Q8.18.6 Explain why it is not possible to say whether or not the minimum sea ice extent

Figure 8.17 A polar bear (*Ursus maritimus*).

Table 8.31 The extent of Arctic sea ice (area of ocean with at least 15% sea ice)

Date	Extent of sea ice ($\times 10^6$ km^2)	
	Average 1981–2010	2015
1 January	13.647	13.005
15 January	14.332	13.507
1 February	14.964	14.093
15 February	15.270	14.459
1 March	15.504	14.469
15 March	15.487	14.356
1 April	15.261	14.320
15 April	14.752	14.039
1 May	14.057	13.342
15 May	13.386	12.664
1 June	12.607	11.733
15 June	11.899	11.083
1 July	10.853	10.084
15 July	9.675	8.758
1 August	8.190	6.924
15 August	7.252	5.756
1 September	6.499	4.594
15 September	6.277	4.447
1 October	6.866	5.088
15 October	8.180	6.464
1 November	9.641	8.614
15 November	10.597	9.715
1 December	11.734	10.863
15 December	12.762	11.961

Data extracted from Ch*arctic* Interactive Sea Ice Graph, National Snow & Ice Data Center, University of Colorado at Boulder. http://nsidc.org/arcticseaicenews/charctic-interactive-sea-ice-graph/ Accessed 17.9.2016.

recorded in 2015 was lower than at any time between 1981 and 2010 from the data in Table 8.31.

Q8.18.7 How does a reduction in the area of sea ice affect polar bears?

Reference/Further Reading

Stirling, I., & Derocher, A. E. (1993). Possible impacts of climatic warming on polar bears. *Arctic*, *46*, 240–245.

CHAPTER 9

CONSERVATION BIOLOGY

This chapter contains exercises concerned with the rational and sustainable use of the world's biological resources.

Bald eagle (*Haliaeetus leucocephalus*)

Examining Ecology. DOI: http://dx.doi.org/10.1016/B978-0-12-809354-2.00009-9

INTENDED LEARNING OUTCOMES

On completion of this chapter you should be able to:

- Explain the importance of tropical forests as a resource.

- Explain the risk posed to animals by road drainage systems.

- Explain how mesh regulation helps to conserve fish stocks.

- Compare the profits made from game ranching with those made from traditional cattle ranching.

- Analyse the economics of wildlife resource utilisation.

- Explain how the biology of a species may affect its long-term survival.

- Use mathematical methods to study the relationship between area and perimeter length in protected areas.

- Identify the challenges associated with the use of reintroduction as a conservation tool.

- Recognise the range of threats to the future survival of a species.

- Discuss the role of zoos in conservation.

- Discuss the ethical issues arising from a variety of ecological studies and ecosystem management strategies.

INTRODUCTION

CONSERVATION

Biological conservation has been defined by Spellerberg and Hardes (1992) as having the aim of 'maintaining the diversity of living organisms, their habitats and the interrelationships between organisms and their environment.' It may involve pollution control, wildlife management, the captive breeding of rare species, the restoration of ecosystems or any of a large number of possible interventions, including changes to laws and policies affecting the environment.

THE IDENTIFICATION OF THREATENED AND ENDANGERED SPECIES

The International Union for the Conservation of Nature and Natural Resources (IUCN) was founded in 1948. It is a partnership of states, government agencies and nongovernmental organisations of over 1,000 members and almost

11,000 volunteer scientists spread across over 160 countries. The IUCN seeks to assist in the conservation of biological diversity and ensure the responsible and equitable use of the world's natural resources. The IUCN produces the Red List: a list of species based on their extinction risk. This uses a series of eight categories: Extinct, Extinct in the Wild, Critically Endangered, Endangered, Vulnerable, Near Threatened, Least Concern and Data Deficient. A species is considered to be threatened if it is classified as Critically Endangered, Endangered or Vulnerable (Fig. 9.1).

IN SITU VERSUS EX SITU CONSERVATION

Conservation efforts may be divided into those that occur in the natural environment (*in situ*) and those that occur 'off-site'(*ex situ*). Protecting an area of African savannah by creating a national park is an example of *in situ* conservation. Maintaining a captive-bred population of cheetahs (*Acinonyx jubatus*) in zoos is an example of *ex situ* conservation. *In situ* conservation measures are generally preferred to *ex situ* measures. However, in many cases *ex situ* measures make important

Figure 9.1 The golden lion tamarin (*Leontopithecus rosalia*) is classified as endangered (EN) by the IUCN.

contributions to the conservation of endangered species especially where wild populations are severely threatened by human activities such as poaching and habitat destruction.

ISLAND BIOGEOGRAPHY AND NATURE RESERVE DESIGN

Nature reserves have been likened to islands as they are increasingly being surrounded by land used by people for other purposes such as farming, housing and industry. In 1967, Robert MacArthur and Edward Wilson attempted to explain the biogeography of islands in their book *The Theory of Island Biogeography* (MacArthur and Wilson, 1967). Subsequently, the principles developed in the study of island biogeography have been used to analyse the effectiveness of nature reserves of different sizes, shapes and spatial arrangements in retaining species and protecting biodiversity.

SPORT HUNTING AND GAME RANCHING

In many parts of the world sport hunting and wildlife conservation survive side by side. Well-managed wildlife populations may produce a sufficient number of surplus animals to support a sport hunting industry. In some African countries (e.g., South Africa, Namibia and Zimbabwe) sport hunting provides an important incentive for people to support conservation measures. A relatively small number of animals are killed by wealthy foreign tourists who pay high hunting fees (especially for iconic species such as rhinoceros and elephant). Some of this income is used to fund clinics and schools for local people, although much of the profit from hunting is returned to hunting companies, airlines, hotels and other tourist facilities.

Experiments with game ranching began in Africa in the 1960s (Dasmann, 1964). Niche separation in wild ungulates means that, for a given area, a much greater biomass of meat can be produced from wild animals than from cattle because the wild species make better use of the available vegetation. However, care must be taken to manage the production of meat from game ranching so as to prevent parasites and diseases from wild animals infecting humans.

ZOOS

Modern zoos have a number of conservation functions. They educate the public about endangered species and threatened habitats; they provide opportunities for scientists to conduct research on the behaviour, welfare, nutrition, reproduction and other aspects of the biology of rare species which may benefit conservation efforts; they engage in the captive breeding of rare species and, in some cases, reintroduce these species back into the wild (Fig. 9.2).

One of the best known examples of a species that was captive bred and then reintroduced into the wild is that of the Arabian oryx (*Oryx leucoryx*). The Fauna Preservation Society (now Fauna and Flora International) launched 'Operation Oryx' in 1961. They captured three wild oryx and obtained others from zoos and private collections in the Middle East. Eventually, a captive population was established at Phoenix Zoo in Arizona and Los Angeles Zoo in California. By 1964 there were 13 oryx in captivity in the USA (Tudge, 1991). The Arabian oryx was first returned to the New Shaumari Reserve in Jordan in 1978. Later oryx were released in the Jiddat al-Harasis plateau in Oman. The first five animals arrived in 1980 and by 1996 there were 450 free-ranging animals.

Other captive-breeding and reintroduction projects undertaken with the help of zoos have involved *Partula* snails (returned to French Polynesia), the black-footed ferret (*Mustela nigripes*) in the United States, and the Hawaiian goose or nēnē (*Branta sandvicensis*).

Figure 9.2 The entrance to Dudley Zoo in England. When it opened in the 1937 it was considered to be the most modern in Europe. It contains many concrete enclosures designed by Berthold Lubetkin – including the entrance – which are now protected as listed buildings.

Although zoos have been involved in a number of reintroduction projects around the world, zoos do not claim that this is their primary function. The reality is that most of the animals kept in zoos were born in a zoo and very few will ever be released to the wild.

SEED BANKS AND FROZEN ZOOS

A seed bank is a place where plants are stored in seed form as a means of ensuring that they do not become extinct in the future. The Millennium Seed Bank Partnership is based in Kew Gardens in England. It is the largest *ex situ* plant conservation programme in the world. It works with a network of partners in 80 countries. So far it has successfully banked over 13% of the wild plant species in the world. By 2020 it plans to have saved 25% of those plant species with bankable seeds: some 75,000 species. The project focuses particularly on those species that are most useful to humans and those whose survival is particularly threatened by human activity such as climate change.

The long-term storage of seeds requires that they are dried until they contain only a very small quantity of water. Most seeds are resistant to desiccation and remain viable when dried and stored at low temperatures. When there is a decline in seed viability they need to be planted and allowed to germinate so that fresh seeds can be produced for storage.

A similar approach has been taken with rare animals. A number of 'frozen zoos' now exist where the sperm and ova of rare species are kept as a genetic resource for the future. The largest facility of this type in the world is located at San Diego Zoo's Institute for Conservation Research. The Frozen Zoo ® contains over 10,000 living cell cultures, oocytes, sperm, and embryos representing nearly 1,000 taxa. This material has the potential to produce offspring using techniques such as artificial insemination and embryo transfer.

EVOLUTIONARILY SIGNIFICANT UNITS

An evolutionarily significant unit (ESU) is essentially the smallest part of a species considered worth saving. The concept has become important in recent years because resources for conservation are finite and we clearly cannot save everything so conservationists must prioritise some taxa over others (Moritz, 1994).

THE ROLE OF THE LAW

Wildlife law is extremely important in supporting conservation efforts within individual countries and at an international level (Rees, 2017). The taking, hunting and sale of rare species is unlawful under national legislation in most, if not all, countries (e.g., the Endangered Species Act of 1973 in the USA; the Wildlife and Countryside Act 1981 in Great Britain; the Wildlife (Protection) Act 1972 in India). At an international level, many treaties have been agreed that protect important habitats and species that cross international boundaries during their migrations (e.g., the UN Convention on Biological Diversity 1992; the Convention on Wetlands of International Importance especially as Waterfowl Habitat 1971). International trade in rare species of animals and plants is regulated by the Convention on International Trade in Endangered Species of Wild Fauna and Flora, 1973 (CITES). The International Whaling Commission introduced a moratorium on commercial whaling in 1986 and has been important in preventing the extinction of many cetacean species.

The following exercises are concerned with the conservation of plants and animals, their sustainable use, the design of nature reserves and the role of zoos in conservation.

References

Dasmann, F. (1964). *African game ranching*. New York: Pergamon Press.

MacArthur, R. H., & Wilson, E. O. (1967). *The theory of island biogeography*. Princeton, NJ: Princeton University Press.

Moritz, C. (1994). Defining 'Evolutionarily Significant Units' for conservation. *Trends in Ecology and Evolution, 9*, 373–375.

Rees, P. A. (2017). *The laws protecting animals and ecosystems*. Chichester: Wiley-Blackwell.

Spellerberg, I. F., & Hardes, S. R. (1992). *Biological conservation*. Cambridge: Cambridge University Press.

Tudge, C. (1991). *Last animals at the zoo. How mass extinction can be stopped*. London: Hutchinson Radius.

COUNTING THREATENED SPECIES: THE IUCN RED LIST

The International Union for the Conservation of Nature and Natural Resources (IUCN) maintains a global list of threatened species in a 'Red List'. The IUCN recognised almost 1.4 million described species of animals in 2016. The conservation status of many of these species had not been assessed at this time. Only about 43,000 of the over 67,000 described vertebrates species had been evaluated, including all of the described species of mammals and birds. In 2016, the 1,208 species of threatened mammals included the snow leopard (*Uncia uncia*) (Fig. 9.3).

Q9.1.1 Complete Table 9.1 by calculating the subtotal for vertebrates and the total including invertebrates.

Q9.1.2 Calculate the total number of threatened species (CR, EN and VU) as a percentage of the number of species described for 1996–98 and 2016.

Q9.1.3 Which taxon has shown the greatest apparent increase in the number of species that have become threatened between the 1996–98 and the 2016 analysis?

Q9.1.4 Explain what other information we would need to be sure that the apparent changes in the number of threatened species recorded here reflect real changes in their status.

Figure 9.3 Snow leopard (*Uncia uncia*).

Table 9.1 Numbers of species described and classified as threatened by the IUCN in 1996–98 and 2015

Taxon	Described species	Number of threatened species (IUCN categories CR, EN and VU)[a]	
		1996–98	**2016**
Mammals	5536	1096	1208
Birds	10,424	1107	1375
Reptiles	10,450	253	989
Amphibians	7538	124	2063
Fishes	33,300	734	2343
Subtotal			
Invertebrates	1,305,250	1891	4383
Total			

Adapted from IUCN (2016).
[a]CR = critically endangered, EN = endangered, VU = vulnerable.

Reference/Further Reading

IUCN. (2016). IUCN Red List Version 2016-2. Table 1: Numbers of threatened species by major groups of organisms (1996–2016) <http://cmsdocs.s3.amazonaws.com/summarystats/2016-2_Summary_Stats_Page_Documents/2016_2_RL_Stats_Table_1.pdf>. Accessed 10.10.16.

THE BIOLOGY OF THE GIANT PANDA

Figure 9.4 Giant panda (*Ailuropoda melanoleuca*).

The giant panda (*Ailuropoda melanoleuca*) is considered to be a national treasure in China and is an iconic species for the conservation movement (Fig. 9.4). A stylised image of a giant panda has been the logo of the World Wildlife Fund (WWF) since it was founded in 1961. Few pandas live in zoos outside China and those that do are effectively rented from the Chinese government. In 2010 Ueno Zoo in Japan agreed to rent a pair of pandas from China for $1m (£640,000) (BBC, 2010). Conservation efforts to protect the panda in the wild have been so successful that it has recently been reclassified by the IUCN from 'endangered' (EN) to 'vulnerable' (VU).

Figure 9.5 The giant panda's diet is chiefly bamboo (*Sinarundinaria*).

Table 9.2 A summary of some aspects of the biology of the giant panda

Food	Chiefly bamboo (*Sinarundinaria*) (Fig. 9.5). Will take small mammals, small birds and fishes. Require about 25 kg of bamboo per day. May feed for up to 11 hours per day. Digestion of bamboo is very inefficient.
Distribution	Highlands of western China from Szechwan, Sheni and Kansu to the high Tibetan plains of Tsinghai.
Habitat	Bamboo forests on open hillsides. These are being replaced by cultivation.
Abundance	Swaisgood, Wang, and Wei (2016) suggest a wild population of around 2,000 individuals. A small number of animals are held in a captive-breeding programme in China and a few animals in zoos outside China.
Behaviour	Solitary, except during the mating season. Shy and difficult to observe in the wild.
Reproduction	Female is only fertile for four days per year. Pregnancy is delayed. Development occurs only in the last four months. Gestation is about 140 days. Female gives birth to one cub. Young weight only about 100 g at birth. Sexual maturity is reached at between 4 and 10 years.

Use the information in Table 9.2 to answer the following questions.

Q9.2.1 What is the greatest single threat to the survival of the giant panda in the wild?

Q9.2.2 Explain why the giant panda is described as a specialist in terms of its feeding behaviour (Fig. 9.5).

Q9.2.3 How has the evolution of its feeding behaviour contributed to the giant panda's demise?

Q9.2.4 What aspects of the giant panda's social behaviour and reproductive biology have made a reduction in its numbers difficult to counteract?

Q9.2.5 What factors are likely to have hampered the study of the population biology of giant pandas in the wild?

Q9.2.6 How can zoos justify the high cost of renting a panda from China?

References/Further Reading

BBC. (2010). Japan to rent pandas from China. <http://news.bbc.co.uk/1/hi/world/asia-pacific/8512979.stm>. Accessed 1.12.2016

Swaisgood, R., Wang, D., & Wei, F. (2016). *Ailuropoda melanoleuca*. The IUCN Red List of Threatened Species 2016: e.T712A45033386. <http://dx.doi.org/10.2305/IUCN.UK.2016-2.RLTS.T712A45033386.en>. Accessed 1.12.2016.

CONSERVING TROPICAL PLANTS: WHY BOTHER?

Ethnobotany is the study of indigenous peoples and how they use wild plants. The data in Table 9.3 show the uses to which the local people put plants in the Tambopata Wildlife Reserve in the Peruvian Amazon.

Q9.3.1 How may this data be used to justify the conservation of tropical rainforests?

Q9.3.2 Which category of use is likely to be the most difficult to authenticate?

Q9.3.3 Explain why it is not possible to calculate the total number of useful plant species in Tambopata from the data in Table 9.3.

Q9.3.4 Use the data in Table 9.4 to justify the concern of conservationists for the loss of vast areas of tropical forest (Fig. 9.6(A) and (B)).

Q9.3.5 On average each of the tree species in the Malaysian forest was represented by 7.4 individuals. Calculate the number of trees per hectare.

Seed banks may be used to store plants for long periods of time in a relatively small amount of space.

Table 9.3 A summary of uses of plants in the Tambopata Wildlife Reserve, Peru

Use	Total number of plant species[a]
Materials (construction, fuel, etc.)	85
Food (fruits, seeds, vegetables, etc.)	42
Medicines	153

Derived from data in Spellerberg & Hardes (1992).
[a]Some species have more than one use.

Table 9.4 The number of tree species recorded in 23 ha of forest in Malaysia and the British Isles

	Tropical forest Malaysia	Temperate forest British Isles
Number of tree species	375	8

Figure 9.6 (A) Temperate forest; (B) tropical forest.

Q9.3.6 Suggest reasons why some plants cannot be stored in seed banks.

Q9.3.7 Why do seeds stored in seed banks need to be germinated periodically?

Reference/Further Reading

Spellerberg, I. F., & Hardes, S. R. (1992). *Biological conservation*. Cambridge: Cambridge University Press.

EXERCISE 9.4

GULLYPOTS: A THREAT TO SMALL VERTEBRATES

Rainwater draining from roads runs into sewers via drains at the edge of the road. In some areas these drains incorporate a gullypot which traps sediment and other materials to prevent them from entering the sewer. Small animals, such as toads (Fig. 9.7) and newts, may be carried in rainwater into the drain and become trapped in the gullypot (Fig. 9.8). Traditional designs do not provide any means by which animals can escape so drains of this type are an important cause of mortality for some species.

Perth and Kinross Ranger Service in Scotland conducted a survey of animals found in gullypots

Figure 9.7 Common toad (*Bufo bufo*).

Figure 9.8 Rainwater drain.

Table 9.5 The fate of animals found in 607 gullypots in Perth and Kinross

Taxon	Alive	Dead	Total	Mortality (%)
Toads	716	92		
Frogs	136	21		
Newts	48	4		
Mammals	0	101		
Total				
%				

Data adapted from Anon. (2012).

in 2012. They examined 607 gullypots, of which 416 contained amphibians and small mammals (Table 9.5).

Q9.4.1 Complete Table 9.5.

Q9.4.2 Which taxon exhibited the highest mortality rate?

Q9.4.3 Which taxon exhibited the lowest mortality rate?

Q9.4.4 Suggest a reason why no mammals survived in the gullypots.

Q9.4.5 Is it possible to conclude from these data that rainwater drains are a greater threat to toads than they are to frogs purely from the information on the numbers of individuals trapped? Explain your answer.

Q9.4.6 Suggest two changes to the design of road drains which could reduce the number of animals trapped.

Reference/Further Reading

Anon. (2012). *Amphibians in Drains Project 2012*, Perth & Kinross Ranger Service. <http://www.arguk.org/info-advice/survey-and-monitoring/220-amphibians-in-drains-project-2012-perth-and-kinross-ranger-service/file>. Accessed 19.10.2016.

BLACK RHINOCEROS POACHING IN TANZANIA

The black rhinoceros (*Diceros bicornis*) is classified by the IUCN as 'critically endangered' (CR) with numbers in the wild reaching a low of just 2,410 in 1995 (Emslie, 2012). A small, heavily-guarded population exists in Ngorongoro Crater in northern Tanzania (Fig. 9.9). A ranger station is located on the floor of the crater and there is a permanent ranger presence to ensure the protection of the crater wildlife, especially the rhinoceros (Fig. 9.10). The following extract has been taken from a recent paper on the historical ecology of this area (Oates and Rees, 2013).

> *Baumann described the shooting of five rhinoceros by his party and stated that 'several others' had also shot rhinoceros during their 5-day visit to the crater (Organ*

& Fosbrooke, 1963). Several studies have suggested that poaching was the primary cause of recent rhinoceros decline in the crater (Moehlman et al., 1996, Mkenda & Butchart, 2000, Mills et al., 2003, 2006). Poaching began in the late 1950s and 1960s when the Masai killed at least 19 rhinoceros in the period immediately after the inauguration of the NCA, probably in response to the severe drought (1959–60), which caused significant mortalities in Masai herds with a resultant increase in poaching to generate income (Fosbrooke, 1972). The offer of a reward for information on poachers in 1961 reduced poaching dramatically (Grzimek, 1964, Goddard, 1967b, Fosbrooke, 1972). However,

Figure 9.9 Ngorongoro Crater, Tanzania.

poaching began again in the early 1970s (Makacha et al. 1979, Mills et al. 2006) and continued until the mid-1980s (Kiwia, 1989a). Borner (1981) estimated that at least 25 rhinoceros were killed in the NCA by 'non-Masai' poachers for commercial gain. Between 1963 and 1965, 82% of crater rhinoceros were adults (total population = 61; Klingel & Klingel, 1966) but this had declined to just 28% (total population = 27) by 1978 (Makacha et al., 1979). This may be evidence for heavy poaching, as commercial poachers select adult rhinoceros with large horns (Borner, 1981). Increased ranger patrols in 1980 reduced poaching, however a subsequent deterioration in ranger activities allowed it to increase. The use of cash rewards for information leading to the conviction of poachers was reestablished in the late 1980s and resulted in another decline in poaching (Homewood & Rodgers, 1991). Runyoro et al. (1995) recorded that there had been no rhinoceros poached in the crater since 1988. However, in 1995, at least four were taken (Moehlman et al., 1996). This led to increased security and no further poaching was reported (Mills et al., 2006). Brett (2001) stated that poaching was still

Figure 9.10 (A) Black rhinoceros (*Diceros bicornis*) in Ngorongoro Crater; (B) The author (second from left) with a botanist from the College of African Wildlife Management (left) and two armed game wardens in Ngorongoro Crater conducting a large mammal census.

a problem for rhinoceros, but Mills et al. (2006) claimed that the population was the most secure in Tanzania.

Runyoro et al. (1995) concluded that poaching was not a significant threat to species other than rhinoceros, while Homewood and Rodgers (1991) believed that poaching and the poor relationship between the NCAA and the Masai threatened the future of the elephant population.

[In this extract NCA means the Ngorongoro Conservation Area and NCAA means Ngorongoro Conservation Area Authority. Readers interested in the the studies to which reference is made in the extract should refer to the original paper by Oates and Rees (2013).]

Q9.5.1 Explain the link between the severe drought of 1959–60 and an increase in the poaching of black rhinoceros.

Q9.5.2 What evidence is there of heavy commercial poaching of black rhinoceros in the 1970s?

Q9.5.3 List strategies which have reduced the poaching of black rhinoceros in the crater.

Q9.5.4 The Austrian explorer Oscar Baumann visited the crater in 1892. How can the killing of five rhinoceros by Baumann's party be justified?

References/Further Reading

Emslie, R. (2012). *Diceros bicornis*. The IUCN Red List of Threatened Species 2012: e.T6557A16980917. <http://dx.doi.org/10.2305/IUCN.UK.2012.RLTS. T6557A16980917.en>. Accessed 1.12.2016.

Oates, L., & Rees, P. A. (2013). The historical ecology of the large mammal populations of Ngorongoro Crater, Tanzania, east Africa. *Mammal Review, 43*, 124–141.

CONSERVING FISH STOCKS BY MESH REGULATION

Figure 9.11 A fishing vessel returning to the port of Reykjavik, Iceland.

Modern fishing fleets are capable of removing fish from the oceans with great efficiency, threatening the long-term future of fish stocks (Fig. 9.11). A common practice in the protection of fisheries is the regulation of the mesh size of nets (Fig. 9.12). The data in Table 9.6 show the effects of increasing the mesh size from 80 mm to 100 mm on a haddock fishery.

Q9.6.1 Calculate the total number of fish caught and the total weight:

    a. before regulation;

    b. after regulation.

Q9.6.2 Which mesh size produces the largest catch in terms of weight?

Q9.6.3 What happened to the 2-year-old fish after regulation?

Q9.6.4 Calculate the percentage increase in the number of 3-year-old fish caught after regulation.

Q9.6.5 Draw two histograms (one above the other) showing the differences in the age structure of the catch before and after regulation.

Figure 9.12 Fishing nets.

Table 9.6 Haddock catch before and after net regulation

Age (years)	Before regulation		After regulation	
	Number $\times 10^6$	Weight (metric tons $\times 10^3$)	Number $\times 10^6$	Weight (metric tons $\times 10^3$)
2	14.0	4.89	–	–
3	5.0	3.05	12.6	7.59
4	1.9	1.88	4.7	4.73
5	0.7	0.97	1.7	2.39
6	0.4	0.81	0.6	1.02
7	–	–	0.4	0.92

Based on Jeves & Margham (n.d.).

Q9.6.6 Explain how mesh regulation has caused this difference.

References/Further Reading

Anderson, L. (1985). Potential economic benefits from gear restrictions and licence limitation in fisheries regulation. *Land Economics, 61,* 409–418.

Jeves, T.M., & Margham, J.P. (Eds) (n.d.). Scottish Ministry of Agriculture, Fisheries and Food Publicity Material. *'A' level biology teacher's notes. Booklet 1—applied biology.* Liverpool: Liverpool Polytechnic.

Waters, J. R. (1991). Restricted access vs. open access methods of management: Toward more effective regulation of fishing effort. *Marine Fisheries Review, 53,* 1–10.

EXERCISE 9.7

GAME RANCHING IN ZIMBABWE

Figure 9.13 Eland (*Taurotragus oryx*).

In the 1960s ecologists began to investigate the viability of ranching game in Africa instead of cattle (Fig. 9.13). This exercise considers the types of animals killed and the profitability of this type of farming on the Hendersen Ranch in Southern Rhodesia in the mid-1960s, based on a 50 square mile study area. Southern Rhodesia became Zimbabwe in 1980.

Q9.7.1 Calculate the total yield (lbs) for each species using Table 9.7.

Q9.7.2 Calculate the mean value per pound of carcass for each species and determine the relative value of each species by completing Table 9.7.

Q9.7.3 Calculate the total meat yield and gross value for all species per square mile.

Q9.7.4 Rank the species from the highest to lowest value and then draw bar charts showing:

a. the recommended crop for each species (highest first);

b. the total meat yield for each species (highest first).

Q9.7.5 Discuss any differences between the relative importance of each species in the two graphs.

Q9.7.6 Compare the net profits from beef and game ranching using the data in Table 9.8.

Table 9.7 Game populations and meat yield from the Hendersen Ranch

Species	Estimated number	Recommended crop	Carcass weight (lbs)	Total meat yield (lbs)	Gross value (£)	Mean value/ lb
A	B	C	D	E (C × D)	F	G (F/E)
Impala	2100	525	65		2133	
Zebra	730	146	255		1168	
Steenbuck	200	40	12		36	
Warthog	170	85	70		297	
Kudu	160	48	225		540	
Wildebeest	160	32	260		416	
Giraffe	90	15	1000		600	
Duiker	80	28	20		42	
Waterbuck	35	7	200		70	
Buffalo	30	5	570		119	
Eland	10	2	600		60	
Klipspringer	10	3	14		3	
Bush pig	10	5	70		17	

Adapted from Dasmann (1964).

Table 9.8 A comparison of profits from beef and game ranching

Type of ranching	Gross profit per year (£)	Expenses per year (£)	Net profit per year (£)
Beef	5198	4692	
Game	5500	2300	

Reference/Further Reading

Dasmann, R. F. (1964). *African game ranching*. Oxford: Pergamon Press.

USE IT OR LOSE IT: SPORT HUNTING AS CONSERVATION

Figure 9.14 Lion (*Panthera leo*) are popular trophies among sport hunters.

Tourism has long been an important source of income for African countries, especially those with substantial wildlife populations like Kenya. The problem with tourism as a justification for conservation is that local people are often more likely to benefit from poaching than from protecting the animals for the benefit of tourists. The Zimbabwe government has developed a project called CAMPFIRE: Communal Areas Management Programme for Indigenous Resources. This project involves the generation of income from tourism and sport hunting (Fig. 9.14).

Q9.8.1 Complete Table 9.9 by calculating:

  a. the total income from the CAMPFIRE project between 1989 and 1992;

b. the percentage of this income obtained from each source;

c. the total allocated;

d. the percentage of the total allocated to each sector of expenditure.

Q9.8.2 What is the most important source of income?

Q9.8.3 Who benefits most from this income in cash terms?

Q9.8.4 Calculate the percentage of the total income from trophy fees contributed by each species in Table 9.10.

Table 9.9 Income and allocation of revenue from CAMPFIRE 1989–1992

	Zimbabwe dollars (Z$)	Percentage of total income
Income		
Sport hunting	10,307,342	
Tourism	163,677	
PAC[a] hides and ivory	243,614	
Other	739,905	
Total income =		
Allocation		**Percentage of total allocated**
District councils	1,339,302	
Wildlife management	2,532,843	
Ward/village/house	5,459,554	
Other	297,588	
Total allocated =		

Adapted from Bond (1994).

[a]Problem animal control: the sale of ivory and hides from animals destroyed as crop raiders.

Table 9.10 The value of wildlife trophy fees to CAMPFIRE in 1992

Species	Total shot	Total income Z$	Percentage of total income
Elephant	56	1,996,400	
Elephant PAC	5	74,065	
Buffalo (males)	230	540,270	
Leopard	39	145,353	
Sable antelope	27	94,014	
Lion (males)	15	56,745	
Other	–	307,199	

Adapted from Bond (1994).

Reference/Further Reading

Bond, I. (1994). Importance of elephant sport hunting to CAMPFIRE revenue in Zimbabwe. *TRAFFIC Bulletin*, *14*, 117–119.

Q9.8.5 Which species contributes the majority of the income in CAMPFIRE?

Q9.8.6 What is the average trophy fee paid to kill one individual of this species?

Q9.8.7 The elephant population of Zimbabwe was estimated to be 76,000 in 1994. Would a ban on elephant hunting at that time have done the elephants and the rural communities more harm than good?

HOW DOES SHAPE AFFECT A PROTECTED AREA?

When animals move beyond the boundaries of a protected area, such as a nature reserve, they often lose the protection they enjoy inside it. The shape, size and perimeter length of a nature reserve affect the loss of animals. This exercise considers the effectiveness of different shapes and sizes of reserves in retaining animals using a theoretical model. Table 9.11 shows the relationship between the radius, area and perimeter of circular reserves of different sizes.

Q9.9.1 Consider two reserves both of $100\,km^2$, one circular and one square. Show by calculation which has the shortest boundary and should therefore be most efficient at retaining individual animals. (Note: perimeter of a circle $= 2\pi r$; $r = \sqrt{A / \pi}$, where $r =$ radius of circle, $A =$ area of circle.)

Q9.9.2 Using the data in Table 9.11, draw a single graph showing how the radius of a circular reserve affects:

a. its area; and

b. its perimeter.

Q9.9.3 With reference to your two graph lines, briefly describe the relationship between radius and:

a. area; and

b. perimeter.

Q9.9.4 Should circular reserves be large or small for maximum retention of animals? Use evidence from Table 9.11 to justify your answer.

Q9.9.5 Which of the following designs would be most appropriate for the maximum retention of animals:

Table 9.11 The relationship between the radius, area and perimeter of circular reserves

Radius (r)	Area (πr^2)	Perimeter ($2\pi r$)	Perimeter length/ unit area ($km\,km^{-2}$)
1	3.14	6.29	2.00
2	12.57	12.57	1.00
3	28.29	18.86	0.67
4	50.29	25.14	0.50
5	78.57	31.43	0.40
6	113.14	37.71	0.33
7	154.00	44.00	0.29
8	201.14	50.29	0.25
9	254.57	56.57	0.22
10	314.29	62.86	0.20

a. a single circular reserve of area $100\,km^2$; or

b. four circular reserves each of $25\,km^2$ (total area $= 100\,km^2$)? Explain your answer by calculating total perimeter length per unit area for each option ($km\,km^{-2}$).

Q9.9.6 Examine the three designs for nature reserves in Fig. 9.15. Rank these designs from the best to the worst in terms of their ability to retain animals and provide an explanation for your ranking.

Q9.9.7 Bearing in mind your answers to the previous questions, what three factors are important in the design and location of nature reserves?

Q9.9.8 Fig. 9.16 shows nature reserves of various shapes and sizes. Explain why nature reserve design A is better at retaining organisms than is design B.

Q9.9.9 Suggest why the arrangement of nature reserves in Fig. 9.16D is better at retaining organisms than that in Fig.9.16C.

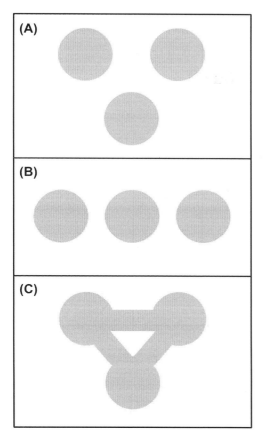

Figure 9.15 Three possible arrangements of a group of three hypothetical nature reserves.

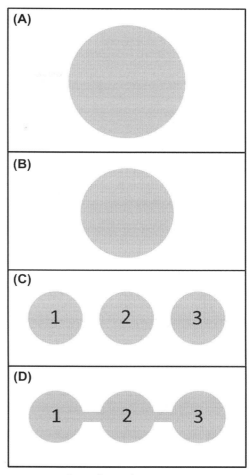

Figure 9.16 Four hypothetical nature reserves.

Q9.9.10 The grey areas in Figs. 9.16 C and D are nature reserves consisting of forest. A fire breaks out in area 2 of Fig. 9.16C and another in area 2 in Fig.9.16D. The areas outside the nature reserves are grassland containing little vegetation. Contrast the likely effects of these fires on the two groups of nature reserves.

Reference/Further Reading

Margules, C., Higgs, A. J., & Rafe, R. W. (1982). Modern biogeographic theory: Are there any lessons for nature reserve design? *Biological Conservation*, *24*, 115–128.

ANIMAL REINTRODUCTIONS

You are the Chief Ecologist at an environmental consultancy and have been asked to plan a project to reintroduce European eagle owls (*Bubo bubo*) into an area that was formerly part of their habitat but from which they have been absent for several hundred years (Fig. 9.17). The area concerned contains land used for agriculture, commercial forestry, tourism and salmon fishing.

Q9.10.1 Suggest ecological studies that need to be conducted in the reintroduction area prior to the reintroduction.

Q9.10.2 Suggest the types of stakeholder that should be consulted about the intention to release these birds.

Q9.10.3 List some possible benefits and some possible harms that might result if this species is released.

Q9.10.4 The birds that will be reintroduced will be taken as chicks from wild populations in areas where the species is relatively common. What legal constraints may cause difficulties for you here?

Q9.10.5 Assuming that it is possible to obtain birds for release what genetic considerations must be addressed in selecting the source population?

Q9.10.6 The chicks will be kept in an aviary and artificially fed prior to release. Why should the keepers provide the food from behind a wooden screen?

Figure 9.17 European eagle owl (*Bubo bubo*).

Q9.10.7 Prior to release, what legal constraints may be encountered?

Q9.10.8 Once the birds have been released what aspects of their biology should be monitored?

References/Further Reading

Rees, P. A. (2001). Is there a legal obligation to reintroduce animal species into their former habitats? *Oryx, 35*, 216–223.

Seddon, P. J., Armstrong, D. P., & Maloney, R. F. (2007). Developing the science of reintroduction biology. *Conservation Biology, 21*, 303–312.

EXERCISE 9.11

CAUSES OF MANATEE MORTALITIES IN FLORIDA

Manatees are herbivorous aquatic mammals that feed opportunistically on a wide range of submerged, floating, and emergent plant species. The American manatee (*Trichechus manatus*) is listed by the IUCN as 'vulnerable' (VU) but the Florida subspecies (*T. m. latirostris*) is listed as 'endangered' (EN) in part because of increasing risks from water traffic in the future.

Q9.11.1 Draw a pie chart showing the number of manatees that died from each cause of death in 2015 using the data in Table 9.12.

Q9.11.2 What percentage of deaths recorded were natural in 2014?

Q9.11.3 Draw a bar chart of the percentage of all recorded deaths that were caused by watercraft between 2004 and 2015.

Q9.11.4 Suggest measures that could be taken to reduce manatee deaths caused by human activity.

Reference/Further Reading

FFWCC. (2016). *Manatee Yearly Mortality Summaries.* Florida Fish and Wildlife Conservation Commission, Marine Mammal Pathology Laboratory. <http://myfwc.com/research/manatee/rescue-mortality-response/mortality-statistics/> Accessed 19.10.16.

Table 9.12 Causes of death in manatees in Florida

Year	Watercraft	Flood gate/ canal lock	Other human	Perinatal	Cold stress	Natural	Undetermined	Unrecovered	Total
2004	69	3	4	72	50	24	51	3	276
2005	79	6	8	89	31	89	90	4	396
2006	92	3	6	70	22	81	116	27	417
2007	73	2	5	59	18	82	66	12	317
2008	90	3	6	101	27	34	69	7	337
2009	97	5	7	114	56	37	103	10	429
2010	83	1	5	97	282	23	208	67	766
2011	88	2	4	78	114	40	115	12	453
2012	82	12	8	70	30	58	124	8	392
2013	73	5	12	129	40	197	274	100	830
2014	69	3	9	100	26	29	118	17	371
2015	86	5	11	91	18	44	119	31	405

Compiled and adapted from FFWCC (2016).

EXERCISE 9.12

THE CONSERVATION ROLE OF ZOOS

Figure 9.18 Haughton House was built in 1898 in Dublin Zoo, Republic of Ireland, in memory of Samuel Haughton, the Royal Zoological Society of Ireland's secretary and a Professor of geology at Trinity College. It originally contained ten animal 'dens', a tearoom and a lecture theatre and is now used as a Learning and Discovery Centre. All zoos in the European Union are required to have an education function.

Zoos have been used as status symbols by royalty and other powerful individuals, to promote civic pride, for scientific enquiry and for entertainment. Since the middle of the 20th century they have been important in the captive breeding of species and modern zoos are now expected to have a conservation function (Fig. 9.18). In the Member States of the European Union zoos are required by the EU Zoos Directive to have a role in the conservation of biological diversity. This may be fulfilled in a variety of ways.

COUNCIL DIRECTIVE 1999/22/EC of 29 March 1999 relating to the keeping of wild animals in zoos

Article 1

Aim

The objectives of this Directive are to protect wild fauna and to conserve biodiversity by providing for the adoption of measures by Member States for the licensing and inspection of zoos in the Community, thereby strengthening the role of zoos in the conservation of biodiversity.

Article 2

Definition

For the purpose of this Directive, 'zoos' means all permanent establishments where animals of wild species are kept for exhibition to the public for 7 or more days a year, with the exception of circuses, pet shops and establishments which Member States exempt from the requirements of this Directive on the grounds that they do not exhibit a significant number of animals or species to the public and that the exemption will not jeopardise the objectives of this Directive.

Article 3

Requirements applicable to zoos

Member States shall take measures under Articles 4, 5, 6 and 7 to ensure all zoos implement the following conservation measures:

- *participating in research from which conservation benefits accrue to the species, and/or training in relevant conservation skills, and/or the exchange of information relating to species conservation and/ or, where appropriate, captive breeding, repopulation or reintroduction of species into the wild,*
- *promoting public education and awareness in relation to the conservation of biodiversity, particularly by providing information about the species exhibited and their natural habitats,*
- *accommodating their animals under conditions which aim to satisfy the biological and conservation requirements of the individual species, inter alia, by providing species specific enrichment of the enclosures; and maintaining a high standard of animal husbandry with a developed programme of preventive and curative veterinary care and nutrition,*
- *preventing the escape of animals in order to avoid possible ecological threats to indigenous species and preventing intrusion of outside pests and vermin,*
- *keeping of up-to-date records of the zoo's collection appropriate to the species recorded.*

Q9.12.1 You own a private zoo in an EU Member State which contains a wide range of exotic animals. Members of the public are not allowed entry to the zoo but you open it to friends and relatives on four days each year. Do you need to comply with the provisions of the Zoos Directive?

Q9.12.2 Zoo B is located in an EU Member State. It has no education programme and poor signage relating to the identification of the animals kept and their conservation. However, it is involved with the captive-breeding of rare species. Has Zoo B complied with the requirements of the Zoos Directive? Explain your answer.

Q9.12.3 Are zoos covered by the Zoos Directive required by law to undertake research on the animals they keep?

Q9.12.4 Why is it important for zoos to keep up-to-date records of their animals?

Q9.12.5 Zoos normally keep records for each individual animal in their collection and keep a record of the total number of each species. For which types of animals would it be very difficult to do this?

Q9.12.6 Explain why the Directive does not require all zoos to participate in projects whose aim is to reintroduce species to the wild.

References/Further Reading

Council Directive 1999/22/EC of 29 March 1999 relating to the keeping of wild animals in zoos.

Rees, P. A. (2011). *An introduction to zoo biology and management*. Chichester: Wiley-Blackwell.

ZOOS AND EVOLUTIONARILY SIGNIFICANT UNITS (ESUS)

An evolutionary significant unit (ESU) may be thought of as a demographically isolated population whose probability of extinction over the timescale of interest (perhaps 100 years) is not substantially affected by natural immigration from other populations. However, definitions of ESUs vary. Most definitions require a degree of geographical separation but the increasing use of DNA analysis has identified previously unknown genetic diversity within some species to the point where it has been possible to separate out new species from within groups of organisms previously considered to be a single species.

The orangutan was previously considered to be a single species (*Pongo pygmaeus*) but is now believed to be two: *P. pygmaeus* (the Bornean orangutan) and *P. abelii* (the Sumatran orangutan). This has caused problems for captive breeding because zoos produced hybrids from these two species in the past that are now no use for breeding. Asian elephants (*Elephas maximus*) may consist of three subspecies. However, there are so few individuals of breeding age in zoos that captive breeding programmes have ignored these subspecies and concentrated solely on saving the species.

Table 9.13 indicates those species of felids for which zoos operate captive breeding programmes in Europe (European Endangered Species programmes (EEPs)) and in North America (Species Survival Plan programs (SSPs)) (Fig. 9.19).

Table 9.13 Felids whose captive populations are managed in EEPs and SSPs

Vernacular name	Scientific name	European Endangered species Programme (EEP)	Species Survival Plan Program (SSP)
Cheetah	*Acinonyx jubatus*	●	●
Fishing cat	*Prionailurus viverrinus*	●	●
Pallas cat	*Otocolobus manul*	●	●
Sand cat	*Felis margarita*	●	●
Black-footed cat	*Felis nigripes*	●	●
Clouded leopard	*Neofelis nebulosa*	●	●
Snow leopard	*Uncia uncia*	●	●
Lion	*Panthera leo*		●
Asian lion	*Panthera leo persica*	●	
Siberian tiger	*Panthera tigris altaica*	●	●
Indochinese tiger	*Panthera tigris corbetti*		●
Sumatran tiger	*Panthera tigris sumatrae*	●	●
Sri Lanka leopard	*Panthera pardus kotiya*	●	
Amur leopard	*Panthera pardus orientalis*	●	
Persian leopard	*Panthera pardus saxicolor*	●	
North Chinese leopard	*Panthera pardus japonensis*	●	
Jaguar	*Panthera onca*	●	
Asian golden cat	*Catopuma temminckii*	●	
Ocelot	*Leopardus pardalis*		
Margay	*Leopardus wiedii*	●	
Oncilla	*Leopardus tigrinus*		●
Geoffroy's cat	*Oncifelis geoffroyi*	●	

Adapted from Rees (2011).

Table 9.14 Elephants held by zoos in Europe and North America in 2006

Region	Species					
	African elephant (*Loxodonata africana*)			Asian elephant (*Elephas maximus*)		
	Bulls	Cows	Sex ratio	Bulls	Cows	Sex ratio
Europe	40	126		51	188	
North America	22	125		23	101	

Adapted from Rees (2009).

Figure 9.19 Amur leopard (*Panthera pardus orientalis*).

Figure 9.20 These Asian elephants (*Elephas maximus*) are part of an EEP for the species.

Q9.13.1 What evidence is there in the table that the definition of an ESU is being applied by zoos inconsistently between species?

Q9.13.2 List those species of felid in Table 9.13 in which the ESU is below the level of species.

Q9.13.3 Suggest justifications for treating the taxa on your list differently from the other felid taxa with respect to their organisation into breeding programmes.

Table 9.14 shows the numbers of elephants held by zoos in North America and Europe in 2006.

Many of these elephants were part of captive breeding programmes (Fig. 9.20).

Q9.13.4 Captive breeding programmes for these two elephant species exist in North America and in Europe. Some authorities recognise subspecies within each of these species but the number of suggested subspecies varies between experts. Suggest three reasons why breeding programmes have not been established for these subspecies.

Q9.13.5 Calculate the sex ratios (male:female) for each species and for each region.

Q9.13.6 Suggest reasons why the ratios are not 1:1.

Q9.13.7 How would you expect these ratios to be different if they were calculated now?

References/Further Reading

Rees, P. A. (2009). The sizes of elephant groups in zoos: Implications for elephant welfare. *Journal of Applied Animal Welfare Science, 12*, 44–60.

Rees, P. A. (2011). *An introduction to zoo biology and management*. Chichester: Wiley-Blackwell.

RETURNING GREY WOLVES TO THEIR FORMER HABITATS IN THE USA

Grey wolves (*Canis lupus*) were systematically removed by hunting and trapping from very large areas of the United States during the early parts of the 20th century (Fig. 9.21). Large predators such as wolves and coyotes (*C. latrans*) were considered to be dangerous pests and eradicating them was seen as worthwhile and morally acceptable to the majority of Americans. President Roosevelt enjoyed hunting and in 1905 he wrote an article describing his participation in a wolf hunt (Roosevelt, 1905) (Fig.9.22). In the 1930s the American government and the Smithsonian were involved in studies aimed at reducing the impact of wolves on livestock (Fig. 9.23).

By the end of the last century the US Fish and Wildlife Service began making attempts to return wolves to their former habitats. The following extract is taken from a paper about reintroduction projects and the legal problems that those undertaking such projects may encounter (Rees, 2001).

In a federal court ruling in December 1997 Downes J. declared that a programme to reintroduce wolves Canis lupus *to Yellowstone National Park and central Idaho was illegal* (Wyoming Farm Bureau Federation v Babbitt, *1997). He ordered that the wolves should be removed, but this order was stayed pending an appeal by conservation organisations. The judge found that assigning the translocated wolves 'experimental and nonessential' status could potentially harm wolves that may naturally migrate from Montana and Canada into the recovery areas, thereby losing their protection under the Endangered Species Act 1973 (ESA).*

Figure 9.21 A grey wolf (*Canis lupus*).

Figure 9.22 President Theodore Roosevelt (second from left) hunting with John Abernathy, a professional wolf hunter, holding a wolf (or possibly a coyote) by the jaw. The hunt is described in Roosevelt (1905). Library of Congress Reproduction Number: LC-DIG-stereo-1s02192.

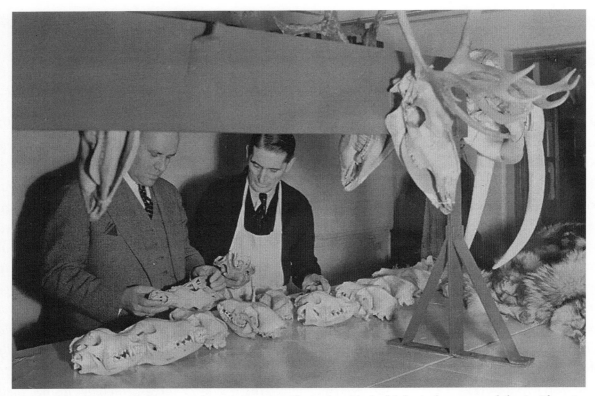

Figure 9.23 Stanley P. Young (left) engaged in a joint scientific study with the biological survey and the Smithsonian Institution in 1939 analysing the skulls and other bones of wolves as a character study with a view to lessening their depredations on livestock in the western regions of the USA. Young published a number of books on American predators including *The War on the Wolf*. Library of Congress Reproduction Number: LC-DIG-hec-26594.

The case was brought by the Idaho, Montana and Wyoming Farms Bureaus against the Department of the Interior (which oversees the endangered species restoration programmes managed by the Fish and Wildlife Service (FWS)). In spite of the existence of a Wolf Compensation Fund set up by the organisation Defenders of Wildlife that compensates ranchers at full market value for verifiable livestock losses, the ranching community is determined to have the wolves removed.

Under the ESA wolves cannot be killed legally in the US. Conferring 'experimental status' on the translocated wolves under section 10(j) of the ESA allowed ranchers

legally to kill any wolves found taking livestock on private land. The judge decided, however, that the effect of assigning these wolves 'experimental status' was to make the introductions illegal under the ESA because they were introduced within the range of nonexperimental (native) populations.

In a second case, brought by the National Audubon Society, attempts have been made to force the FWS to restore full ESA protection to Idaho's wolves, claiming that when experimental animals and endangered populations overlap, the 'experimental' animals revert to 'endangered'.

The American experience with wolf reintroduction programmes shows us the importance of clear legal definitions. The ESA extends the debate on reintroductions by creating a category of animals that are deemed to be 'nonessential experimental populations' within a protected native species. The translocated wolves came originally from Canada. The species was virtually exterminated from the lower 48 states of the US by an intensive government-funded predator-eradication programme in the early decades of the twentieth century. The species has therefore only been absent from the recovery areas for less than 100 years. By any sensible definition the introduced wolves were native species, and their classification as 'experimental' was only instituted to allow ranchers to protect their livestock. If there had been no public objections to the reintroductions these wolves would undoubtedly have been classified as endangered (native) species.

Eventually common sense prevailed. In January 2000, after a long legal battle, the 10th Circuit Court of Appeals overturned the original 1997 decision and determined that an overly technical interpretation of the ESA was inappropriate (Wyoming Farm Bureau Federation v Babbitt 01/13/2000). The wolves have been allowed to stay and are now an important tourist attraction. From 1995 to the end of 1999 approximately 50,000 visitors observed the Druid Peak pack in Lamar Valley, Yellowstone National Park, making them arguably the most viewed wolf pack in the world (Smith et al., 1999).

In the United States, the Endangered Species Act of 1973 allows for experimental populations to be treated differently from other populations of the same species:

Endangered Species Act of 1973

Sec.10(j) EXPERIMENTAL POPULA-TIONS.—(1) For purposes of this subsection, the term "experimental population" means any population (including any offspring arising solely therefrom) authorised by the Secretary for release under paragraph (2), but only when, and at such times as, the population is wholly separate geographically from nonexperimental populations of the same species.

(2)(A) The Secretary may authorise the release (and the related transportation) of any population (including eggs, propagules, or individuals) of an endangered species or a threatened species outside the current range of such species if the Secretary determines that such release will further the conservation of such species.

(B) Before authorising the release of any population under subparagraph (A), the Secretary shall by regulation identify the population and determine, on the basis of the best available information, whether or not such population is essential to the continued existence of an endangered species or a threatened species.

Q9.14.1 Justify the wholesale destruction of populations of large predatory mammals that occurred in the United States in the early parts of the 20th century.

Q9.14.2 How have conservationists justified efforts to reintroduce wolves into parts of the United States?

Q9.14.3 Explain how the designation of reintroduced wolves as an 'experimental population' was used by the ranchers in the Yellowstone area in an attempt to have them removed.

Q9.14.4 Under what circumstances could ranchers legally kill the reintroduced wolves?

Q9.14.5 Why was the original decision of the court to have the wolves removed stayed?

Q9.14.6 What benefits do the wolves reintroduced into Yellowstone National Park bring to the area?

References/Further Reading

Endangered Species Act of 1973.

Rees, P. A. (2001). Is there a legal obligation to reintroduce animal species into their former habitats? *Oryx*, *35*, 216–223.

Roosevelt, T. (1905). A Wolf Hunt in Oklahoma. *Scribner's Magazine*, XXXVIII(5), 513–532. November 1905.

EXERCISE 9.15

DO GOOD FENCES MAKE GOOD NEIGHBOURS? ARE WILDLIFE FENCES PART OF THE PROBLEM OR PART OF THE SOLUTION?

There is an old saying that 'good fences make good neighbours.' In many parts of the world the wildlife authorities are resorting to fencing protected areas. Elsewhere, fences are being erected to restrict the movement of human migrants across international borders (Linnell, 2016).

A dog fence has existed across parts of southern Australia since the 1880s. The current fence stretches from the Nullarbor Plain in South Australia almost to the Queensland coast. In the state of South Australia, the fence was established in law by the Dog Fence Act 1946. The purpose of the fence is to exclude dingoes (*Canis familiaris dingo*) from agricultural areas to protect livestock (Fig. 9.24).

Q9.15.1 Suggest two benefits of fencing a protected area.

Q9.15.2 Suggest two disadvantages of erecting wildlife fences.

Q9.15.3 Why are small holes – too small for an adult dingo to pass through – jeopardising the integrity of the dog fence in Australia.

Crowther et al. (2016) assign the dingo the scientific name *Canis dingo* and believe that it deserves to be protected. When dingoes first arrived in Australia they were genetically isolated but more recently they have hybridised with domestic dogs, making their taxonomic status unclear. This is partly because there are no detailed descriptions available of the original specimens against which to compare modern individuals.

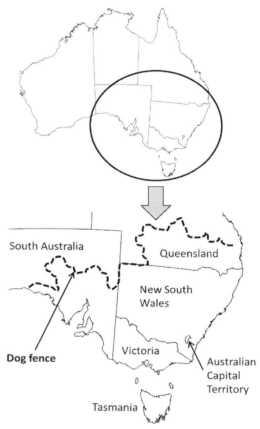

Figure 9.24 The dog fence in southeast Australia is intended to keep dingoes away from livestock areas.

Q9.15.4 Suggest arguments for and against giving dingoes legal protection.

References/Further Reading

Crowther, M. S., Fillios, M., Colman, N., & Letnic, M. (2014). An updated description of the Australian dingo (*Canis dingo* Meyer, 1793). *Journal of Zoology, 293*, 192–203.

Linnell, J. D. C. (2016). Border controls: Refugee fences fragment wildlife. *Nature, 529*, 156.

ETHICS IN ECOLOGICAL RESEARCH AND ECOSYSTEM MANAGEMENT

Many early ecological studies would not have been possible without killing animals and plants and, in some cases, destroying ecosystems. Conservation projects often involve capturing and moving animals and, sometimes, taking them into captivity (Fig. 9.25). Nowadays, ecologists are much more aware of the welfare needs of animals and the need to minimise environmental damage than they were in the past. It is important that all ecologists consider the legal and ethical restrictions that constrain their work. Most academic journals which publish studies on animals require the authors to follow strict ethical guidelines.

Discuss the ethical issues that arise in each of the following projects:

Q9.16.1 A study of the feeding ecology of minke whales (*Balaenoptera acutorostrata*) which requires the killing of whales to examine their stomach contents.

Q9.16.2 The felling of 100 hectares of woodland in order to study the effect of forestry operations on runoff and nutrient cycles.

Q9.16.3 The attachment of a radio collar to a lion (*Panthera leo*) in the Serengeti National Park in order to track its movements.

Q9.16.4 The capture, marking and release of a species of small mammal (e.g., a vole) in order to calculate its population size.

Q9.16.5 The cropping of all of the vegetation in a metre square in order to determine the biomass of material present.

Figure 9.25 European beaver (*Castor fiber*). The Royal Zoological Society of Scotland, the Scottish Wildlife Trust and Forestry Commission Scotland have reintroduced beavers into Scotland after being absent since the 16th century.

Q9.16.6 The capture and taking into captivity of a population of a species of raptor in order to establish a captive breeding programme.

Q9.16.7 The release of captive bred animals into the wild as part of a reintroduction programme.

Q9.16.8 Clipping the toes of small mammals as a method of marking them for population studies.

Q9.16.9 Culling a population of African elephants (*Loxodonta africana*) to reduce elephant damage to *Acacia* woodland.

Q9.16.10 Eradicating feral cats (*Felis catus*) on an island where they are threatening the survival of populations of rare ground-nesting seabirds.

Reference/Further Reading

Farnsworth, E. J., & Rosovsky, J. (1993). The ethics of ecological field experimentation. *Conservation Biology, 7*, 463–472.

CHAPTER 10

STATISTICS

This chapter is concerned with the use of statistical methods to analyse ecological data.

Lichens are important indicators of air quality

Examining Ecology. DOI: http://dx.doi.org/10.1016/B978-0-12-809354-2.00010-5

INTENDED LEARNING OUTCOMES

On completion of this chapter you should be able to:

- Calculate measures of central tendency.

- Calculate measures of dispersion.

- Explain the difference between a population and a sample.

- Distinguish between continuous and discontinuous variables.

- Construct graphs appropriate to different types of data.

- Interpret graphs with liner and logarithmic scales.

- Recognise a normal distribution.

- Recognise a skewed distribution.

- Formulate a null hypothesis.

- Distinguish between one- and two-tailed hypotheses.

- Explain the difference between a type I and a type II error.

- Calculate simple probabilities.

- Perform a variety of statistical tests.

- Determine whether or not the results of a statistical test are statistically significant.

INTRODUCTION

Statistics is the branch of mathematics concerned with the description of the properties of samples of measurements and with drawing inferences from numerical data, based on probability theory.

VARIABLES, POPULATIONS, AND SAMPLES

Variables

A variable is a number, quantity or characteristic that changes in value over time or in different situations, for example, relative humidity, the number of females in a population, or the amount of time an animal spends sleeping.

Variables may be categorised as independent or dependent variables, on the basis of their relationship to one another. An independent variable (IV) may assume different values regardless of changes in the value of other variables. A dependent variable (DV) takes on different values in response to changes in an independent variable. In most graphs the IV is plotted against the x-axis and the DV is plotted against the y-axis. However, sometimes this is reversed to provide a better pictorial representation of the data being plotted.

Populations and Samples

In statistics, a population is the complete set of values of a particular variable in a given situation, e.g., the heights of all of the one year old tigers in India. It is sometimes referred to as the parent population.

A sample is a small portion or small number of something whose characteristics are taken to represent the whole. For example, a sample of voles taken from a population of voles; a sample of weights of blackbirds taken from a population (of weights) of blackbirds.

A population possess characteristics referred to as parameters, for example, a mean (μ) and a standard deviation (σ). A sample taken from this population has characteristics called statistics including a mean ($\overline{x}$) and a standard deviation (s). Note that the symbols used for the same measure are different.

If the samples taken are representative of the population as a whole it is possible to make inferences about the population by analysing these samples using appropriate statistical methods.

DESCRIPTIVE STATISTICS

The results of most studies are likely to produce values for variables that are amenable, at the very least, to examination using descriptive statistics. This may include simple graphical summaries showing the spread of data or the calculation of measures of central tendency (a 'middle' or 'typical' value, such as a mean) along with an indication of dispersion (the spread of values around this middle value, such as a standard deviation). The discussion of measures of central tendency and dispersion that follows is concerned with samples rather than populations.

Graphs

A graph should be self-explanatory. It should have a title which explains its purpose without requiring the reader to refer to explanatory text, with the exception of very complex graphs. Each axis should have a title and include information about the units used.

Bar charts may be used where the variable being measured is discontinuous. For example, each column in a bar chart might represent the percentage of birds in a survey belonging to various species (species K, L, M, N, etc.). Each species is unrelated to any of the others so the order in which the columns are displayed does not really matter and there should be a gap between each column. It is often useful to draw bars in rank order of height (e.g., tallest first).

Histograms are used to display data that is continuous, for example, the length of individual animals all belonging to the same species. Here we would use a scale for the x-axis of our graph in which the shortest length class appeared at the left side of the x-axis and the longest length class at the right, with various intermediate classes between them. There should be no gap between the columns. Determining an appropriate range of values for classes is largely a matter of judgement and experience.

Should Points be Joined?

When the points in a graph are joined by a line there is an implication that there are missing values that could have been plotted – if sufficient measurements had been made – between those which actually occur in the graph. This may or may not be true. The process of 'adding' these new data points is referred to as 'interpolation'. If all of the plotted data points lie on a perfectly straight line, interpolation is straightforward. In some cases they may lie on a curve whose path is predictable. In such cases it may even be possible to extrapolate the line beyond the data points plotted in order to make predictions outside of the values of the variables studied. In many cases it does not make sense to interpolate new data points because we have no way of knowing what might have happened between the data points that we have.

Some data are collected by sampling two variables simultaneously. If these data are plotted as a scatter diagram the best way of representing the relationship between the variables may be to add a line-of-best-fit (trend line). This may be a straight line or a curve. It may pass through some of the points or none of them. The process of fitting such a line to data is called 'regression analysis'.

Linear and Logarithmic Scales

A linear scale is one which is separated into equal divisions for equal values so, for example, the distance between 1 and 2 is the same as the distance between 3 and 4 or 5 and 6. This is the type of scale used on most graphs. A logarithmic (log) scale is one which plots the logarithm (usually to the base 10, and written $\log_{10}$) of the values instead of the values themselves. For example the $\log_{10}$ of 10 is 1, the $\log_{10}$ of 100 is 2, and the $\log_{10}$ is 1000 is 3.

The use of a log scale allows differences between low values to be distinguished more easily when the scale extends over a very wide range of values. It has the effect of making some curves appear as straight lines, or at least, straighter than if they had been plotted on a linear scale. Log scales are often used in survivorship curves.

Measures of Central Tendency

A measure of central tendency is a number that indicates the 'middle' of a set of values. How this 'middle' is calculated depends upon the particular measure of central tendency used.

Mean

The mean is also known as the arithmetic average. This is the sum of all of the values in the data set divided by the number of values:

$$\overline{x} = \frac{\sum x}{n}$$

where,

x = a particular value from the data set;

$\sum x$ = the sum of all of these values;

n = the total number of values in the data set.

The mean of the values 2, 5, 6, 3 = 16/4 = 4. Although the mean is easy to calculate, and to understand, it is not always the most appropriate measure to calculate. For example, some data sets contain values which are exceptionally high or exceptionally low compared with the rest of the data (outliers). Such outlying values can be disregarded by calculating the median.

Median

The median is the middle value when all of the data are ranked from the lowest value to the highest. The values 9, 3, 6, 4, 2 become 2, 3, 4, 6, 9 when ranked so the median is 4. If there is an even number of values the median is the mean of the two values either side of the middle. When the values 12, 3, 1, 4, 7, 2 are ranked they become 1, 2, 3, 4, 7, 12. The median of the values is (3 + 4)/2 = 3.5.

The median is a useful measure of central tendency where the data set contains outliers because it prevents them from affecting the calculation. For example, if the values were 1, 3, 6, 9, 11 the median would be 6. If the values were 1, 3, 6, 9, 128 the median would still be 6.

Mode

This is the value within a set of data which occurs most frequently. The mode of the values 1, 5, 7,

2, 5, 3 is 5, because it occurs twice (i.e., more often than any of the others).

If we wanted to know how many offspring a particular mammal species most often produces – the commonest litter size – calculating a mean would most likely produce a value which was not a whole number. The median could also generate a value which was misleading because few individuals – and possibly no individuals, if the sample size was an even number thereby producing a median with a fractional component – may produce a litter of this size.

Consider the following data:

1, 2, 2, 3, 3, 3, 4

Here, the mean is 2.57, the median is 3, and the mode is also 3. Some distributions have no mode, while others are bimodal, i.e., the distribution has two modes, e.g., 2, 3, 4, 4, 4, 5, 6, 7, 9, 9, 9, 10, 11. The calculation of the mean or median would not identify a bimodal distribution as only a single value is produced.

Measures of Dispersion

Measures of dispersion indicate the distribution of values within the data set. A measure of central tendency should be accompanied by a measure of dispersion because two sets of data could have identical means even though the distributions were not identical. For example, consider the following sets of data:

Data set K – 5, 5, 5, 5, 5

Data set L – 2, 3, 5, 6, 9

The mean of both data sets is 5, but clearly the distribution of values in each is different. In these particular cases, the median is also 5, and in data set K the mode is 5.

The simplest measure of dispersion is the range.

Range

This is the difference between the highest and lowest values in a set of data, e.g., in the data set 4, 8, 9, 11, 18, the range is 14 (18–4). This takes no account of extremely large or extremely small values which might make the dispersion of the values appear much greater than it would if any outliers were to be removed. The range of the values 2, 4, 6, 7, 100 is 100–2 = 98. The range of 2, 83, 88, 95, 100 is also $100 - 2 = 98$.

Variance (s^2)

The variance is a measure of the dispersion of values around the mean and is calculated using the equation below.

$$s^2 = \frac{\sum (x - \bar{x})^2}{n - 1}$$

where,

x = a particular value from the data set;

$\bar{x}$ = the mean of all values in the data set;

n = the total number of values in the data set.

For the values, 2, 3, 5, 4, 1, the mean is 15/5 = 3.

Each value of x should now be subtracted from the mean and squared. The squared values should then be added together:

x	$x-\bar{x}$	$(x-\bar{x})^2$
2	2–3 = −1	1
3	3–3 = 0	0
5	5–3 = 2	4
4	4–3 = 1	1
1	1–3 = −2	4
	Total	10

$$s^2 = \frac{10}{5 - 1} = 10/4 = 2.5$$

The variance is the square of the standard deviation.

Standard Deviation (s)

The standard deviation is an alternative measure of the dispersion of values around the mean. It is the square root of the variance. In a normal distribution over 99% of all values fall between three standards deviations either side of the mean.

$$s = \sqrt{\frac{\sum (x - \bar{x})^2}{n - 1}}$$

where,

x = a particular value from the data set;

$\bar{x}$ = the mean of all values in the data set;

n = the total number of values in the data set.

The calculation is identical to that for the variance except that the square root must be taken of the value obtained. So, for the data in the example for the calculation of the variance above,

$$s = \sqrt{\frac{10}{5 - 1}} = \sqrt{2.5} = 1.581$$

The Normal Distribution

The best known distribution found in biology is the 'normal' or Gaussian distribution, which occurs in variables such as height, weight, wingspan and length of pregnancy. It takes the form of a smooth, symmetrical, bell-shaped curve, where the mean, median and mode of the population correspond to the highest point on the curve. Some 68.29% of all of the values in a normal distribution fall between one standard deviation of the mean; 95.44% occur within two standard deviations of the mean and almost all values (99.74%) lie within three standard deviations of the mean. Many variables assume an approximately normal distribution, especially when sample size is large.

A skewed distribution is one whose apex is positioned to the right or left of its centre. A positively skewed distribution has a long 'tail' to the right of the apex and the mean is also located to the right of the apex. A negatively skewed distribution has a long 'tail' to the left of the apex and the mean is also located to the left of the apex.

INFERENTIAL STATISTICS

Inferential statistics are concerned with making inferences or predictions about a population from observations and analyses that have been made on a sample. Before describing a range of different statistical methods we should first consider the process of formulating testable hypotheses.

Hypotheses and Hypothesis Testing

An hypothesis is a statement or proposition which is assumed to be true for the sake of argument or as the basis for experimentation or investigation of the evidence; it is a provisional explanation. The formulation of hypotheses is an essential step in the testing of ideas in science.

Some studies do not state a hypothesis because they may be attempting to establish a fact or a set of facts. For example, what is the average size of social groups of wild lions and how much variation is there in this? Such a study could result in the calculation of a mean and a standard deviation, and the production of a graph showing the distribution of the frequencies of different group sizes. Other studies need a clearly defined hypothesis to test. This may be designed to examine particular types of questions:

1. Is variable C greater than variable D? For example, do male baboons on average exhibit more aggression than female baboons?

2. Is there an association between variable K and variable L? For example, does the depth of the water table affect the frequency of germination of the seeds of some plant species?

Statistical methods are required to test such questions. In statistical testing two different hypotheses are defined. These hypotheses are referred to as the null hypothesis and the alternative hypothesis.

A null hypothesis (H_0) is a hypothesis that a scientist attempts to disprove (reject or refute) in the course of a particular scientific study. It is generally paired with an alternative hypothesis (H_1). In statistics often the purpose of a test is to try to show that the null hypothesis is likely to be wrong; to reject or refute the null hypothesis. For example, if we were studying temperatures in two locations (M and N) we could formulate the following hypotheses:

H_1 = There is a difference in the mean maximum daily temperature at sites M and N.

H_0 = There is *no* difference in the mean maximum daily temperature at sites M and N.

In attempting to establish that H_1 is likely to be true we must obtain evidence from a statistical test that allows us to reject H_0. Statistical tests such as chi-squared tests, and t-tests allow us to test whether or not apparent differences between sets of data are likely to have occurred by chance.

Probability and Statistical Significance

If the results of a study are statistically significant this means that they are unlikely to have occurred by chance. How do we determine whether or not an event has occurred by chance? If we toss a coin the chance (probability) that it will come down heads is 0.5 (1/2) and the chance that it will come down tails is also 0.5 (1/2) as there is no other possibility and the probabilities of all possible events must add up to 1.0.

The results of statistical analyses produce a test statistic (e.g., a t value) which may, for example, indicate the difference between the means of two sets of data. By looking this value up in the appropriate table the probability that this value would be obtained by chance may be determined. Such tables are widely available in statistics textbooks and on the Internet.

A P value is associated with a test statistic and is defined as the probability of observing a value of the test statistic greater than or equal to the one actually observed (assuming that the test statistic is distributed as it would be under the null hypothesis). The P value is the actual probability determined by the test. For example, $P < 0.05$ means that if the same experiment or study was undertaken 100 times the results obtained would occur by chance on fewer than five occasions. In other words we can be 95% certain that our results have been caused by a real effect and not by chance. On the other hand, $P > 0.05$ means that if the same experiment or study was undertaken 100 times the results obtained would occur by chance on more than five occasions. Similar statements could be made for other levels of probability.

The alpha (α) value is the threshold value against which P values are measured: a number between 0 and 1, e.g., 0.05 (5%), 0.01 (1%). This value is specified before the test is performed. So, for example we may set this value as 0.05. This would mean that we would reject the null hypothesis if our P value was less than this.

If $\alpha = 0.05$ and the calculated P value is 0.004 we reject the null hypothesis.

If $\alpha = 0.05$ and the calculated P value is 0.125 we accept the null hypothesis.

So, when we say that something is statistically significant this means that a given result is unlikely to have occurred by chance assuming the null hypothesis is actually correct (i.e., there is no difference between two sets of data, there is no relationship between two variables etc.).

One-Tailed and Two-Tailed Tests

A statistical test is said to be 'two-tailed' if we do not care about the direction of the difference between two sets of data and 'one-tailed' if we do. So if we are testing to see if A is larger than B we use a one-tailed test. Similarly, if we were testing if B is larger than A we would use a one-tailed test. If we are only interested in if A and B are different (i.e., A > B *or* A < B) because we have no a priori reason for thinking that there is a particular direction to the difference, we would use a two-tailed test. It is important to decide which type of test you need to perform before you calculate your statistic because the critical value with which you must compare it will depend upon whether you are using a one-tailed or a two-tailed test.

Degrees of Freedom

The degree of freedom (*df*) is essentially the number of values in a calculation that are free to vary. If the mean of a set of data is 6 and we know that it has been calculated from the four values 2, 10 and 4 and an unknown value (*x*), we can calculate *x* from the other values (i.e., 8). In this case, the number of degrees of freedom is three, because if we know three of the values and their mean we can calculate the value of the missing number because its value is fixed.

When applying a statistical test the number of degrees of freedom is the number of values in calculating the statistic that are free to vary. We need to know this in order to look up the critical values in the appropriate place in a statistical table.

Type I and Type II Errors

If an inappropriate level of significance is used for a statistical test an error may be made in the interpretation of the results. A type I error occurs when a null hypothesis is rejected when it should have been accepted. This may result in the researcher concluding that there is an effect or relationship when there is not: a false positive. A type II error occurs when a null hypothesis is accepted when it should have been rejected. This may result in a researcher concluding that there is no effect or relationship when, in fact, one exists: a false negative. For example, in a correlation analysis, concluding that there was a correlation between variables X and Y when there was not would be a type I error. Conversely, concluding that there was no correlation when, in fact, a correlation existed would be a type II error.

Statistical Tests
Testing for Differences Between Sets of Data

Independent *t*-Test   The independent *t*-test, or Student's *t*-test, is used to determine if two sets of data have equivalent means. It assumes the data are normally distributed and the two sets of data do not have to contain the same number of observations. For example, we could use this test if we wanted to examine whether there was a difference between the time taken for male and female chimpanzees to obtain food from a puzzle feeder. The number of each sex could be the same or it could be different. Student's *t* is calculated as:

$$t = \frac{\overline{x}_A - \overline{x}_B}{\sqrt{\dfrac{s_A^2}{n_A} + \dfrac{s_B^2}{n_B}}}$$

where,

$\bar{x}_A$ = mean of data set A;

$\bar{x}_B$ = mean of data set B;

s_A^2 = variance of data set A;

s_B^2 = variance of data set B;

n_A = number of values in data set A;

n_B = number of values in data set B.

$$df = (n_A + n_B) - 2$$

The value of t for the data in the following table is calculated as follows:

	A	$\bar{x}_A$–A	$(\bar{x}_A–A)^2$	B	$\bar{x}_B$–B	$(\bar{x}_B–B)^2$
	8	10.67–8 = 2.67	7.13	6	6.4–6 = 0.4	0.16
	12	10.67–12 = − 1.33	1.77	5	6.4 – 5 = 1.4	1.96
	10	10.67–10 = 0.67	0.45	7	6.4–7 = − 0.6	0.36
	9	10.67–9 = 1.67	2.79	6	6.4–6 = 0.4	0.16
	13	10.67–13 = − 2.33	5.43	8	6.4–8 = − 1.6	2.56
	12	10.67–12 = − 1.33	1.77			
Total	64	–	19.34			5.2
	$\bar{x}_A = 64/6 = 10.67$			$\bar{x}_B = 32/5 = 6.4$		

$\bar{x}_A = 10.67$

$\bar{x}_B = 6.4$

$s_A^2 = \dfrac{19.34}{6 - 1} = 3.868$

$s_B^2 = \dfrac{5.2}{5 - 1} = 1.3$

$n_A = 6$

$n_B = 5$

$t = \dfrac{\bar{x}_A - \bar{x}_B}{\sqrt{\dfrac{s_A^2}{n_A} + \dfrac{s_B^2}{n_B}}}$

$= \dfrac{10.67 - 6.4}{\sqrt{\dfrac{3.868}{6} + \dfrac{1.3}{5}}}$

$t = 4.484$

$df = (6 + 5) - 2 = 9.$

As we have no reason to expect one sex to be faster than the other at using the puzzle feeder we must perform a two-tailed test because we are interested in a difference in either direction, i.e., males faster than females, or females faster than males. The critical value for the the 5% level of significance is 2.262 (with 9 degrees of freedom). Our calculated value exceeds this ($t = 4.484$). It also exceeds the critical value for the 1% level of significance (3.250). We can therefore say that we are more than 99% certain that the difference observed between the two sexes did not occur by chance. Alternatively, the probability of this difference occurring by chance is less than 1% ($P < 0.01$).

Testing for Associations Between Sets of Data

Correlation and Regression Correlation analysis is a mathematical method that generates a single value (correlation coefficient) to represent the relationship between two variables. This value ranges from +1 (a perfect positive correlation) to −1 (a perfect negative correlation). A value of 0 indicates no correlation. The higher the absolute value of the correlation (i.e., the value disregarding the sign) the stronger the relationship between the variables. A positive correlation between variables X and Y means that as X increases, so too does Y. A negative correlation between these variables means that as X increases Y decreases and vice versa.

Whether or not a correlation coefficient is statistically significant depends upon the sample size. Imagine two points on a graph. By definition, they can be joined by a straight line. If this line is anything other than perfectly horizontal, the correlation coefficient will be either $+1$ or -1, i.e., there is a perfect positive correlation or a perfect negative correlation. This has no meaning because the sample size is just two, the minimum number of points required for a straight line, and by definition a line which passes through these points must go through both and can only produce a value of $+1$ or -1 or zero if the line is horizontal (in which case the calculation cannot be performed because it results in a division by zero error).

Even where correlations are high and statistically significant, it does not follow that a change in one variable necessarily results in a change in the other, i.e., the correlation does not prove 'cause and effect'. It may be that both variables are linked to a change in a third variable which has not been measured.

Product-Moment Correlation Coefficient (Pearson r)   This statistic is used to measure the strength of the linear relationship between two variables, where both are measured on an interval or a ratio scale, and is calculated as:

$$r = \frac{\sum[(x - \bar{x})(y - \bar{y})]}{\sqrt{[\sum(x - \bar{x})^2 \sum(y - \bar{y})^2]}}$$

where,

x = individual value from data set x;

y = individual value from data set y;

$\bar{x}$ = mean of data set x;

$\bar{y}$ = mean of data set y;

n = number of pairs of data;

$df = n - 2$.

The value of r may be thought of as a measure of how far the points on a graph of one variable against the other fall from a straight line drawn through these points. The closer they fall to the line, the stronger the correlation and the nearer the value of r will be to 1. If the points fall exactly on a straight line the value of r will be either $+1$ (a prefect positive correlation, where both variables increase together) or -1 (a perfect negative correlation, where one variable increases as the other decreases). If $r = 0$ this does not mean there is no correlation between the two variables but there is no *linear* relationship.

The following table illustrates the method of calculating r for the two sets of data x and y.

x	$\bar{x} - x$	$(\bar{x} - x)^2$	y	$\bar{y} - y$	$(\bar{y} - y)^2$	$(\bar{x} - x) \times (\bar{y} - y)$
5	1.17	1.36	7	−1.33	1.78	−1.56
7	−0.83	0.69	9	−3.33	11.11	2.78
3	3.17	10.03	1	4.67	21.78	14.78
8	−1.83	3.36	6	−0.33	0.11	0.61
5	1.17	1.36	4	1.67	2.78	1.94
9	− 2.83	8.03	7	−1.33	1.78	3.78
37		$\Sigma = 24.83$	34		$\Sigma = 39.33$	$\Sigma = 22.33$
37/6			34/6			
$\bar{x} = 6.17$			$\bar{y} = 5.67$			

$$r = \frac{22.33}{\sqrt{24.83 \times 39.33}} = 0.7146$$

If we are looking for a correlation in either direction, i.e., positive or negative, we would apply a two-tailed test. If we had reason to believe that there should be a positive correlation we would apply a one-tailed test. Similarly, if we had reason to believe the data were negatively correlated we would also apply a one-tailed test.

If we apply a two-tailed test to these data, the critical value of r at the 5% level of significance for six pairs of data is 0.811. For a one-tailed test, the critical value at 5% is 0.729. In both cases the correlation is not significant ($P > 0.05$).

Linear Regression Analysis The purpose of linear regression analysis is to fit a straight line (a line-of-best-fit) to a series of points on a graph which represents the relationship between two variables (x and y). This line allows us to predict the value of one variable from the other.

The equation of a straight line is:

$$y = bx + a$$

In this equation,

$$a = \bar{y} - b\bar{x}$$

where,

$\bar{x} =$ mean of data set x;

$\bar{y} =$ mean of data set y.

The value of b (the gradient of the line) is calculated as:

$$b = \frac{\Sigma(x - \bar{x})(y - \bar{y})}{\Sigma(x - \bar{x})^2}$$

Using the data from the example for the correlation coefficient, the regression line is calculated as follows:

$\bar{x} - x$	$\bar{y} - y$	$(\bar{x} - x)(\bar{y} - y)$
1.17	−1.33	−1.5561
−0.83	−3.33	2.7639
3.17	4.67	14.8039
−1.83	−0.33	0.6039
1.17	1.67	1.9539
−2.83	−1.33	3.7639
	Σ	22.3334

b = 22.3334/24.83 = 0.899452

a = 5.67 − 0.899452 × 6.17 = 0.1202

y = 0.8895x + 0.1204

Care must be taken not to infer that there is necessarily a causal relationship between the variables. The method described here applies only to *linear* relationships.

Chi-Squared Tests (χ^2)   The chi-squared (χ^2) statistic may be used to compare the counts of categorical responses between two or more independent groups. For example, it may be used to test if the numbers of males and females in a population conforms to a 1:1 ratio or if an animal uses all 10 equal-sized zones in a zoo enclosure equally. A χ^2 test may only be used on raw counts and not on percentages, proportions, means, or other calculated values. The basic formula used to calculate chi-squared is:

$$\chi^2 = \sum \frac{(O_i - E_i)^2}{E_i}$$

where,

O_i = observed value;

E_i = expected value;

df = number of comparisons – 1.

This type of test may be used whenever we wish to examine a hypothesis which specifies the frequency with which the observations should fall into certain classificatory groups. This is called a goodness-of-fit test. Consider the following example.

An ecologist sampled the population of a species of vole in a woodland. A total of 25 males and 35 females were captured. Do these numbers differ from a 1:1 ratio? The calculation of χ^2 is performed as follows:

	Males	Females
Observed values (O)	25	35
Expected values (E)	$(25 + 35)/2 = 30$	$(25 + 35)/2 = 30$
$\dfrac{(O - E)^2}{E}$	$\dfrac{(25 - 30)^2}{30}$	$\dfrac{(35 - 30)^2}{30}$
χ^2	0.833	0.833
		1.666

This value must now be looked up in χ^2 tables. For this to be statistically significant at the 5% level our calculated value must exceed 3.148 (with one degree of freedom). In this case it does not so there is no significant difference between the observed and expected values ($P > 0.05$).

The 2 × 2 Contingency Table   A 2 × 2 contingency table may be used to test for independence between two categorical variables. The data to be tested should be arranged in two columns and two rows as indicated below:

		Variable A		
		Present	Absent	Totals
Variable B	Present	a	b	$a + b$
	Absent	c	d	$c + d$
	Totals	$a + c$	$b + d$	$a + b + c + d = n$

Chi-squared is calculated as:

$$\chi^2 = \frac{n(|ad - bc| - 0.5n)^2}{(a + b)(c + d)(a + c)(b + d)}$$

$$df = 1$$

This formula includes Yates' correction. This is applied when some of the expected values are rather small, generally taken to mean less than 5, and can do no harm when numbers are large. This method could be used to test for an association between, for example, the presence of species A and species B in the same quadrat:

		Species A		
		Present	Absent	Totals
Species B	Present	36	16	52
	Absent	22	34	56
	Totals	58	50	108

STATISTICS

$$\chi^2 = \frac{n(|ad - bc| - 0.5n)^2}{(a + b)(c + d)(a + c)(b + d)}$$

$$= \frac{108(|(36 \times 34) - (16 \times 22)| - (0.5 \times 108))^2}{52 \times 56 \times 58 \times 50}$$

$$= 72265392/8444800$$

$$\chi^2 = 8.557$$

$$df = 1$$

The critical value of χ^2 of the 5% level of significance ($df = 1$) is 2.71. The calculated value of 8.557 exceeds this. It also exceeds the the critical value at the 1% level (5.412). There is, therefore, a statistically significant association between the two variables ($P < 0.01$).

CALCULATING PROBABILITIES

The likelihood that an event will occur can be expressed as a fraction, a percentage, or a number between zero (impossible) and 1 (certain). The sum of the probabilities of all possible outcomes is always 1 (or 100%).

$$\text{The probability of an outcome} = \frac{\textit{The number of ways this outcome can occur}}{\textit{The total possible outcomes}}$$

The probability that events X *and* Y both occur is equal to the probability that event X occurs multiplied by the probability that event Y occurs. This is known as the multiplication rule. The probability that event K *or* event M occurs is equal to the probability that event K occurs plus the probability that event M occurs. This is known as the addition rule.

This following exercises are intended to provide practice in using various methods in statistics to analyse the types of data produced by ecological studies.

Reference

van Emden, H. (2008). *Statistics for Terrified Biologists*. Oxford: Blackwell Publishing Ltd.

SEWAGE SLUDGE PARASITES

During the process of sewage (wastewater) treatment the process of settlement produces sludge in which a number of organisms are evident. This sludge is frequently treated in anaerobic digesters. The ova of the nematode worms *Ascaris* and *Toxocara* occur in anaerobically digested wastewater sludge. Table 10.1 shows the number of ova found in samples of sludge and the number that eventually embryonated.

Q10.1.1 Calculate the probability of an embryo developing from:

a. an *Ascaris* ovum;

b. a *Toxocara* ovum.

Q10.1.2 Eggs of the tapeworm *Taenia* sink to the bottom of sewage in settlement tanks and are therefore removed in the sludge. In an experiment on the effects of settlement time on egg removal using a 475 mm column of raw sludge the following results were obtained (Table 10.2).

Table 10.1 The numbers of ova observed and embryonated in two genera of nematodes

Parasite	Total number of ova observed	Number of ova embryonated
Ascaris spp.	33	21
Toxocara spp.	64	34

Table 10.2 The effect of settling time on *Taenia* egg removal

Settling period (minutes) (x)	Percentage removal of eggs (y)
15	51
30	65
60	81
120	98

a. draw a scatter diagram showing the relationship between egg removal and settling period;

b. draw the regression line of y on x, showing the formula of the line;

c. using the formula calculate the percentage removal you would expect after 45 minutes.

EXERCISE 10.2

OZONE LEVELS IN A CITY

The following measurements were made of ozone levels (in Dobson Units, DU) at two street level locations within a city (Table 10.3).

Q10.2.1 For the ozone levels at each location calculate:

    a. the mean;

    b. the standard deviation;

    c. the median.

Q10.2.2 Use an independent t-test test to determine whether or not there is a statistically significant difference between the ozone levels recorded at these two sites.

    a. You have no reason to believe that the ozone level is higher at one site than the other. Would you perform a one-tailed test or a two-tailed test?

    b. State your null hypothesis (H_o).

Table 10.3 Ozone levels at two locations in a city (Dobson Units)

Location A	Location B
400	389
365	399
450	375
388	403
391	366
420	385
405	406
398	388
405	378
410	
420	

    c. Calculate the value of t.

    d. Using statistical tables determine whether or not your t value is statistically significant (state the level).

    e. Should you reject or accept the null hypothesis?

LEAD POLLUTION IN GULLS

Samples of herring gulls (*Larus argentatus*) that were washed up on the beach near a factory at Northcliff and others that were shot in a colony living 20 miles south of the factory at Wellbeach were analysed for lead content. The results of seawater sampling at Northcliff are provided in Table 10.4 and analyses of lead levels in the brains of the gulls sampled are presented in Table 10.5.

Table 10.4 Lead levels (ppm) in samples of seawater

10	13	11	24
45	27	25	32
25	7	29	45
60	51	14	29
30	22	8	45
6	27	2	51
43	22	28	12
33	42	26	21

Table 10.5 Lead levels (ppm) in two samples of herring gulls

Sample 1 Northcliff	Sample 2 Wellbeach
112	56
87	78
78	49
110	79
67	66
103	51
98	85
77	56
105	67
88	79
64	101
79	45
90	77
75	89
101	

Q10.3.1 Draw a bar chart of the distribution of lead levels in the seawater samples listed in Table 10.4. Use class intervals of 10ppm (i.e. 0–10, 11–20 etc.).

Q10.3.2 For the data on lead levels in the sea, calculate:

  a. the mean;

  b. the mode;

  c. the median;

  d. the standard deviation;

  e. the range.

Q10.3.3 Use an independent *t*-test test to determine whether or not there is a statistically significant difference between the lead levels in gulls sampled at the two sites.

  a. You believe that lead levels in gulls should be higher in sample 1 than in sample 2 because these animals were taken in an area where lead pollution was suspected. Would you perform a one-tailed test or a two-tailed test?

  b. State your null hypothesis (H_o).

  c. Calculate the value of *t*.

  d. Using statistical, tables determine whether or not your *t* value is statistically significant (state the level).

  e. Should you reject or accept the null hypothesis?

Q10.3.4 Describe precautions that must be taken in sampling the water and the gulls in order to insure that any statistical analyses undertaken are valid.

EXERCISE 10.4

BODY LENGTH AND MASS IN HUMPBACK WHALES

The following data show the relationship between body length and mass in 10 humpback whales (*Megaptera novaeangliae*) (Table 10.6).

Q10.4.1 Examine the relationship between length and mass by calculating the correlation coefficient (r) between these two variables. You

Table 10.6 The body length and mass of 10 humpback whales

Length (metres) x	Mass (tonnes) y
12.5	26.1
16.5	32.0
15.7	29.0
15.8	32.1
14.2	29.3
12.0	14.0
15.9	30.0
13.6	28.0
14.9	31.0
16.0	33.0

believe that mass should be proportional to length.

a. Should you perform one-tailed or a two-tailed test?

b. State your null hypothesis (H_0).

c. Calculate the value of r.

d. Using statistical tables determine whether or not your r value is statistically significant (state the level).

e. Should you reject or accept the null hypothesis?

Q10.4.2 Draw a scatter diagram showing the relationship between length and mass.

Q10.4.3 Calculate the position of the regression line (y on x) and draw this on your graph, giving the formula of the line.

EXERCISE 10.5

SEALS AND DISEASE

Common seals (*Phoca vitulina*) and grey seals (*Halichoerus grypus*) were tested for the presence of phocine distemper virus (PDV). In total 108 males and 92 females were tested. The results of tests for the presence of the virus are provided in Table 10.7

Q10.5.1 Use a chi-squared test to determine whether or not the ratio of males to females used in the study significantly differs from 1:1.

   a. State your null hypothesis (H_o).

   b. Calculate the value of chi-squared.

Table 10.7 The presence of phocine distemper virus in common and grey seals

	Grey seal	Common seal
Virus present	17	34
Virus absent	83	66

c. Using statistical tables determine whether or not your chi-squared value is statistically significant (state the level).

d. Should you reject or accept the null hypothesis?

Q10.5.2 Use a 2 × 2 chi-squared contingency table to determine whether or not PDV preferentially attacks either of the seal species.

   a. State your null hypothesis (H_o).

   b. Calculate the value of chi-squared.

c. Using statistical tables determine whether or not your chi-squared value is statistically significant (state the level).

d. Should you reject or accept the null hypothesis?

EXERCISE 10.6

BOD AND *TUBIFEX*

Biological Oxygen Demand (BOD) is a measure of the amount of organic matter present in a water sample: the higher the BOD, the lower the amount of oxygen present (because microbes in the water use it in respiration).

The following data show the relationship between BOD and the number of *Tubifex* worms present in samples taken at various points along a river (Table 10.8).

Table 10.8 BOD measurements and the number of *Tubifex* worms recorded at 10 sites along a river

Site	BOD ppm (x)	No. of *Tubifex* worms (y)
1	10	98
2	9	87
3	6	63
4	7	69
5	6	55
6	10	105
7	3	21
8	2	15
9	5	45
10	1	4

Q10.6.1 Examine the relationship between BOD and the number of *Tubifex* worms by calculating the correlation coefficient (r) between these two variables. You expect *Tubifex* numbers to be higher in areas of low oxygen.

a. Should you perform a one-tailed or a two-tailed test?

b. State your null hypothesis (H_0).

c. Calculate the value of r.

d. Using statistical tables determine whether or not your r value is statistically significant (state the level).

e. Should you reject or accept the null hypothesis?

Q10.6.2 Draw a scatter diagram showing the relationship between BOD and the number of *Tubifex* worms.

Q10.6.3 Calculate the position of the regression line (y on x) and draw this on your graph, giving the formula of the line.

WEATHER MEASUREMENTS

The maximum daily temperatures (°C) at two weather stations in a desert in July are recorded in Table 10.9.

Q10.7.1 For the temperatures at each location calculate:

   a. the mean;

   b. the standard deviation;

   c. the median.

Q10.7.2 Use an independent t-test test to determine whether or not there is a statistically significant difference between the maximum daily temperatures recorded at the two weather stations.

   a. You have no reason to believe that the maximum daily temperature is higher at one station than the other. Would you perform a one-tailed test or a two-tailed test?

   b. State your null hypothesis (H_0).

Table 10.9 Maximum daily temperature (°C) at two weather stations

Site A	Site B
35	29
36	26
35	25
29	24
27	30
29	34
33	35
35	27
30	28
29	
26	

   c. Calculate the value of t.

   d. Using statistical tables determine whether or not your t value is statistically significant (state the level).

   e. Should you reject or accept the null hypothesis?

EXERCISE 10.8

LICHEN DIVERSITY AND AIR POLLUTION

The following data show the relationship between distance from a city centre (in miles) along a straight line transect and lichen diversity (number of species) on stone walls and trees (Table 10.10).

Q10.8.1 Examine the relationship between distance from the city centre and the number of lichen species by calculating the correlation coefficient (*r*) between these two variables. You expect lichen diversity to be lower in areas of high air pollution.

a. Should you perform a one-tailed or a two-tailed test?

b. State your null hypothesis (H$_o$).

c. Calculate the value of *r*.

d. Using statistical tables determine whether or not your *r* value is statistically significant (state the level).

e. Should you reject or accept the null hypothesis?

Table 10.10 The number of lichen species recorded at various distances from a city centre

Miles from city centre (miles) (*x*)	Lichen diversity (number of lichen species) (*y*)
1	4
2	8
3	14
4	30
5	29
6	38
7	47
8	51
9	67
10	65

Q10.8.2 Draw a scatter diagram showing the relationship between distance from the city centre and the number of lichen species.

Q10.8.3 Calculate the position of the regression line (*y* on *x*) and draw this on your graph, giving the formula of the line.

CHAPTER 11

MULTIPLE CHOICE QUESTIONS

This chapter is a series of tests consisting of multiple choice questions on the topics covered by Chapters 1 to 10. Answers are provided in Chapter 12.

A tusk from an African elephant (*Loxodonta africana*), Ngorongoro Crater, Tanzania.

Examining Ecology. DOI: http://dx.doi.org/10.1016/B978-0-12-809354-2.00011-7

Chapter 1 Biodiversity and Taxonomy

1. Which of the following represents the correct sequence of taxonomic ranks used to classify animals (from the smallest to the largest)?

   a. species, genus, family, order, class, phylum

   b. species, genus, order, family, class, phylum

   c. species, genus, class, family, order, phylum

   d. species, genus, family, class, order, phylum

2. Which of the following lists of taxa is correctly ranked from the smallest (least species) to the largest (most species)?

   a. mammals, birds, insects, fishes

   b. mammals, birds, fishes, insects

   c. insects, fishes, birds, mammals

   d. birds, mammals, fishes, insects

3. Which of the following statements about taxonomy is *not* true?

   a. The system of assigning two names to a species is known as the binomial system of nomenclature

   b. In the species *Panthera leo persica* the last name indicates a subspecies

   c. The scientific name of an organism is also known as its vernacular name

   d. The original organism from which a species was first formally described is called the holotype

4. Which of the following statements about *Canis lupus* and *Canis latrans* is *not* true?

   a. The two species are in different genera

   b. Both species are in the same order

   c. Both species are in the same class

   d. The two species are in the same family

5. When is it legitimate to abbreviate the name of the genus in the scientific name of an animal (e.g., to abbreviate *Acinonynx jubatus* to *A. jubatus*)?

   a. Only when the scientific name has previously been written out in full

   b. Only when the name of the genus has previously been written out in full when referring to a different species of the same genus

   c. When either the scientific name has previously been written out in full or when the name of the genus has previously been written out in full when referring to a different species of the same genus

   d. Never

6. New species may be discovered by which of the following means?

   a. By analysing the DNA of museum specimens that appear, on morphological evidence, to be the same species

   b. Discovering previously undescribed organisms in the wild

   c. Assigning animals to new species when they had previously been classified as subspecies

   d. All of the above

7. In the scientific (binomial) name of an animal, the name indicating the species may begin with a capital (upper case) letter:

   a. only if the name is derived from that of a country (e.g., *Africana*)

   b. only if the name is derived from the name of a person (e.g., *Smithii*)

   c. if the name is derived from that of a country or a person's name

   d. under no circumstances

8. The national bird of the United States of America is known as the bald eagle. This name is its:

   a. scientific or vernacular name

   b. vernacular or common name

   c. binomial or scientific name

   d. common or binomial name

9. The abbreviation spp. means:

   a. all species of a higher taxon

   b. subspecies

   c. species

   d. all species in a family

10. The suffix 'idae' in the name of a zoological taxon (e.g., Picidae) indicates:

    a. a genus

    b. an order

    c. a family

    d. a class

11. A key which identifies a species by asking a series of questions to which there are only two possible answers is known as a:

    a. branching key

    b. diverging key

    c. bifurcating key

    d. dichotomous key

12. Which of the following statements about the animal known as *Panthera tigris altaica* is *not* true?

    a. The term *altaica* is the name of the subspecies to which the animal belongs

    b. The term *Panthera* is the name of the genus to which the animal belongs

    c. The term *tigris* is the name of the species

    d. The three names taken together are known as the binomial name

13. Which of the following statements about animal names is *not* true?

    a. The subspecies which was the first to be discovered has the species name repeated as its subspecific name, e.g., *Panthera leo leo*

    b. Each genus of animals contains more than one species

    c. Occasionally a species' scientific name is changed

    d. The vernacular name of a species may vary within and between countries

14. Studies of biodiversity (and other subjects) conducted by using the general public to collect data are referred to as:

    a. citizen science projects

    b. popular science projects

    c. public science projects

    d. people science projects

15. In communities which consist of a relatively large number of species the distribution of relative species abundance is almost always:

    a. bimodal

    b. normal

    c. lognormal

    d. negatively skewed

16. Carl Linnaeus, the creator of the binomial system of nomenclature, was a:

    a. German zoologist

    b. Swedish physician

    c. Swiss botanist

    d. Dutch biologist

17. The animal phylum that contains the most species is the:

    a. Chordata

    b. Mollusca

    c. Annelida

    d. Arthropoda

18. The technique known as 'knock-down fogging' is most likely to be used to collect samples of:

    a. plants

    b. insects

    c. small mammals

    d. seeds

19. How many species of insects have been identified?

    a. 10,000

    b. 100,000

    c. 1,000,000

    d. 10,000,000

20. The organisation known as the IUCN is:

    a. the International Union for the Conservation of Nature

    b. the International Union for the Conservation of Natural Resources

    c. the International Union for Conserving Nature

    d. the International Union for the Conservation of Nature and Natural Resources

Chapter 2 Abiotic Factors and Ecophysiology

1. A dimictic lake is one in which:

    a. the water mixes from the bottom to the top during two mixing periods each year

    b. no thermal stratification occurs at any time of the year

    c. the water mixes from the bottom to the top during a single mixing each year

    d. the water is exceptionally shallow

2. In a thermally stratified lake, the thermocline is:

    a. located above the epilimnion

    b. located below the hypolimnion

    c. a thin layer of water in which the temperature changes with depth less rapidly than in the layer above or the layer below

    d. a thin layer of water in which the temperature changes with depth more rapidly than in the layer above or the layer below

3. Asian elephants spend more time throwing soil over their bodies (dusting) on hot days than on cold days. The purpose of this behaviour is:

    a. to keep the sun off the skin to prevent overheating

    b. to keep parasites from biting the skin

    c. to keep the skin in good condition

    d. impossible to establish from this information alone

4. Which of the following lists contains only the abiotic components of an ecosystem?

    a. water, air, gravel, fungi, roads, buildings

    b. highways, water, air, rocks, ferns, bridges

    c. air, water, rocks, buildings, highways, soil

    d. buildings, air, highways, pylons, gravel, water

5. Which of the following physiological events would *not* occur when a seal dives?

    a. Increased blood flow to the brain cortex

    b. Increased blood flow to the liver

    c. Reduced heart rate

    d. Reduced blood flow to the diaphragm

6. Birds such as gulls, petrels and fulmars are able to maintain osmotic balance by excreting sodium in their:

    a. nasal gland secretions and urine

    b. nasal gland secretions only

    c. urine only

    d. faeces and urine only

7. Which of the following statements is *not* true for a camel deprived of water?

    a. Water lost during the hottest hours of the day would be lower than a fully hydrated camel

    b. Maximum rectal temperature would be higher in late afternoon than that of a fully hydrated camel

    c. Minimum rectal temperature would be higher at night than that of a fully hydrated camel

    d. It stores heat energy during the day and loses it at night

8. In which of the following experimental conditions would green plants be likely to achieve the highest photosynthetic rate?

    a. Condition K – 0.03% CO_2, low light intensity

    b. Condition L – 0.13% CO_2, high light intensity

    c. Condition M – 0.03% CO_2, high light intensity

    d. Condition N – 0.13% CO_2, low light intensity

9. Sodium concentration in urine should be measured as:

    a. $mmol\ L^{-1}$

    b. $cm^2\ L^{-1}$

    c. mmol

    d. $L\ mmol^{-1}$

10. A delay in the development of an animal in response to a recurring period of adverse environmental conditions is known as:

    a. hibernation

    b. aestivation

c. gestation

d. diapause

11. Which of the following biotopes has a climate that is hot with a distinct dry season?

   a. taiga

   b. tundra

   c. tropical forest

   d. savannah

12. *Grapholita* larvae (caterpillars) enter diapause in response to:

   a. a reduction in day length

   b. a fall in temperature

   c. a change in humidity

   d. exposure to frost

13. Which of the following is *not* characteristic of hibernation in black bears?

   a. Lowered heart rate

   b. Increased metabolic rate

   c. Lowered oxygen consumption

   d. Reduced urine output

14. If a temperature sensor is lowered into a thermally stratified lake, in which order will it pass through the various layers?

   a. Thermocline, epiliminion, hypolimnion

   b. Epilimnion, hypolimnion, thermocline

   c. Hypolimnion, thermocline, epilimnion

   d. Epilimnion, thermocline, hypolimnion

15. Which of the following types of animals are homiotherms?

   a. Mammals only

   b. Mammals and reptiles only

c. Birds and reptiles only

d. Birds and mammals only

16. When an animal such as a woodlouse moves in an undirected random manner it is exhibiting a behaviour known as a:

   a. reflex

   b. taxis

   c. kinesis

   d. tropism

17. The physiological responses of the body that occur when a seal dives are collectively known as the:

   a. diving response

   b. submergence reflex

   c. diving reflex

   d. survival reflex

18. The source of the carbon used by terrestrial photosynthetic plants to make carbohydrates is:

   a. carbon dioxide in the atmosphere

   b. the soil water

   c. carbon monoxide in the atmosphere

   d. calcium carbonate in the soil

19. Which of the following is *not* an adaptation that allows animals to avoid adverse weather conditions?

   a. homoeostasis

   b. hibernation

   c. diapause

   d. aestivation

20. When the water table is 20 cm below the surface what response would you expect to find in the seeds of *Juncus effusus* (a rush)?

    a. Germination rate will be higher than that at 5 cm below the surface

    b. Germination will be slightly lower than that at 5 cm below the surface

    c. There will be no difference between the germination rate compared with that when the water table reaches the soil surface

    d. Seeds will not germinate

Chapter 3 Ecosystems, Energy and Nutrients

Test 1

1. The source of energy in ecosystems is:

    a. always the sun

    b. always chemicals in the environment

    c. sometimes the sun and sometimes chemicals in the environment

    d. the heat produced by respiration

2. Entropy is a measure of disorder or randomness. Nature tends towards producing situations of high entropy (high disorder) as complex organic molecules are broken down into their constituent parts when organisms die and decay. In plant cells, the process of photosynthesis causes:

    a. an increase in entropy

    b. a decrease in entropy

    c. no change in entropy

    d. an increase in entropy followed by a decrease in entropy

3. The bacteria and fungi found in the soil are collectively known as:

    a. decomposers

    b. detritivores

    c. chemotrophs

    d. primary producers

4. At each stage in a food chain energy is lost as the heat of:

    a. assimilation

    b. digestion

    c. respiration

    d. excretion

5. The efficiency with which an animal converts food into usable energy is known as its:

    a. digestive efficiency

    b. feeding efficiency

    c. ingestion efficiency

    d. assimilation efficiency

6. A secondary producer is also called:

    a. a primary consumer

    b. a secondary consumer

    c. a tertiary consumer

    d. none of the above

7. The primary productivity of a grassland ecosystem is best measured by calculating the:

    a. dry biomass of green plants per unit area per unit time

    b. wet biomass of green plants per unit area per unit time

c. dry biomass of animals per unit area per unit time

d. wet biomass of animals per unit area per unit time

8. An inverted pyramid of numbers would be seen in which of the following food chains:

a. host → parasites → hyperparasites

b. phytoplantkton → zooplankton → fishes → seals

c. grasses → gazelles → cheetahs

d. Acacia trees → elephants

9. Which of the following statements is *not* true?

a. All green plants are autotrophs

b. All chemotrophic bacteria are autotrophs

c. All green plants are phototrophs

d. All autotrophs are phototrophs

10. Which of the following statements is *not* true?

a. All pyramids of numbers are wider at the bottom than at the top

b. All pyramids of energy are narrower at the top than at the bottom

c. All pyramids of biomass are wider at the bottom than at the top

d. In parasitic food chains the pyramid of numbers is inverted

11. Which of the following elements is central to the structure of chlorophyll?

a. zinc

b. iron

c. magnesium

d. nitrogen

12. In which parts of the electromagnetic spectrum does chlorophyll absorb light most strongly?

a. blue and red

b. green and red

c. blue and green

d. green and yellow

13. Which of the following is a pair of terms that do *not* mean the same thing?

a. secondary producer and primary consumer

b. trophic level 1 and primary producer

c. decomposer and detritivore

d. tertiary producer and secondary consumer

14. Which of the following statements is true?

a. Net primary production = gross primary production – respiration

b. Gross primary production = net primary production – respiration

c. Net primary production = gross secondary production – respiration

d. Net secondary production = gross secondary production + respiration

15. Which of the following statements is *not* true?

a. Secondary production is the production of chemical energy by animals

b. Primary production is the production of chemical energy from light

c. All heterotrophs are animals

d. Net primary production may be measured as units of mass per unit area per unit time interval

16. Consider the following food chain:

    phytoplankton → zooplankton → fish → seal → polar bear

    Which of the following statements about this food chain is *not* true?

    a. There must be more zooplankton than seals and fewer polar bears than fish in this food chain

    b. The fish is a secondary consumer

    c. The ultimate source of the energy for this food chain is sunlight

    d. As energy moves along this food chain some energy is lost as the heat of respiration between the phytoplankton and the polar bear at three points only

17. In ecosystems, complexity of food webs tends to lead to:

    a. instability

    b. stability

    c. extinction

    d. speciation

18. In a woodland, the living roots of the trees are part of the:

    a. secondary production

    b. tertiary production

    c. detritus

    d. primary production

19. Which of the following terms is *not* used for the predator at the top of a food chain?

    a. apical predator

    b. climax predator

    c. alpha predator

    d. apex predator

20. In the water cycle, which of the following equations is true?

    a. transpiration = evaporation–evapotranspiration

    b. evapotranspiration = evaporation + transpiration

    c. evaporation = transpiration–evapotranspiration

    d. transpiration + evapotranspiration = evaporation

Test 2

1. Which of the following statements about the 'ten per cent rule' in relation to energy flow in food chains is true?

    a. It states that when energy passes between organisms in a food chain each trophic level loses 10% of its energy in respiration

    b. It suggests that only 10% of the energy present in each trophic level is passed on to the next trophic level, but it is an overgeneralisation

    c. It is generally attributed to the British ecologist Charles Elton

    d. It suggests that 10% of the energy present in each trophic level is lost to decomposers, but it is an overgeneralisation

2. With reference to the pyramid of numbers below, which of the following statements is true?

    a. A = herbivores and D = carnivores

    b. A = decomposers and B = primary producers

    c. E = top carnivores and C = decomposers

    d. B = energy from the sun and C = herbivores

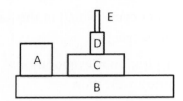

3. In the diagram of energy flow in a food chain below, which of the following statements is true?

   a. C represents herbivores and G represents decomposers

   b. A represents energy from the sun and E represents decomposers

   c. G represents energy loss via respiration and F represents decomposers

   d. B represents green plants and D represents top carnivores

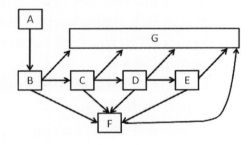

4. Which of the following statements about the diagram below of a pyramid of numbers is *not* true?

   a. The carnivores are more common than the primary producers

   b. The pyramid could represent a parasitic food chain

   c. A could represent trees, B could represent herbivorous insects and C could represent bacteria living in the insects

d. A could represent grasses, B could represent gazelles and C could represent lions

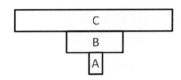

5. When a stream's energy resources are derived from within they are referred to as:

   a. autochthonous

   b. allochthonous

   c. autotrophic

   d. allopatric

6. Nitrogen is *not* an important component of:

   a. proteins

   b. nucleic acids

   c. carbohydrates

   d. enzymes

7. Which of the following elements are commonly found in general plant fertilisers?

   a. nitrogen, phosphorus and potassium

   b. magnesium, zinc and manganese

   c. iron, nitrogen and zinc

   d. phosphorus, zinc and magnesium

8. The process by which nutrients are washed out of the top layers of the soil is called:

   a. leaching

   b. denitrification

   c. decomposition

   d. leaking

9. Green plants absorb nitrogen through their roots mostly as:

    a. ammonium ions

    b. nitrite ions

    c. nitrate ions

    d. all of the above

10. The concept of a 'pyramid of numbers' was first suggested by:

    a. Ernst Haeckel

    b. Eugene Odum

    c. Robert MacArthur

    d. Charles Elton

11. Which of the following statements about the nitrogen cycle is *not* true?

    a. Lightning fixes atmospheric nitrogen

    b. Nitrates are converted back to atmospheric nitrogen by denitrification

    c. Nitrifying bacteria convert nitrate to nitrite

    d. Organic nitrogen in dead organisms is converted to ammonium by ammonification

12. In waterlogged soil, denitrifying bacteria:

    a. absorb nitrites from the soil

    b. break down nitrates and release nitrogen to the air

    c. absorb nitrogen from the air

    d. convert ammonia to nitrates

13. When large quantities of plant nutrients are washed into freshwater ecosystems excessive algal growth occurs. This process is called:

    a. ammonification

    b. nitrification

    c. acidification

    d. eutrophication

14. Which of the following ranges of soil pH are likely to result in the optimal uptake of nutrients by plants?

    a. 5.5–6.5

    b. 6.5–7.5

    c. 7.5–8.5

    d. 5.5–7.5

15. Which of the following statements relating to plant nutrients in soil is *not* true?

    a. Molybdenum is less available to plants in acidic soils

    b. Nitrogen and potassium are optimally available in soils that are slightly acidic

    c. When pH > 7.5 phosphorus reacts with calcium and magnesium increasing the availability of phosphorus to plants

    d. Sulphur is optimally available to plants in slightly acidic soils

16. NH_4^+ is the symbol for:

    a. a nitrate ion

    b. a nitrite ion

    c. an ammonium ion

    d. nitrous oxide

17. One of the benefits of using sewage sludge cake as a source of nitrogen for agricultural crops rather than liquid chemical fertilisers is that:

    a. sewage sludge contains low levels of organic matter compared with chemical fertilisers

b. the nitrogen in the sewage sludge is in a more soluble form than that in chemical fertilisers

c. all of the nitrogen in sewage sludge is released quickly and is immediately available for uptake by the crop

d. the nitrogen in sewage sludge is released slowly and will build up in the soil with repeated applications

18. In agricultural production systems, net yield return is:

   a. another name for the benefit–cost ratio

   b. a measure of the value of the increased yield by adding fertiliser

   c. a measure of the yield produced by an unfertilised crop

   d. a measure of the yield produced by a fertilised crop

19. Which of the following is likely to be the most important nitrogen sink in a large lake?

   a. the fish

   b. the outflow

   c. the weeds

   d. sedimentation

20. In the diagram below 'A' represents:

   Nitrite $\xrightarrow{A}$ Nitrate
   NO_2^- $\qquad$ NO_3^-

   a. *Nitrobacter*

   b. *Nitrosomonas*

   c. *Azotobacter*

   d. lightening

Chapter 4 Determining Abundance and Distribution

1. During the second trapping of a population of animals that was marked and trapped on an earlier occasion an ecologist found 20 marked animals in the sample captured. In order to estimate the population size using the Lincoln Index the ecologist needs to know:

   a. only the number of animals marked during the first trapping occasion

   b. only the number of animals not marked during the first trapping occasion

   c. no further information

   d. the number marked on the first trapping occasion and the total number captured on the second occasion

2. If marked animals die and disappear between the first and second trapping when attempting to calculate the population size using the Lincoln Index:

   a. there will be no effect on the estimate obtained

   b. the population estimate obtained may be either too high or too low

   c. the population estimate obtained would be too high

   d. the population estimate obtained would be too low

3. Which of the following formulae represents the Lincoln Index? (n_1 = number caught on first occasion (and then marked); n_2 = total number caught on second occasion (marked and unmarked); r = number caught on both occasions (recaptures)):

   a. $r/(n_1 \times n_2)$

   b. $(n_1 \times n_2)/r$

   c. $(n_1 + n_2)/r$

   d. $(r \times n_2)/n_1$

4. The number of plants growing in an area divided by the size of the area gives its:

   a. density

   b. population size

   c. dispersion

   d. intensity

5. An ecologist samples a population of plant species A 100 times using a one metre square quadrat. In total, 67 of the quadrats contained no plants. The other 23 quadrats contained between 23 and 45 individuals of species A. The distribution of species A is best described as:

   a. random

   b. clumped

   c. regular

   d. normal

6. Which of the following statements about quadrats is *not* true?

   a. They may be of any size

   b. They may be used to estimate the density of ground-nesting birds' nests

   c. They should be located at random or in a regular pattern

   d. They cannot be used to estimate the size of snail populations

7. If a plant population is sampled by counting the number of individuals in each of 100 quadrats, which of the following statements about the mean number of individuals per quadrat and variance for the population is true if the population is distributed at random?

   a. variance = mean

   b. variance < mean

   c. variance > mean

   d. variance = mean^2

8. The UTM location of a white-beaked dolphin was recorded as 26 V 507544 mE 7168741 mN at 10.35 and 26 V 507463 m east 7169004 m north at 10.50. In which general direction is the dolphin moving?

   a. north

   b. northwest

   c. southeast

   d. northeast

9. Which of the following assumptions are made by the Lincoln Index? Between the first trapping (when animals are marked) and the second trapping:

   i. No immigration occurs

   ii. No emigration occurs

   iii. No births occur

   iv. No deaths occur

   v. No marks are lost

a. i, ii and iii only

b. ii, iv and v only

c. i, ii, iii, iv and v

d. v only

10. A calendar of catches is most likely to be used to study:

    a. predator–prey relationships

    b. intraspecific competition

    c. changes in the size and composition of an animal population over time

    d. plant population dynamics

11. The mapping system shown in the diagram below records the presence of an animal species in 10-km squares. Which of the following statements about this map is true?

    a. Only eight animals have been recorded in this area

    b. Thirty-two per cent of the area is inhabited by the species

    c. At least one individual of this species has been recorded in each of the eight 10-km squares indicated by a black circle

d. The species is found throughout each of the eight 10-km squares indicated by a black circle

12. The UTM (Universal Transverse Mercator) coordinate system divides the Earth into blocks. Each block is divided up by lines running north–south and east–west at intervals of:

    a. 1 m

    b. 5 m

    c. 10 m

    d. 100 m

13. On an exposed rocky shore different algal species occur in bands running more-or-less parallel to the shoreline, depending upon the amount of time they can survive being exposed to the air. This banding pattern is known as:

    a. stratification

    b. zonation

    c. banding

    d. lamination

14. Which of the following sampling methods does *not* involve the use of a transect?

    a. The population density of deer in a park is estimated by walking in a straight line for 2 kms and counting the deer observed 500 m either side of the line

    b. The distribution of flowering plants was recorded by assessing percentage cover of each species within quadrats placed at 20 m intervals in a straight line across a salt marsh

    c. The density of elephants in an area of savannah was estimated by counting animals observed 1 km either side of the flight path of a light aircraft flown due north for 50 kms

d. The distribution and abundance of various species of snails were determined in an area of deciduous woodland by examining 50 randomly located quadrats

15. Individual animals of the same species were removed from a population on four occasions to estimate the total population by removal trapping. The estimated total population is calculated by drawing a scatter diagram, plotting the number captured on each sample day on the y-axis against the cumulative total previously removed on the x-axis. Which of the following statements is true?

a. The animals trapped on each occasion should be returned to the population after counting them

b. A line of best fit drawn through the points crosses the x-axis at the estimated population size

c. A line of best fit drawn through the points crosses the y-axis at the estimated population size

d. A line joining all of the dots crosses the x-axis at the estimated population size

16. Which of the following statements is most likely to be true for historical records of plant distributions from the 19th century?

a. They are an important source of accurate records made by professional botanists

b. They are likely to cover large areas as a result of rigorous field surveys

c. They were conducted using comprehensive and accurate identification guides

d. They are likely to be prone to error because they were largely made by amateurs

17. Large mammals were surveyed by flying a light aircraft for a single day over savannah in straight lines and counting the animals seen 500 m either side of the path of the aircraft in sections 10 km long. This method allows you to calculate, for each species identified:

a. an accurate total number for the whole area

b. an estimate of its density

c. trends in population size

d. patterns of migration

18. National records of the distribution of species in the United Kingdom are kept by the:

a. Biological Records Centre

b. Ecological Records Centre

c. Species Mapping Centre

d. County recorders

19. To estimate the position of a whale from a ship at sea you would need to know:

a. the position of the ship and the bearing from the ship to the whale

b. the bearing to the whale and the distance from the ship to the whale

c. the position of the ship, the distance from the ship to the whale and the bearing from the ship to the whale

d. some other combination of measurements

20. Which of the following is true about the effect of quadrat sampling on the estimate obtained for the density of a species of plant in a field?

   a. Taking a small number of quadrat samples will always result in an underestimate of the density

   b. The more quadrats sampled the more accurate the estimate is likely to be up to a point but taking further samples will not change the estimate very much thereafter

   c. Taking a small number of quadrats samples will always result in an overestimate of the density

   d. The number of quadrat samples taken has no effect on the estimate of density

Chapter 5 Population Growth

1. Which of the following is *not* a density-dependent factor which could influence the size of a population of herbivores?

   a. frost

   b. competition

   c. disease

   d. predation

2. A logistic curve levels out (reaches its asymptote) when:

   a. birth rate > death rate

   b. death rate > birth rate

   c. birth rate = 0

   d. birth rate = death rate

3. J-shaped growth is also called:

   a. exponential growth

   b. linear growth

   c. boom and bust growth

   d. logistic growth

4. Which of the following organisms could be described as a *K*-strategist?

   a. a fruit fly

   b. an elephant

   c. a bacterium

   d. a virus

5. A table showing the death rates of different age classes of individuals in a population of red deer is called:

   a. a death table

   b. a mortality table

   c. a life table

   d. a survival table

6. The asymptote of a curve showing population growth occurs when the population has reached:

   a. the carrying capacity of the environment

   b. its maximum growth rate

   c. its minimum growth rate

   d. its optimum growth rate

7. The term fecundity refers to a population's:

   a. mortality rate

   b. growth rate

   c. migration rate

   d. reproductive rate

8. A density-independent factor affecting population growth has:

   a. a greater effect when population density is high than when it is low

   b. a smaller effect when population density is high than when it is low

   c. a similar effect regardless of density

   d. no effect on population size

9. Which of the following could *not* be described as a population?

   a. All of the lions in the Serengeti National Park

   b. All of the birds in New York State

   c. All of the blue whales in the ocean

   d. All of the red deer in Scotland

10. The size of a population at a particular point in time is determined by its:

    a. birth rate, death rate, immigration rate and emigration rate

    b. death rate, immigration rate, birth rate and density

    c. immigration rate, density, birth rate and emigration rate

    d. birth rate, death rate, emigration rate and density

11. Which of the following sequences describes the growth of a population which exceeds the carrying capacity of the environment? (BR = birth rate; DR = death rate):

    a. BR < DR, BR > DR, BR = DR

    b. BR > DR, BR < DR, BR = DR

    c. BR > DR, BR = DR, BR < DR

    d. BR = DR, BR > DR, BR < DR

12. When a species invades an area which it has not occupied previously, where the environment is favourable, and it contains no predators or competitors, its initial population growth is likely to be:

    a. stable

    b. exponential

    c. logistic

    d. J-shaped

13. Which of the following sequences of numbers represents J-shaped growth? (Each sequence represents the size of a population at fixed time intervals).

    a. 2, 4, 8, 16, 32, 64, 128, 256, 512, 1024

    b. 17, 25, 36, 54, 109, 135, 227, 401, 732

    c. 2, 4, 8, 16, 23, 64, 128, 125, 129, 127, 130

    d. 2, 4, 8, 16, 32, 64, 128, 93, 75, 40, 23

14. Which of the following could be described as an *r*-strategist?

    a. locust

    b. *Tyrannosaurus rex*

    c. blue whale

    d. *Homo sapiens*

15. The continuous version of the logistic model of population growth is described by the differential equation:

$$\frac{dN}{dt} = \frac{rN(K - N)}{K}$$

In this equation, the rate of maximum population growth (intrinsic rate of natural increase) is denoted by:

    a. *r*

    b. *K*

    c. *dN*

    d. *rN*

16. Which type of life survivorship curve is exhibited by parasites?

    a. Type I

    b. Type II

    c. Type III

    d. Type IV

17. Which of the following is *not* a characteristic of a *K*-strategist?

    a. It has a very large number of young

    b. It has a long life span

    c. It is large in size

    d. Individuals reproduce more than once during their lifetime

18. To construct a static life table an ecologist should determine:

    a. the age of all of the animals present in a population at a particular point in time

    b. the age at which each animal dies from a cohort of individuals that were all born at the same time (e.g., in the same year)

    c. the number of offspring produced in a population each year

    d. the number of deaths that occur in the population each year

19. In a life table, the abbreviation l_x refers to:

    a. the number of deaths in an age class

    b. the number of survivors in an age class

    c. the expectation of further life in an age class

    d. the number of births in an age class

20. Population growth may be studied using a Leslie matrix. In the matrix below, the probability of an individual from year 0 surviving to year 1 is given by:

$$\begin{bmatrix} n_{0t+1} \\ n_{1t+1} \\ n_{2t+1} \end{bmatrix} = \begin{bmatrix} f_0 & f_1 & f_2 \\ p_0 & 0 & 0 \\ 0 & p_1 & 0 \end{bmatrix} \times \begin{bmatrix} n_{0t} \\ n_{1t} \\ n_{2t} \end{bmatrix}$$

    a. p_1

    b. n_{1t}

    c. p_0

    d. n_{0t}

Chapter 6 Species Interactions

1. A species that has a disproportionately large effect on the community in which it occurs is known as a:

    a. polymorphic species

    b. pioneer species

    c. superpredator

    d. keystone species

2. The contention that 'complete competitors cannot coexist' is known as the:

    a. competitive exclusion principle

    b. competitive expulsion principle

    c. competitive coexistence principle

    d. coexistence and competition principle

3. Which of the following statements about niches is *not* true?

    a. The niche was first described by Ernst Haeckel

    b. Charles Elton classified niches based on feeding habits

    c. Grinnell defined an organism's niche in terms of the habitat it occupies

    d. McArthur and Levins discussed the niche in terms of resource utilisation

4. Hutchinson's concept of the niche may be described in terms of:

    a. a 3-dimensional hyperspace

    b. a 2-dimensional megavolume

    c. an *n*-dimensional space

    d. an *n*-dimensional hypervolume

5. Which of the following statements about niches is *not* true?

    a. Niche breadth in tropical forest species is generally large

    b. The realised niche of a species is smaller than its fundamental niche

    c. Invasive species may expand rapidly if they are introduced into an ecosystem with a suitable vacant niche

    d. Niche separation is the same as niche partitioning

6. A scientist planted wheat seeds at five different densities and then calculated the germination rate and the biomass of plant material produced per plant at each density. The scientist was most likely to be studying:

    a. interspecific competition

    b. intraspecific competition

    c. growth curves

    d. productivity

7. In African grasslands, antelopes occupy a similar ecological niche to that occupied by kangaroos in Australian grasslands. This is an example of:

    a. divergent evolution

    b. convergent evolution

    c. stabilising selection

    d. disruptive selection

8. Which of the following statements about predator–prey interactions model by the Lotka-Volterra equation is *not* true?

    a. The peaks in numbers of predators and prey do not occur at exactly the same time

    b. The peak in predator numbers precedes the peak in prey numbers

    c. The troughs in numbers do not occur at exactly the same time

    d. The peaks in prey numbers precede the peaks in predator numbers

9. A guild is a:

    a. group of taxa that exploit the same resources

    b. another name for a niche

    c. a group of prey species that all have the same predator

    d. a group of colonial insects

10. Cormorants and shags are seabirds that appear to avoid competition by:

    a. feeding at different times

    b. feeding at different depths in the sea

c. living in different habitats

d. having completely different diets

11. In what type of birds did MacArthur study niche separation in coniferous forests in the United States?

   a. titmice

   b. raptors

   c. woodpeckers

   d. warblers

12. The use of a natural predator or parasite to control the numbers of a pest species is known as:

   a. natural control

   b. ecological control

   c. biological control

   d. pest control

13. Elton and Nicholson conducted a famous predator–prey study of which two species in the wild?

   a. foxes and rabbits

   b. lynx and hares

   c. bobcat and rabbits

   d. foxes and hares

14. Georgii Gause conducted studies of competition using two species of:

   a. *Paramecium*

   b. flour beetles

   c. fruit flies

   d. woodlice

15. In Brown and Davidson's study of competition between ants and rodents, which of the following statements is *not* true?

   a. When ants were removed rodent numbers increased

   b. When rodents were removed the number of ant colonies increased

   c. When ants and rodents were removed seed density was higher than when both were present

   d. Rodents and ants feed on seeds of the same size

16. In the Serengeti, which of the following animals is a major carrier of distemper virus?

   a. wildebeest

   b. lions

   c. domestic dogs

   d. rhinoceros

17. Phocine distemper virus is a disease of:

   a. cattle

   b. badgers

   c. seals

   d. dogs

18. Eastern red oak seedlings survive under forest stands while loblolly pine seedlings do not because:

   a. oak seedlings are better at competing for soil nutrients than pine seedlings

   b. pine seedlings are attacked by insect species that do not attack oak seedlings

c. pine seedling only do well in dry conditions

d. light conditions are too low for the pine seedlings to thrive

19. In eastern Australia, in which of the following soil types should rabbits build their warrens to avoid attack by foxes?

a. sand dunes

b. desert loams

c. stoney banks

d. loams

20. In which decade did Hutchinson publish his definition of the niche?

a. 1950s

b. 1960s

c. 1970s

d. 1980s

Chapter 7 Behavioural Ecology and Ecological Genetics

1. ESS is an abbreviation for:

a. ecologically stable strategy

b. evolutionarily static strategy

c. ecologically strategic state

d. evolutionarily stable strategy

2. In dung flies an ESS would dictate that males would copulate with females for a period of time:

a. equal to that spent searching for new females

b. that would result in the optimum number of eggs being fertilised for the time spent copulating

c. greater than the mean amount of time spent searching for new females

d. sufficient to fertilise 100% of their eggs

3. In order to determine the location of an animal that is wearing a radio-tracking collar we would require information about the direction from which the signal is coming derived from:

a. two locations recorded simultaneously or recorded a short time apart

b. two recordings made at different times from a single location

c. at least three different locations

d. one recording made from a single location

4. Feeding territory size in raptors that feed on small mammals is likely to be:

a. negatively correlated with prey density

b. directly proportional to prey density

c. unrelated to prey density

d. positively correlated with prey density

5. An index of association calculated between two individual animals in a group may range in value from:

a. 1 to 10

b. 0 to 1

c. 1 to 100

d. 0 to 10

6. Which of the following pairs of characteristics is likely to be possessed by a desert lizard species that adopts a 'sit-and-wait' feeding strategy?

   a. small brain and high endurance capacity

   b. streamlined and limited endurance capacity

   c. limited endurance capacity and stocky build

   d. limited learning ability and highly active

7. The primary purpose of roaring contests in red deer is to:

   a. allow rival males to locate each other

   b. allow rival males to assess each other's fighting ability

   c. attract females

   d. deter subadult stags from stealing hinds

8. An elephant was recorded feeding during 65 out of 98 scan samples. On 3 of the 98 occasions when recordings were made the elephant was out of site. The percentage of time spent feeding should be calculated as:

   a. $(65/98) \times 100$

   b. $(65 - 3)/98 \times 100$

   c. $(65/98) \times (100 - 3)$

   d. $65/(98 - 3) \times 100$

9. The feeding strategy of an animal, in terms of the costs and benefits of chasing large and small prey at high and low speeds, is best assessed by calculating:

   a. a game matrix

   b. an activity budget

   c. an index of association

   d. a correlation coefficient

10. Which of the following statements about activity budgets is *not* true?

    a. An ethogram should be constructed before recording begins

    b. Behaviour categories should be mutually exclusive

    c. All recordings should be made at five minute intervals

    d. Scan sampling involves making simultaneous recordings for a group of animals

11. Which of the following statements about Cynthia Moss's study of the elephants of Amboseli National Park is true?

    a. Each of the elephants in a family was assigned a name at random and a three character identification code

    b. Each of the elephants in a family was assigned a name beginning with the same letter and a unique three character identification code

    c. Each elephant in a family was assigned a three character identification code but no name

    d. Each of the elephants in a family was assigned a name beginning with the same letter but no code

12. Add the missing term to the sentence below:

    The ………… population size is defined as the size of an ideal population that would lose genetic variation by genetic drift at the same rate.

    a. affective

    b. nominal

    c. effective

    d. selective

13. Plants able to grow successfully on soil contaminated by heavy metals are known as:

    a. heavy metal resistant

    b. heavy metal tolerant

    c. heavy metal selected

    d. heavy metal lenient

14. The detailed work on resource utilisation and adaptive radiation in Galapagos finches was undertaken by:

    a. David Lack

    b. Charles Darwin

    c. Robert FitzRoy

    d. Robert MacArthur

15. The peppered moth (*Biston betularia*) appears light grey in clean environments and black in environments where trees and buildings are dark in colour as a result of air pollution. This is an example of:

    a. convergent evolution

    b. industrial melanism

    c. industrial albinism

    d. stabilising selection

16. Inbreeding of heterogyzotes results in:

    a. increased heterozygosity in future generations

    b. hybrid vigour in future generations

    c. increased homozygosity in future generations

    d. increased mutation in future generations

17. If a population has experienced a genetic bottleneck in the past this may result in:

    a. reduced genetic diversity

    b. increased genetic diversity

    c. increased mutation rate

    d. reduced biodiversity

18. The last two individuals possessing the gene for a particular character are killed by disease, thereby removing this gene from the population. This process is known as:

    a. genomic drift

    b. genetic extinction

    c. genetic elimination

    d. genetic drift

19. Which of the following statements about Charles Darwin is *not* true?

    a. During his visit to the Galapagos Islands he carefully labelled specimens with the name of the island on which they were collected

    b. He was ship's naturalist on *HMS Beagle*

    c. He donated a large number of specimens to the Zoological Society of London

    d. He graduated in divinity

20. *Agrostis capillaris* is a grass species that has evolved so that some populations are able to grow on soil contaminated with zinc. If a transect is constructed across a boundary between zinc-contaminated soil and uncontaminated soil we would expect to find:

    a. a gradual change in tolerance at the boundary

    b. a rapid change in tolerance at the boundary

    c. no marked change in tolerance at the boundary

    d. no predictable pattern of change at the boundary

Chapter 8 Environmental Pollution and Perturbations

1. If a chemical compound A has a lower LD_{50} than chemical compound B, which of the following statement is true?

    a. Compound A has higher toxicity than compound B

    b. Compound A has a lower toxicity than compound B

    c. There is no difference in the toxicity of compounds A and B

    d. It is not possible to say from the LD_{50} values which compound is the more toxic

2. When cold air becomes trapped under warm air, for example, on a hot day following a cold night, the dispersion of atmospheric pollutants may be poor. This condition is known as:

    a. a temperature conversion

    b. a temperature reversal

    c. a temperature transposition

    d. a temperature inversion

3. When organic matter of human origin enters a lake, the lake may quickly become enriched and suffer serious damage due to eutrophication. This is process is usually referred to as:

    a. traditional eutrophication

    b. cultural eutrophication

    c. organic eutrophication

    d. enriched eutrophication

4. Which of the following pairs of abbreviations are not both breakdown products of the pesticide DDT?

    a. DDE and DDD

    b. DDC and DDE

    c. DDD and DDC

    d. DDF and DDD

5. When the temperature of a body of freshwater increases, which of the following is least likely to be true?

    a. Oxygen concentration falls

    b. The heart rate of *Daphnia* decreases

    c. The internal body temperature of aquatic homiotherms is unaffected

    d. Light penetration will remain unchanged

6. Which of the following sectors is *not* a source of greenhouse gas emissions?

    a. agriculture

    b. energy supply

    c. waste management

    d. none – they all produce greenhouse gas emissions

7. Which of the following statements about rivers and river catchments is *not* true?

    a. Sediment discharge increases when a catchment is deforested

    b. Total sediment yields are higher in urban areas than rural areas of similar size

    c. Nutrient cycles are not disrupted when catchments are deforested

    d. During summer, storms are likely to add heat to rivers in urban areas

8. Which of the following statements about DDT is *not* true?

    a. DDT is likely to be present in higher concentrations in top predators than in herbivores

    b. DDT causes a reduction in eggshell thickness in birds of prey

    c. DDT is an organophosphate insecticide

    d. DDT was widely used in tropical areas in an attempt to reduce the incidence and spread of malaria

9. A lichen is an association between:

    a. a green plant and a green alga

    b. a green alga and a fungus

    c. a fungus and a protozoan

    d. a moss and a fungus

10. Which of the following statements about *Tubifex* is *not* true?

    a. It is an annelid

    b. Its blood contains haemoglobin

    c. It is an indicator of high oxygen levels in water

    d. It is unlikely to found in high numbers where mayflies are common

11. Which of the following isotopes has the longest half-life?

    a. iodine-131

    b. plutonium-339

    c. strontium-90

    d. uranium-238

12. Which of the following statements about the relationships between *Daphnia*, water temperature and oxygen concentration in water is *not* true?

    a. Heart rate in *Daphnia* is positively correlated with temperature

    b. Oxygen concentration in water is positively correlated with water temperature

    c. *Daphnia* is poikilothermic

    d. Oxygen is more soluble in cold than in warm water

13. A high density of which of the following taxa would indicate water polluted with organic material?

    a. *Asellus* (hoglouse)

    b. *Gammarus* (shrimp)

    c. *Baetis* (mayfly)

    d. *Polycelis* (flatworm)

14. Acid rain causes a particular problem for aquatic molluscs because:

    a. it prevents osmoregulation

    b. it cause eutrophication

    c. it reacts with oxygen making it difficult for them to obtain sufficient oxygen for respiration

    d. it breaks down the calcium carbonate they need to make their shells

15. A biodiversity index measures:

    a. the number of types of species present only

    b. the number of individual organisms present only

c. the number of types of species present and the evenness of the distribution of individuals between these species

d. the evenness of the distribution of individuals between the species present only

16. An ecologist measured the invertebrate biodiversity of pond A using Simpson's index and the invertebrate biodiversity of pond B using Menhinick's index. The values obtained were 0.73 for pond A and 0.65 for pond B. Which of the following statements is true?

a. The biodiversity of invertebrates is higher in pond A than in pond B

b. The biodiversity of invertebrates is higher in pond B than in pond A

c. It is not possible to compare values obtained from different indices

d. Simpson's index should not be used to measure invertebrate biodiversity

17. Which of the following is *not* a greenhouse gas?

a. CH_4

b. CO_2

c. O_3

d. N_2

18. Acid rain has a pH of less than:

a. 7.0

b. 4.5

c. 5.6

d. 3.5

19. Global warming causing polar ice loss is having a negative impact on polar bear populations because:

a. it makes it difficult for them to find sufficient food

b. it reduces the populations of mammalian prey species

c. it is causing a reduction in the penguin populations on which they feed

d. the higher temperatures are causing a reduction in their birth rate

20. Which of the following will not cause eutrophication if released into a lake?

a. milk

b. sewage

c. chemical fertiliser

d. mercury

Chapter 9 Conservation Biology

1. What is the acronym used for the management programme for natural resources in Zimbabwe?

a. CROSSFIRE

b. CAMPFIRE

c. WILDFIRE

d. BUSHFIRE

2. Which area of ecology has been important in the theoretical study of the design of nature reserves?

a. population dynamics

b. island biogeography

c. production ecology

d. ecological genetics

3. As the radius of a theoretical circular nature reserve increases, the perimeter length per unit area:

   a. increases

   b. decreases

   c. is constant

   d. changes unpredictably

4. Grey wolves (*Canis lupus*) reintroduced to Yellowstone National Park were given a special status under the Endangered Species Act of 1973. These wolves were considered to be:

   a. a restoration population

   b. a reestablishment population

   c. a trial population

   d. an experimental population

5. In the European Union, Member States are required to have a conservation function. Which of the following functions is mandatory?

   a. Education

   b. Research from which conservation benefits accrue to the species

   c. Captive-breeding

   d. Reintroduction of species into the wild

6. Experiments were conducted on game ranching in Africa in the 1960s by:

   a. Raymond Dasmann

   b. Eugene Odum

   c. Robert MacArthur

   d. Charles Elton

7. The study of indigenous peoples and how they use wild plants is known as:

   a. anthrobotany

   b. ecobotany

   c. ethnobotany

   d. homobotany

8. Which of the following changes to fish catches would you expect to occur after mesh regulation that increased the size of the holes in fishing nets?

   a. Fewer very young fish caught

   b. More older fish caught

   c. No very young fish and more older fish caught

   d. A greater weight of young fish caught

9. Which of the following statements about game ranching in Southern Rhodesia in the 1960s is *not* true?

   a. Only ungulate species were cropped

   b. It was more profitable that traditional cattle ranching

   c. Giraffe were taken as part of the crop

   d. Elephants were taken as part of the crop

10. Which of the following statements about giant pandas (*Ailuropoda melanoleuca*) is *not* true?

    a. Gestation is about 140 days

    b. The female is only fertile for 12 days per year

    c. The female normally gives birth to a single cub

    d. Newborn cubs are small and very poorly developed

11. In relation to the income generated from sport hunting in Zimbabwe, PAC stands for:

    a. preliminary antelope count

    b. potential animal cull

    c. problem animal control

    d. problem animal count

12. Red foxes (*Vulpes vulpes*) are not native to Australia but were taken there by the British so that the colonists could continue to engage in the traditional sport of fox hunting. This release of red foxes is most accurately described as:

    a. a reintroduction

    b. population supplementation

    c. a translocation

    d. an introduction

13. Oregon Zoo takes native adult silverspot butterflies from the wild each year, induces them to lay eggs, overwinters the larvae in refrigerators and then feeds them before releasing them back into the wild to areas where populations of the species already exist. This activity is best described as:

    a. an introduction programme

    b. a reintroduction programme

    c. a population supplementation programme

    d. a translocation programme

14. The list of threatened species maintained by the International Union for the Conservation of Nature and Natural Resources (IUCN) is known as the:

    a. Black List

    b. Red List

    c. Green List

    d. Blue List

15. The IUCN list groups the following categories together and defines them as 'threatened':

    a. CR and EN only

    b. CR, EN and VU

    c. EN and VU only

    d. VU and CR only

16. In 2016, a total of 989 species of reptiles were classified as threatened by the IUCN out of 10,450 identified reptile species. Which of the following statements about calculating the percentage of all identified reptile species which are threatened is *not* true?

    a. The percentage is calculated as $989/10{,}450 \times 100$

    b. The percentage cannot be calculated unless we know that the conservation status of all identified reptile species has been assessed

    c. At least 9.5% of all identified reptile species are threatened

    d. It is possible, but unlikely, that 9,461 of the known reptile species are not threatened

17. A demographically isolated population whose probability of extinction over the timescale of interest (perhaps 100 years) is not substantially affected by natural immigration from other populations is called an:

    a. ECU

    b. ESS

    c. ESU

    d. ECS

18. The terms most likely to be used for places where animal gametes and plant propagules are stored in very cold conditions are:

    a. frozen zoo and plant bank

    b. seed bank and frozen zoo

    c. gamete bank and propagule bank

    d. seed store and cool zoo

19. Which of the following statements about 'Operation Oryx' is *not* true?

    a. It involved Phoenix Zoo in Arizona

    b. It was a project initiated by the World Wildlife Fund (WWF)

    c. Some oryx were returned to Jordan

    d. Some oryx were returned to Oman

20. Which of the following are killed in significant numbers by watercraft in Florida?

    a. marine turtles

    b. killer whales

    c. river dolphins

    d. manatees

Chapter 10 Statistics

1. In a normal distribution:

    a. the mean, median and mode are all located in the centre of the distribution

    b. the mean and mode are located in the centre of the distribution

    c. only the mean is located in the centre of the distribution

    d. the median and the mean are located in the centre of the distribution

2. In a normal distribution what percentage of the area under the curve lies within three standard deviations either side of the mean?

    a. 68%

    b. 99.7%

    c. 95.5%

    d. 100%

3. Which of the following is NOT a measure of central tendency?

    a. mean

    b. standard deviation

    c. mode

    d. median

4. The correlation coefficient may have any value between:

    a. 0 and +1

    b. −1 and 0

    c. 0 and 100

    d. −1 and +1

5. The standard deviation of a set of values =

    a. $\sqrt{variance}$

    b. $variance^2$

    c. $variance^3$

    d. $\sqrt{mean}$

6. The mode of the values, 5, 6, 8, 12, 7, 1, 6, 9, 4 is:

    a. 12

    b. 6

    c. 6.4

    d. 7.1

7. The correlation coefficient between two sets of data is 0.9.

    a. This is statistically significant at the 5% level

    b. This is statistically significant at the 1% level

    c. This is not statistically significant

    d. It is not possible to say whether or not this is statistically significant

8. Which measure of central tendency would best represent the following data: 3, 49, 7, 12, 8, 6, 2, 9, 11?

    a. the mean

    b. the mode

    c. the median

    d. the range

9. A population of deer consists of 127 males and 167 females. Which of the following statistical tests would you use to determine whether or not these values differ from a 1:1 ratio of males:females?

    a. independent t-test

    b. dependent t-test

    c. correlation coefficient

    d. chi-squared test

10. An ecologist wants to compare the diameter of limpets found on a beach which has been polluted by oil with those found on one that is unpolluted because she suspects that the oil has reduced their growth rate. She has 200 measurements of limpets from each beach. Which of the following statistical tests should she use?

    a. 2 × 2 chi-squared test

    b. independent t-test

    c. regression analysis

    d. dependent t-test

11. A type I error is:

    a. detecting an effect that is not present

    b. not detecting an effect that is present

    c. the same as a sampling error

    d. the same as a standard error

12. A correlation coefficient of zero between variable A and variable B indicates that:

    a. as A increases B increases

    b. as A increases B decreases

    c. there is no relationship between changes in A and changes in B

    d. none of the above

13. A positively skewed distribution has:

    a. a long 'tail' to the left of the apex

    b. a long 'tail' to the right of the apex

    c. a skew of zero

    d. a mean value to the left of the apex

14. Regression analysis calculates:

    a. the position of a line-of-best fit through a series of points on a graph

    b. the skewness of a distribution

    c. the dispersion of values around the mean

    d. the extent to which the difference between the means of two distributions is statistically significant

15. Which of the following is *not* a continuous variable?

    a. height

    b. mass

c. colour

d. length

16. Which of the following symbols is used for the mean of a population?

    a. σ

    b. μ

    c. $\bar{x}$

    d. s

17. The abbreviation for a null hypothesis is:

    a. NH

    b. H_1

    c. H_0

    d. H_N

18. You have carried out a χ^2 test using a 2×2 contingency table. When looking up the critical value in χ^2 tables how many degrees of freedom would you use?

    a. 1

    b. 2

    c. 3

    d. 4

19. Which of the following pairs of terms are measures of dispersion?

    a. mean and mode

    b. mean and standard deviation

    c. median and standard deviation

    d. variance and standard deviation

20. If the results of a t-test are significant at the 5% level, this may be abbreviated as:

    a. $P > 0.05$

    b. $P < 0.05$

    c. $P < 0.005$

    d. $P < 0.5$

CHAPTER 12

ANSWERS TO EXERCISES AND MULTIPLE CHOICE TESTS

This chapter provides answers to the exercises in Chapters 1 to 10 and the multiple choice questions in Chapter 11.

Grey heron (*Ardea cinerea*)

Examining Ecology. DOI: http://dx.doi.org/10.1016/B978-0-12-809354-2.00012-9

Chapter 1 Biodiversity and Taxonomy

1.1. *Ecology and taxonomy*

Q1.1.1. The definitions provided by Elton and Odum are rather vague. Neither mentions relationships or the study of the distribution and abundance of organisms.

Q1.1.2. Ecology is the study of the relationship between organisms and each other and organisms and the environment. Clearly, the identities of the species present in any ecosystem must be determined if we are to make sense of their interrelationships.

Q1.1.3. The UN Convention on Biological Diversity requires Parties to identify components of biological diversity that are important for the conservation of biodiversity and its sustainable use. This work can only be undertaken by biologists with a sound knowledge of taxonomy, so it follows that a shortage of taxonomists would make the identification of species and their subsequent monitoring very difficult to achieve.

Q1.1.4. Deforestation, overfishing, air, water and land pollution, hunting, poaching, industrialisation, habitat loss, etc.

Q1.1.5. Some species have ranges that extend across many countries and some species migrate from one country to another at certain times of the year. It is important that countries cooperate by creating protected areas that cross international boundaries and protecting the places that animals use in their migrations. A great deal of expertise in taxonomy, ecology and conservation has been developed by scientists,

particularly in North America and Europe, and it is essential that this expertise is shared with other parts of the world. Many ecological challenges are global (e.g., climate change and marine pollution) and it is essential that countries work together to mitigate the damage done to the planet in general and biodiversity in particular. The global protection of our genetic resources is important if we are to conserve species that may be useful in agriculture, medicine and other areas of human activity.

1.2. *What's in a name? vernacular versus scientific names*

Q1.2.1. The binomial (scientific) name is unique, whereas the same vernacular name may be used for more than one species or a particular species may have more than one vernacular name. Vernacular names tend to vary between countries and languages whereas the binomial name of a species is the same in all languages and in all countries.

Q1.2.2. The term 'blackbird' has little meaning from an evolutionary and taxonomic point of view because it has been used for a number of unrelated species that have been classified into different genera.

Q1.2.3. This species is known by a large number of different names in this part of the world and by more than one name in some languages so you would need to use all of them in the questionnaire. It would also be useful to show them a photograph of an ocelot and photographs of several small cat species of similar appearance and ask them to point to any they have seen.

1.3. *The classification of animals*

Q1.3.1. A family tree of the Hominidae

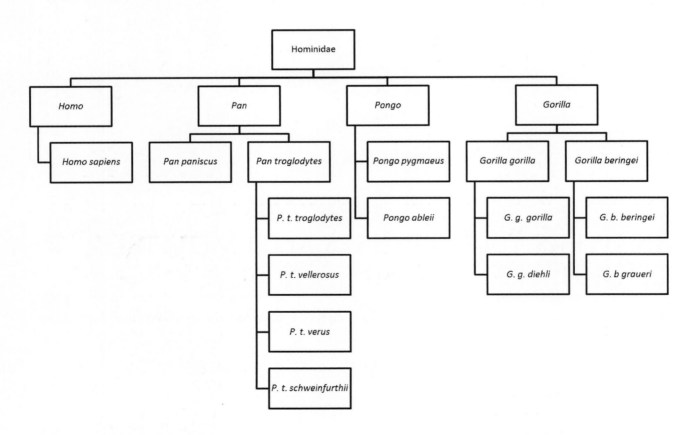

Q1.3.2. They refer to geographical areas, suggesting that the subspecies have evolved as the result of geographical isolation.

Q1.3.3. The subspecies of chimpanzee that was the first to be described was the Central chimpanzee *Pan troglodytes trogylodytes*. This is indicated by the repetition of the specific epithet *trogylodytes*.

Q1.3.4. The name inaccurately suggests a short stature and that the animal is a type (possibly a subspecies) of chimpanzee, which it is not.

Q1.3.5. No, it is not possible to determine from the family tree which species is most closely related to humans. It is clear that humans belong to the genus *Homo* and no other extant species does, but that is as far as it is possible to go. The most effective way to analyse evolutionary relationships is to compare the DNA of the various species.

1.4. Constructing a dichotomous key

Q1.4.1. They are known as dichotomous keys because they split into two branches when each question is answered; each question has only two possible answers.

Q1.4.2. Questions relating to the relative size of a structure are not really appropriate in a key because they assume that the user is in a position to make a comparison. The question 'Does the animal have long wings?' cannot be answered unless the user is examining two animals: one with short wings and another with longer wings. Whether an animal is considered to be large or small, or tall or short requires the user to make a comparison. Conversely, the question 'Does the animal have six legs?' merely requires the user of the key to count legs.

Q1.4.3.

1.	Does it have straight antennae?	Yes – Go to Q2	No – Go to Q3.
2.	Does it have a straight tail?	Yes – *Species A*	No – *Species F.*
3.	Does it have a curved tail?	Yes – *Species B*	No – Go to Q4.
4.	Does it have spots on the last segment?	Yes – Go to Q5	No – *Species E.*
5.	Does it have three spots on the last segment?	Yes – *Species C*	No – *Species D*

Q1.4.4. This allows a direct comparison to be made between the specimen being examined and an image of what a particular species (or other taxon) looks like. If a mistake has been made answering the questions the user may misidentify an organism by following the wrong route through the key. A mistake is easier to identify if drawings or photographs are part of the end point of a key.

1.5. *Global biodiversity: the numbers of recognised species*

Q1.5.1. 1,372,498

Q1.5.2.

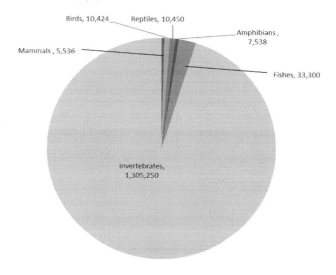

Q1.5.3.

Arachnids as a percentage of all invertebrate species	$102,248/1,305,250 \times 100 = 7.8\%$
Invertebrates as a percentage of all animal species	$1,305,250/1,372,498 \times 100 = 95.1\%$
Fishes as a percentage of all vertebrate species	$33,300/67,248 \times 100 = 49.5\%$
Homiotherms (mammals and birds) as a percentage of all vertebrate species	$(5538 + 10,424)/67,248 \times 100 = 23.7\%$
Insects as a percentage of all animal species	$1,000,000/1,372,498 \times 100 = 72.9\%$
Mammals as a percentage of all vertebrate species	$5538/67,248 \times 100 = 8.2\%$
Homiotherms as a percentage of all animal species	$(5538 + 10,424)/1,372,498 \times 100 = 1.16\%$

Q1.5.4. Invertebrates make up over 95% of all known species of animals. Of these 7.8% are spiders and almost 73% of all animals are insects. Most conservation effort is put into conserving mammals and birds which make up less than 24% of all vertebrate species and only about 1.2% of all animal species. If the purpose of conservation is to conserve biodiversity, efforts should reflect the relative abundance of the various taxa that inhabit the Earth. Many invertebrates perform important ecological functions and provide essential ecosystem services.

1.6. *Diversity in chalcid wasps*

Q1.6.1.

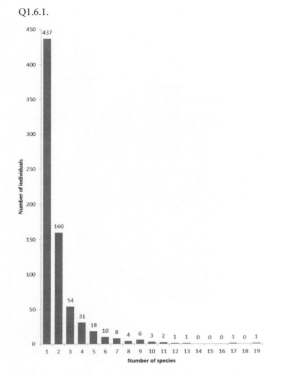

Q1.6.2. 737 species.

Q1.6.3. 1448 individuals.

Q1.6.4. $(19 \times 1) + (17 \times 1) + (13 \times 1) + (12 \times 1) = 61$

Q1.6.5. Some extremely rare species may have been present at such a low density that they did not appear in the sample.

Q1.6.6. The data do support the proposition that 'Common species are rare in nature.'

Q1.6.7. It should be specific to insects/arthropods and degrade quickly. For example, natural pyrethrum diluted in a highly raffinated white oil is specific to arthopods and will degrade photochemically within a few hours.

1.7. *Lognormal distribution of species relative abundance*

Q1.7.1.

Interval	Number of individuals	Log$_2$ number of individuals	Number of species
1	1–2	1	16
2	2–4	1–2	21
3	4–8	2–3	25
4	8–16	3–4	23
5	16–32	4–5	20
6	32–64	5–6	15
7	64–128	6–7	12
8	128–256	7–8	8
9	256–512	8–9	5
10	512–1024	9–10	3

Q1.7.2.

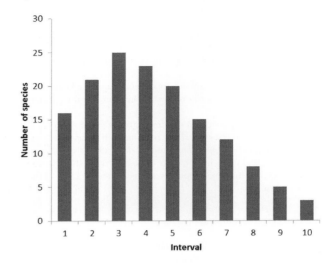

Q1.7.3. Rare species would be most likely to be missed since, by definition, they are represented by very few individuals. Common species could be underrepresented but are highly unlikely to be completely missed.

1.8. *The discovery of new species*

Q1.8.1. Analysis of the amount of variation in the DNA of a species may lead to it being split into two or more species.

Q1.8.2. Occasionally species are declared to be extinct in the wild when in fact they are not. Animals that live in remote places are rarely seen by humans and, if they become very rare, sightings may cease altogether. If a species has not been seen in the wild for many years it may be declared extinct when, in fact, the density of the species is so low that it is highly unlikely to be seen. It could be found by a new expedition to the area.

Q1.8.3. Many of the newly discovered species have been found in forest ecosystem, especially in southeast Asia, Africa and South America.

Q1.8.4. The Santa Cruz giant tortoise, the soprano pipistrelle and the bonobo were all discovered by reexamining existing species.

Q1.8.5. The two phenotypes emit sounds of different frequencies. The 45 kHz phenotype is *Pipistrellus pipistrellus*; the 55 kHz phenotype is *Pipistrellus pygmaeus*.

Q1.8.6. The USA has a large human population which includes many bird-watchers, hunters and ecologists. The country has been extensively explored and its wildlife is well known to scientists and amateur naturalists.

Q1.8.7. The scientific name is *Okapia johnstoni*. It was named after the British Governor of Uganda, Sir Harry Johnston, who was the first person to acquire a specimen of this species for science.

1.9. *Extinctions versus discovery of new species*

Q1.9.1. More susceptible.

Q1.9.2. Birds.

Q1.9.3. Island species have a small geographical range and low population size. Consequently any natural fluctuation in numbers or natural occurrences such as fire or disease may make the population nonviable. Island species are also vulnerable to the effects of introduced species that may compete with them, prey on them or bring in new disease to which they have no defence mechanism. Hunting by humans and habitat destruction may also cause extinctions on islands.

Q1.9.4. Ground-nesting birds that have evolved in the absence of predators are easy prey for introduced predators (e.g., cats, dogs, rats) and humans that may take adults and their eggs for food.

Q1.9.5. 2.4/year × 10 years = 24 species

1.10. The historical ecology of the large mammals of Ngorongoro crater

Q1.10.1. Baumann's account provided very little information. He referred to just four large mammals species – wildebeest, zebras, hippopotamuses and black rhinoceros – and no indication of their numbers. It is inconceivable that there were no other species present considering the diversity of mammals currently living in the crater.

Q1.10.2. Historical accounts vary greatly in their estimates of large mammal population sizes. They can be little more than guesses because they were made when scientific methods of counting animals did not exist. Considering the more modest numbers suggested by recent estimates it seems unlikely that the higher historical estimates can be justified.

Q1.10.3. No, because we would need to know the time of year when each count was conducted and this information is not provided.

Q1.10.4. The vegetation in the Serengeti National Park is not the same as that in the crater. This may account for the absence of giraffes and impalas (and possibly other antelopes) from the crater. The crater floor is surrounded by high walls. Although this is not a barrier as we know that some species move in and out of the crater, it may have affected the dispersal of some species.

Q1.10.5. Some estimates of numbers were made from the air while others were no more than guesses made from observations on the ground. Some were made by lay individuals, some by experienced hunters and others by scientists. In order to examine changes to numbers with time we would ideally need data collected using the same method on a number of different occasions. Such data is not available for the crater.

Q1.10.6. It was too time-consuming to separate them out. Also, these animals often occur as mixed species groups so are difficult to distinguish.

1.11. Pest eradication in New Zealand

Q1.11.1. New Zealand has no indigenous large predators so the native species of animals have not evolved to avoid predators or defend themselves against them.

Q1.11.2. Cats and dogs, as they are not native species and many species in New Zealand have not evolved behaviours to avoid predators.

Q1.11.3. Keep them under control; enclose them in their own gardens; control breeding. Cats can be given collars fitted with bells to warn birds etc. of their approach.

Q1.11.4. Pests have been estimated to do economic damage amounting to $3.3 billion a year.

Q1.11.5. Under s.25(2) of the Hazardous Substances and New Organisms Act 1996 no permit shall be issued to import any species listed in Schedule 2 of the Act: Prohibited new organisms. Prairie dogs are listed in this schedule.

Q1.11.6. No. Under s.25(2) of the Hazardous Substances and New Organisms Act 1996 no permit shall be issued to field test, or release any species listed in Schedule 2 of the Act: Prohibited new organisms. This includes the red squirrel.

1.12. Citizen science: biodiversity and environmental studies

Q1.12.1. They require a great deal of data from a wide geographical area; they require minimal scientific or technical knowledge; recordings are easy to make; the projects are of general topical interest to the public.

Q1.12.2. They reduce research costs; they allow a great deal of data to be collected often in a relatively short time and over a large geographical area; they help to engage the public with ecological issues and science in general.

Q1.12.3. Participants are likely to be unskilled (in relation to the work to be done) so tasks must be kept straightforward or training provided; difficult to ensure consistency in data collection; difficult to control the study when using volunteers (e.g., the geographical distribution of participants in ecological studies may be clumped so whole areas may not be sampled).

Q1.12.4. Common bluebell = A and E. All of the others are Spanish bluebell.

Q1.12.5. May need expensive bird identification guides, binoculars, GPS equipment, access to a computer and the Internet.

Q1.12.6. Local knowledge of biodiversity is lost as fewer people engage with nature. People who have become separated from nature are unlikely to show interest in biodiversity policy.

Q1.12.7. Much conservation and biodiversity education is passive and involves dispensing information in printed form. Citizen science requires participants to engage in data collection and collation (perhaps even rudimentary analysis) and is likely to encourage a broader interest in science and the environment than participation in an education programme.

1.13. The ecological value of ancient woodland

Q1.13.1. Ancient woodland is, by definition, over 400 years old. It is a relatively stable ecosystem which supports a large number of species. This woodland is impossible to replace once lost. As well as having ecological interest, ancient woodland may also have considerable historical importance.

Q1.13.2. The species that are most likely to have been omitted from the list are those that are very small (e.g., microorganisms) and those that live in the soil or as parasites inside other species (animals or plants). Migratory species are also likely to have been missed as their presence is transitory.

Q1.13.3.

a. Although translocating soil will probably preserve the seed bank, soil organisms and soil fertility, it will not be practical (although it is technically possible) to transplant the large trees that will be lost and it will be impossible to capture and translocate all of the fauna. There is no good evidence that translocating soils can produce a better woodland than new planting on nontranslocated soils.

b. Ancient woodland is defined as having existed continuously since 1600. Clearly, planting new trees in a different location will not replace an ecosystem that has developed over at least 400 years. It will take decades before mature trees are present and there is no guarantee that the fauna lost from destroyed ancient woodland will colonise newly planted woodland.

Chapter 2 Abiotic Factors and Ecophysiology

2.1. *Climatic variations*

Q2.1.1.

(A)

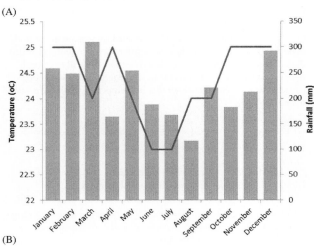

(B)

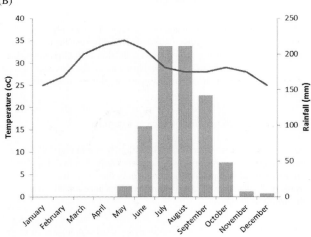

Note: Lines = temperature; bars = rainfall.

Q2.1.2.

	Biotope A	**Biotope B**
Mean monthly temperature (°C)	24.3	29.4
Mean monthly rainfall (mm)	218.3	61.6
Total annual rainfall (mm)	2619	739

Q2.1.3. Biotope A has a high rainfall, more or less evenly distributed throughout the year, with rain in every month. Rainfall in region B is highly seasonal, with no rain between January and April and an annual total that is only around 28% that of region A.

Q2.1.4. The mean monthly rainfall figures mask the fact that no rainfall falls at all in biotope B for four months and two other months have very little rainfall. Most of the rain falls in just four months of the year.

Q2.1.5. Biotope A is tropical forest because the climate is hot and wet throughout the year. Biotope B is savannah because it has distinct wet and dry seasons.

Q2.1.6. In biotope B plant growth would be seasonal. Seeds will germinate at the beginning of the rainy season and established plants will grow only if sufficient soil water is available.

2.2. *Lake stratification*

Q2.2.1.

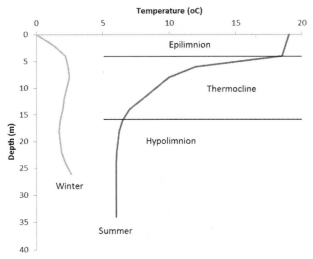

Figure 12.1

Q2.2.2. The scales are unusual because the dependent variable is temperature and this would normally be plotted against the *y*-axis. The graph makes more sense visually if depth is plotted on the *y*-axis even though this is the independent variable. Normally, values increase from the bottom to the top of the *y*-axis. This does not make sense with depth as this obviously increases from the top to the bottom of a lake. Breaking two basic rules used in drawing graphs produces a graph that better reflects reality.

Q2.2.3. See Fig. 12.1.

Q2.2.4. As winter approaches, the upper layer of the lake cools, winds cause increased mixing and the thermocline disappears.

Q2.2.5. Oxygen levels will fall in the water below the thermocline during summer because the layers are stable and there is little mixing of water between the epilimnion and the hypolimnion.

2.3. *Carbon dioxide and photosynthesis*

Q2.3.1.

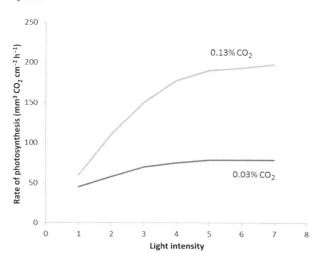

Q2.3.2. When the carbon dioxide concentration is 0.03% the rate of photosynthesis increases as light intensity increases until light intensity reaches 5 units. Thereafter, photosynthetic rate is constant. When carbon dioxide concentration is raised to 0.13% the photosynthetic rate is higher at all light intensities and continues to increase as light intensity increases.

Q2.3.3. These data suggest that an increase in atmospheric carbon dioxide concentration would increase photosynthesis and thus plant production.

Q2.3.4. Atmospheric carbon dioxide has been increased by the burning of fossil fuels, notably coal, gas and oil, by industry and for domestic use.

Q2.3.5. Marine algae use hydrogen carbonate ions as a source of carbon dioxide.

Q2.3.6. Forests are capable of absorbing large quantities of carbon dioxide from the atmosphere, and in so doing, reducing its concentration. The carbon dioxide is used in photosynthesis to make sugars resulting in carbon being locked up as cellulose and other carbon-based molecules in the trees. The carbon is therefore stored as biomass.

2.4. *The effect of water table depth on seed germination*

Q2.4.1.

Water table depth (cm below soil surface)	Number of seedlings germinated from 6000 seeds	Germination (%)
0	540	9.00
5	151	2.52
10	12	0.20
20	0	0.00

Q2.4.2.

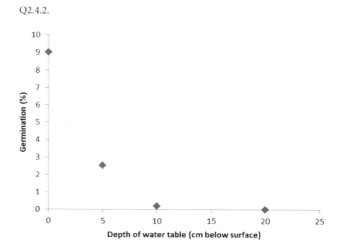

Q2.4.3. A negative correlation. As the depth of the water table increases the germination rate decreases.

Q2.4.4. *Juncus effusus* is a rush and grows in water and waterlogged soil. The seeds will not germinate in dry conditions as such conditions would not be suitable for the seedlings and they would not establish.

2.5. *The effect of oxygen level on midge larvae survival*

Q2.5.1.

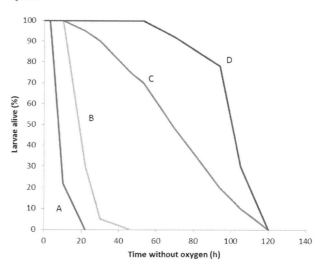

Q2.5.2. All larvae of all species survived for at least 3 hours without oxygen. Thereafter survival varied considerably between species. All individuals of species A were dead after 22 hours without oxygen. Species B survived longer but were all dead after 46 hours. Species C and D survived much longer, with 30% of larvae of species D still alive after 105 hours.

Q2.5.3. Species A and B are likely to be stream-dwellers as they cannot survive for long without oxygen. Water in streams is constantly oxygenated as a result of water movement. Species C and D can survive for long periods without oxygen and are likely to dwell in stagnant water where oxygen levels are low.

Q2.5.4. D would have to be deprived of oxygen for approximately 100 minutes for 50% of the larvae to die.

2.6. *Temperature selection in two fish species*

Q2.6.1.

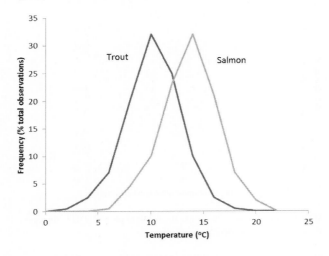

Q2.6.2. Trout survive in the range 2–18°C; salmon survive in the range 6–20°C.

Q2.6.3. Trout prefer 10°C; salmon prefer 14°C.

Q2.6.4. The fish would be most active at the two ends of their preferred temperature ranges.

Q2.6.5. Fish are poikilothermic so their body temperature depends upon the environmental temperature. In order to maintain a body temperature that allows them to remain active they need to be able to move away from unfavourable temperatures and towards favourable temperatures.

2.7. *Adaptation to life underwater: diving in seals*

Q2.7.1.

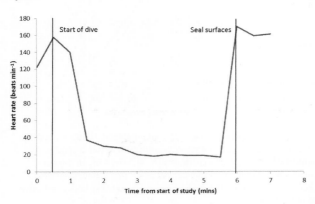

Q2.7.2.

    a. Dive began at 30 seconds.
    b. Dive ended at 6.0 minutes.

Q2.7.3. During a dive the heart beats at a much slower rate than at the surface. It is uses less energy when beating slower so needs less oxygen and glucose and therefore less blood.

Q2.7.4. Breathing is suspended during the dive so the diaphragm does no work and therefore requires very little blood.

Q2.7.5. During the dive blood is diverted to those organs which are most important for its immediate survival, especially the brain and the eyes.

2.8. *Surviving salty environments: maintaining osmotic balance*

Q2.8.1.

Nasal secretion	
Volume (cm^3)	56.3
Sodium (mmol)	43.7
Cloacal secretion	
Volume (cm^3)	75.2
Sodium (mmol)	4.41

Q2.8.2. Nasal secretion removed 43.7 mmol of sodium compared with just 4.14 mmol removed via the cloaca.

Q2.8.3.

Time (min)	Sodium concentration (mmol L^{-1})	
	Nasal secretion	**Cloacal secretion**
15	772.7	48.3
40	752.3	71.2
70	781.7	80.0
100	776.4	60.8
130	794.1	33.9
160	804.9	9.6
175	750.0	13.2

Q2.8.4.

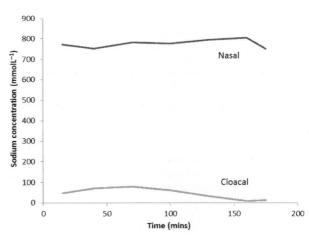

Q2.8.5.

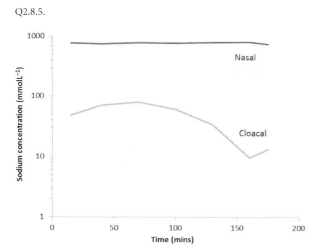

Q2.8.6. The second graph (using a log scale on the vertical axis) makes it easier to detect that the cloacal secretion peaks 70 minutes after ingestion while the concentration of sodium in the nasal secretion remains more or less constant with time.

Q2.8.7. Oceanic birds have the highest concentration of sodium in their nasal secretion, followed by coastal birds and inland birds have the lowest concentration. This probably reflects the extent to which they take in salty water.

2.9. *Living in the desert: temperature tolerance in the camel*

Q2.9.1. The body temperature range in camels is much greater than in the other animals listed.

Q2.9.2. Camels store excess heat when they are deprived of water and hence dehydrated.

Q2.9.3. Water is normally used to lose heat by sweating. As sweat evaporates the body loses the latent heat of vaporisation of the sweat as it changes from a liquid form to a gas. If the body temperature is allowed to rise, the water that would have been lost as sweat to cool it down is saved.

Q2.9.4. Maintenance of a high body temperature reduces the heat gradient between the body and the hot environment. This reduces further heat gain by conduction. A cold body will warm up faster than a warm body in a hot environment.

Q2.9.5. At night, when the air is cool, the body loses heat by transferring it to the environment by radiation and conduction. By allowing its temperature to fall to a low temperature during the night it can store more heat the next day before it needs to sweat.

Q2.9.6. Fur insulates the skin of the camel from the intense heat from the sun. When the fur is shorn the camel loses more water by evaporation to keep cool.

Q2.9.7. A fully hydrated camel would lose 9.1 L in the 10 hottest hours of the day. The dehydrated camel loses only 2.8 L thereby saving 9.1 − 2.8 L = 6.3 L. The heat energy which would have been used to evaporate this has been stored, i.e., 6.3 L × 2426 kJ = 15283.8 kJ.

2.10. *Diapause in* Grapholita

Q2.10.1.

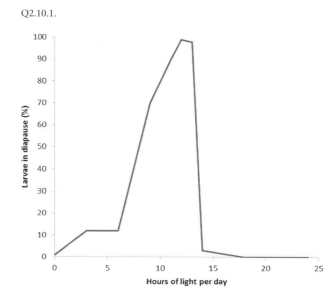

Q2.10.2. Once the minimum number of hours of daylight falls to 12, 99% of larvae are in diapause.

Q2.10.3. This indicates the end of summer and the beginning of autumn (fall) in southern California, i.e., beginning of October.

Q2.10.4. It is clear from the data that *Grapholita* larvae that hatched after about 25 August were exposed to around 13 hours of daylight. The larvae use this day length to predict the coming winter and the associated lower temperatures.

2.11. *The effect of temperature on dusting behaviour in an Asian elephant*

Q2.11.1.

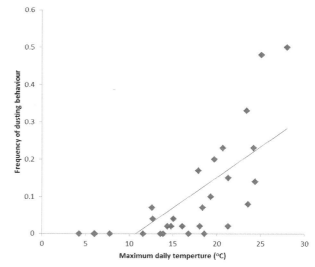

Figure 12.2

Q2.11.2. See Fig. 12.2.

Q2.11.3. The correlation coefficient, $r = 0.717$.

Q2.11.4. The is a positive correlation between temperature and the frequency of dusting behaviour.

Q2.11.5. The lowest temperature at which dusting was observed was 12.6°C.

Q2.11.6. Elephants may dust bathe to protect the skin from burning by sunlight, to act as a screen to prevent the skin becoming hot (i.e., to keep cool), to prevent insects from biting the skin, or for some other reason that is not obvious.

Q2.11.7. No, it is not possible to determine why elephants dust more in warm weather than in cold weather from this study alone. All that can be said is that there is a positive correlation between temperature and the frequency of dusting behaviour. This does not mean that we have established cause and effect (i.e., we cannot say that changes in temperature directly affect dusting frequency) let alone the purpose of the behaviour.

2.12. *Hibernation in the black bear*

Q2.12.1. Metabolism involves the breakdown of food using oxygen in the process of aerobic respiration:

$$C_6H_{12}O_6 + 6O_2 \rightarrow 6CO_2 + 6H_2O + Energy$$

Measuring the quantity of oxygen used indirectly measures the quantity of food (glucose in this case) used in respiration.

Q2.12.2.

(A)

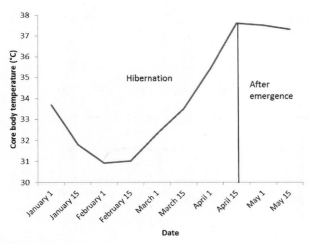

Figure 12.3

(B)

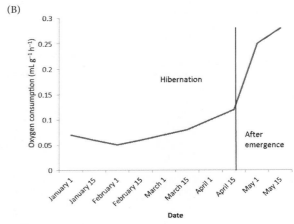

Figure 12.4

Q2.12.3. See Figs. 12.3 and 12.4.

Q2.12.4. $37.6 - 30.9 = 6.7°C$.

Q2.12.5. $0.05/0.28 \times 100 = 17.86\%$.

Q2.12.6. When the two graphs are compared it is clear that, as the bear emerges from hibernation, the core temperature rises and then stabilises but the metabolic rate continues to rise. This suggests that low temperature alone does not cause a reduction in metabolic rate. If it did we would expect the metabolic rate to stabilise as soon as the bear's body temperature had stabilised.

2.13. *Microclimate preferences in woodlice*

Q2.13.1. We would expect to find five woodlice on each side because there would be a 50% chance that any particular woodlouse would be on any particular side.

Q2.13.2.

Chamber number	Light	Dark
1	3	7
2	4	6
3	3	7
4	4	6
5	4	6
6	3	7
7	1	9
8	5	5
9	4	6
10	3	7
Total (observed values)	34	66
Total (expected values)	$\frac{34+66}{2} = 50$	$\frac{34+66}{2} = 50$

For the dark/light choice, $\chi^2 = 10.24$. This is highly significant: $P < 0.005$ ($df = 1$)

ANSWERS TO EXERCISES AND MULTIPLE CHOICE TESTS

Q2.13.3.

Chamber number	Low humidity	high humidity
1	4	6
2	5	5
3	3	7
4	4	6
5	4	6
6	3	7
7	2	8
8	6	4
9	5	5
10	3	7
Total (observed values)	39	61
Total (expected values)	$\dfrac{39 + 61}{2} = 50$	$\dfrac{39 + 61}{2} = 50$

For the dry/humid choice $\chi^2 = 4.84$. This is significant: $P < 0.05$ ($df = 1$)

Q2.13.4. These experiments demonstrate that woodlice prefer dark to light conditions and prefer humid to dry conditions. Dark conditions would help to hide them from predators and humid conditions would prevent desiccation.

Chapter 3 Ecosystems, Energy and Nutrients

3.1. *A steppe ecosystem: biotic and abiotic factors*

Q3.1.1. Biological or biotic components, e.g., the kiang, lichens, and starlings. Physical or abiotic factors, e.g., temperature, snow and rock.

Q3.1.2.

Abiotic factors	Primary producers	Primary consumers	Secondary consumers	Decomposers
Wind	Lily	Gazelles	Buzzards	Soil bacteria
Humidity	Hyacinth	Kiang	Wild cats	Fungi
Fire	Feather grass	Rabbits	Heron	
Temperature	Iris	Ducks	Fox	
Frost	Sage brush	Locusts	Falcons	
Air	Fescue grass	Ground squirrel	Stoat	
Snow	Sedges	Hares	Eagles	
Sunlight	Daisy	Geese	Sand lizards	
Rock	Thistles	*Finches*	*Finches*	
Ice	Tulip	*Starlings*	*Starlings*	
Rain	Lichens	Saiga (antelope)	Crows	
Rivers		Marmot (a rodent)	Wolf	
		Roe deer	Polecat	
		Przewalski's horse	Kestrels	
		Wild boar		

Q3.1.3. The steppe consists of a biological community and a physical environment. It has the three major features of an ecosystem: energy flows between its biological components; nutrients cycle between its biological and physical components and it possesses biological diversity.

Q3.1.4. This type of classification uses evolutionary relationships. Ecologists are interested in the ecological relationships between the organisms because they are concerned with how ecosystems work.

3.2. *Food webs in an English woodland*

Q3.2.1.

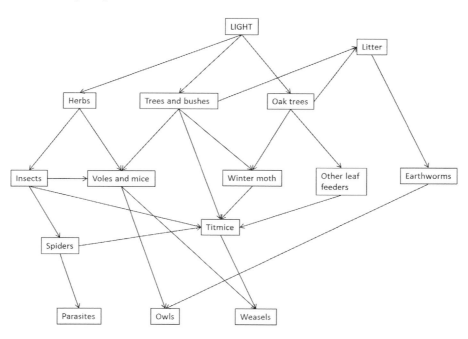

Q3.2.2.

Taxon	Trophic level
Oak tree	Primary producer
Parasites	Tertiary consumer
Owls	Secondary consumer
Weasels	Tertiary consumer
Winter moth	Primary consumer

Q3.2.3. The increase in voles and mice would provide additional food for owls and weasels and their young. This is likely to result in an increase in the numbers of owls and weasels.

Q3.2.4. Feeding relationships could be determined by simple observation of feeding in the wild; by taking organisms into captivity and offering them food choices (probably not suitable in this study); by examining stomach contents, faeces and owl pellets; by using radioactive tracers introduced into the food chain (again, probably not suitable in this study).

Q3.2.5. Decomposers (bacteria and fungi) have been omitted from the food web. This is often the case in food webs because of the difficulty of counting individual organisms and precisely describing their roles.

3.3. *Ecological pyramids*

Q3.3.1.

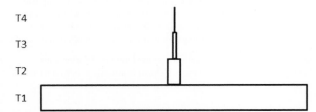

T4

T3

T2

T1

Figure 12.5

Q3.3.2. Decomposers (bacteria and fungi).

Q3.3.3. It would be difficult to draw a pyramid of numbers for Silver Springs to scale because it would not be possible to show the smallest and largest values on the same linear scale (i.e., the width of the smallest value would have to be displayed as the thickness of a single line). We could get around this problem by using a logarithmic scale but this would produce a visual distortion of the relationships between the trophic levels.

Q3.3.4.

Efficiency of transfer from T_3 to T_4	$(67/1602) \times 100 =$	4.18%
Efficiency of transfer from T_2 to T_3	$(1602/(14 \times 10^3)) \times 100 =$	11.44%
Efficiency of transfer from T_1 to T_2	$(14 \times 10^3/(87 \times 10^3)) \times 100 =$	16.09%

Q3.3.5. The efficiency of transfer of energy between one trophic level and the next varies between trophic levels and from one type of ecosystem to another. This depends upon the types of organisms present and the efficiency with which they feed. It also depends upon the relative importance of decomposers and the process of

decomposition in any system. The efficiency of transfer was once thought to be around 10% for each set of trophic levels but in some ecosystems it may be much higher than this.

Q3.3.6. The energy that is not transferred to the next level is lost in the process of respiration (as heat), in faeces and urine, in fur, antlers and other body parts that may be shed (including leaves and branches in plants). The energy in these materials will be utilised by decomposers.

Q3.3.7. Fish → nematodes → bacteria

Q3.3.8. Bacteria

Q3.3.9.

Organism	(A) Number of individuals	(B) Mass of each individual (g)	(C) Total biomass (g)
Bacteria	10^9	10^{-15}	10^{-6}
Nematodes	10^3	0.1	100
Fish	1	1000	1000

Q3.3.10.

(a)

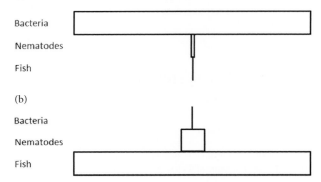

Bacteria

Nematodes

Fish

(b)

Bacteria

Nematodes

Fish

Q3.3.11. A pyramid of energy for this food chain would be similar in shape to the pyramid of biomass because the amount of energy in an organism is proportional to its mass. A single large organism can clearly contain more energy than a large number of small organisms whose combined mass is smaller than that of the large organism.

3.4. *Energy flow in ecosystems*

Q3.4.1. The light energy that is not absorbed by the green plants may be transmitted through its leaves or reflected. Green leaves are green because they reflect green light.

Q3.4.2. The percentage of the total insolation:

a. absorbed by green plants is 50%.
b. converted into net primary production is 0.5%.
c. converted into net secondary production is 0.05%.
d. converted into net tertiary production is 0.01%.

Q3.4.3. The percentage of the gross production used in respiration in

a. green plants is $0.5/1.0 \times 100 = 50\%$.
b. herbivores is $0.45/0.50 \times 100 = 90\%$.
c. carnivores is $0.04/0.05 \times 100 = 80\%$.

Q3.4.4. In respiration energy is lost as heat.

Q3.4.5.

a. Energy is lost from the grazing food chain when plants die and decompose and when they shed leaves, release seeds, fruits, lose bark, etc.
b. Energy is lost to the carnivores in the food chain when herbivores lose skin, fur, antlers and other body parts, in faeces and urine and when they die (unless eaten by scavengers) because the energy in their bodies will pass to the organisms in the decomposer food chain.

3.5. *A comparison of energy budgets in two ecosystems*

Q3.5.1. Silver Springs receives more than 3.5 times the isolation received by the old field site in Michigan due to differences in climate and day length. Florida experiences a longer day length than Michigan because it is located further south and the sun is higher in the sky. Both of these factors increase isolation.

Q3.5.2.

a. The percentage of the total insolation which is used by the green plants (to produce the gross primary production)
Old field: 1.237%
Silver Springs: 1.231%.
b. The percentage of the total insolation that is not used by the green plants
Old field: 98.485%
Silver Springs: 98.732%.
c. The percentage of the total insolation that becomes the net primary production
Old field: 1.051%
Silver Springs: 0.523%.
d. The percentage of the gross primary production that is used in respiration.
Old field: 15.020%
Silver Springs: 57.551%.

Q3.5.3. Ecosystems are inefficient at using light as a source of energy. In both of these systems a little more than 1% of the energy in insolation is used to produce the gross primary production and almost 99% of this energy is unused. Notice that respiratory losses of energy from plants are a much higher percentage of gross primary production in the freshwater system than in the field system.

3.6. *Chemotrophism*

Q3.6.1. Autotrophs. They use chemical as a source of energy rather than the sun.

Q3.6.2. Chemosynthesis.

Q3.6.3.

i	$4FeCO_3 + O_2 + 6H_2O \rightarrow$ $4Fe(OH)_3 + 4CO_2 + Energy$	Iron bacteria (e.g., *Leptothrix*).
ii	$2S + 3O_2 + 2H_2O \rightarrow 2H_2SO_4$ $+ Energy$	Colourless sulphur bacteria (e.g., *Thiobacillus*);
iii	$2NH_3 + 3O_2 \rightarrow 2HNO_2 +$ $2H_2O + Energy$	Nitrifying bacteria (e.g., *Nitrosomonas*);

Q3.6.4. The H_2S is removed by oxidation

Q3.6.5. H_2S is highly corrosive and expensive to remove using chemical means. The use of chemotrophic bacteria to remove this would remove the costs of chemical treatment and reduce running costs of machinery caused by corrosion.

3.7. *Grazing and detritus food chains*

Q3.7.1. Many of the herbivores in these systems are large organisms that consume vast numbers of very small plants, feeding for much of the day. There is therefore little opportunity for plankton and algae to die and decompose.

Q3.7.2. Smaller: most of the herbivores in a forest are insects.

Q3.7.3. Animals cannot digest lignin because they lack the appropriate enzymes. Some fungi and bacteria possess ligninase enzymes and they will breakdown bark on the forest floor. The presence of lignin in plant tissues increases the amount of energy available to detritus food chains.

Q3.7.4.

a. Cattle take 15% of the 37% of net primary production taken by grazers. Grazers other than cattle take 37% − 15% = 22% of NPP.
b. The other herbivores would be rabbits, deer and any other native vertebrate grazers, along with insects, snails and other invertebrates.

Q3.7.5. *Spartina* marsh and grazed meadow.

Q3.7.6. They break dead organisms up into smaller fragments, thereby making a larger surface area available to bacteria and fungi. Material passes through the gut of invertebrates such as earthworms and is released back to the soil partially digested.

3.8. *Energy sources in aquatic ecosystems*

Q3.8.1. 83%.

Q3.8.2. Almost all of the energy in the streams in the deciduous forest came from outside of the streams. This will have been in the form of leaves, branches, etc. that have fallen into the water. In contrast most of the energy in the lake came from within the lake. This will have come from organisms in the lake.

Q3.8.3. No. Almost all of the energy came from within the deciduous forest itself.

Q3.8.4. Most of the energy in the lake in the deciduous forest came from within the lake (83%), while most of the energy in the lake in the coniferous forest came from outside the lake (66%).

Q3.8.5. The foodwebs in any ecosystem are ultimately dependent upon the sources of energy that support that system. Any changes to the energy sources are likely to affect the abundance of organisms. An increase in energy supplied to a system is likely to cause an increase in the populations of organisms that can access that energy. If more energy is supplied to the small invertebrates in an aquatic system this could cause an increase in their number, resulting in an increase in fish numbers etc.

3.9. *Energy budget of a bank vole*

Q3.9.1.

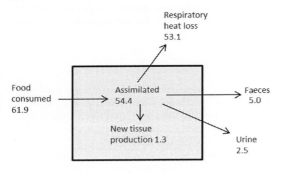

Q3.9.2. The amount of energy assimilated as a percentage of the energy consumed = 54.4/61.9 × 100 = 87.9%.

Q3.9.3. The percentage of the assimilated energy that is:

a. lost as the heat produced by respiration is 53.1/54.4 × 100 = 97.6%.
b. used to produce new tissue is 1.3/54.4 × 100 = 2.4%.

Q3.9.4. The percentage of the consumed energy that is lost in:

a. urine: 2.5/ 61.6 × 100 = 4.0%.
b. faeces: 5/61.6 × 100 = 8.1%.

Q3.9.5. The amount of energy a pregnant female bank vole would need to consume per day assuming that her energy requirements were increased by approximately 60% during pregnancy is:

$$= 61.6 \times 10^3 \, J \times 160 \,/\, 100 = 98.56 \, kJ$$

Q3.9.6. Annual production (J) = number × mean weight (g) × calorific value (Jg^{-1})

$$= 50 \times 23 \times 20,900$$
$$= 24,035,000 \, J \, year^{-1} \text{ or } 24,035 \, KJ \, year^{-1}$$

3.10. *Wildlife biomass in the Serengeti*

Q3.10.1. Lion, cheetah, leopard, spotted hyena, hunting dog, golden jackal and black-backed jackal.

Q3.10.2.

Species	Number	Mean weight (kg)	Total biomass (kg)
Wildebeest	370,000	128	47,360,000
Plains zebra	193,000	164	31,652,000
Lion	1650	120	198,000
Thomson's gazelle	980,000	12	11,760,000
Grant's gazelle	3100	32	99,200
Eland	7200	225	1,620,000
Leopard	500	35	17,500
Cheetah	500	40	20,000

(*Continued*)

(Continued)

Species	Number	Mean weight (kg)	Total biomass (kg)
Topi	26,000	82	2,132,000
Hartebeest	20,000	95	1,900,000
Spotted hyena	6000	55	330,000
Impala	75,000	32	2,400,000
Hunting dog	1100	18	19,800
Waterbuck	3200	131	419,200
Giraffe	8000	716	5,728,000
Black-backed jackal	13,500	6	81,000
Warthog	17,000	40	680,000
Buffalo	38,000	420	15,960,000
Golden jackal	5000	6	30,000

Predators are indicated in italics.

Q3.10.3.

a. The total number of predators = 28,250.
b. The total number of herbivores = 1,740,500.

Q3.10.4. The herbivore numbers must be supported by the vegetation available. Carnivore numbers are largely determined by the availability of food in the form of herbivores. In systems such as this we would always expect to find fewer carnivores than herbivores because each carnivore must eat many herbivores each year in order to survive.

Q3.10.5.

a. The total biomass of predators = 696,300 kg.
b. The total biomass of herbivores = 121,710,400.

Q3.10.6.

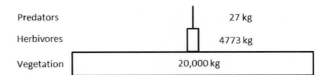

Figure 12.6

Area of Serengeti = 25,500 km²
Biomass of predators km⁻² = 696,300/ 25,500 = 27.31 kg
Biomass of herbivores km⁻² = 121,710,400/ 25,500 = 4,772.96 kg
Biomass of vegetation km⁻² = 2 × 10⁷/1,000 = 20,000 kg

Q3.10.7. Many of the herbivores feed selectively on plants of a particular species or length. The great diversity of plants available and the selectivity of the grazers allows many herbivores to coexist.

Q3.10.8. These two species are very numerous and they migrate to the Maasai Mara in Kenya when the rains cause an increase in grass growth in this area at a time when food is in short supply in the Serengeti. In effect they follow the rains.

3.11. *Assimilation efficiency in the African elephant*

Q3.11.1.

Food offered	Total wet weight (kg)	Dry matter %	Total dry weight (kg)	Variable
Hay	112	83	92.96	
Straw	30	81	24.30	
Carrots	293	11	32.23	
Sprouts	94	13	12.22	
Beetroot	3	9	2.70	
Total food offered			164.41 =	O
Waste (food not eaten)	89	21	18.69 =	W
Food eaten (offered – waste)			145.72 =	I
Dung voided	579	19	110.01 =	E

Q3.11.2.

$$\text{The gross assimilation efficiency} = \frac{145.72 - 110.01}{145.72} \times 100 = 24.5\%$$

Q3.11.3. The water content of food and faeces varies considerably. We need to discount the water content because it is the dry matter in the food that is supplying energy and nutrients.

Q3.11.4. The ratio of food to dung in terms of dry weight is approximately 1.32:1:

1 unit of dung represents approximately 1.32 units of food.

Q3.11.5. This elephant produces $3\,kg \times 0.2 = 0.6\,kg$ dry weight dung per hour, or $0.6 \times 24 = 14.4\,kg$ dry weight per day. Using the ratio of 1.32 food:1 dung, this represents $14.4 \times 1.32 = 19.01\,kg$ of food (dry weight).

3.12. *Capybara farming in Venezuela*

Q3.12.1. In 4.5 years, three litters of six capybara could be raised to their slaughter weight (at age 1.5 years), one litter at a time. In the same period of 4.5 years only one cow could be produced and raised to slaughter weight (at age 4.5 years). It takes the same amount of food to feed these 18 capybara (6×3) as it does to feed a single cow over the 4.5 year period.

Q3.12.2. The mean live weight of a capybara at slaughter is $540\,kg/18 = 30\,kg$.

Q3.12.3. The capybara together produce $60\,kg\,year^{-1}$; the cow produces only $40\,kg\,yr^{-1}$.

Q3.12.4. The capybara have more offspring (reproduce at a faster rate) than cows and are slaughtered at a younger age than cows.

Q3.12.5. $40/60 \times 100 = 66.7\%$.

Q3.12.6. Once an animal is seen as a resource (in this case food) it assumes an economic value. Most societies do not attach any economic value to their wildlife but they do value their food resources.

3.13. *Nitrogen cycle*

Q3.13.1.

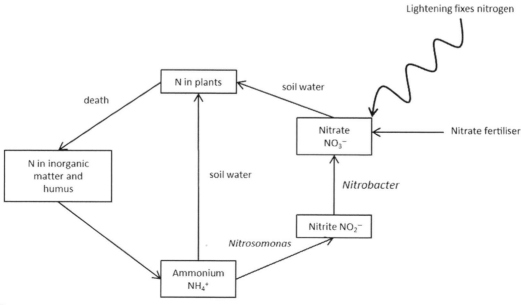

Figure 12.7

Q3.13.2. Nitrate and ammonium.

Q3.13.3. See Fig. 12.7.

Q3.13.4. See Fig. 12.7.

Q3.13.5.

a. Nitrates are leached downwards (away from plant roots) and out of the upper layers of the soil by rainwater.
b. Freshwater ecosystems may be polluted by runoff contaminated by nitrate fertiliser resulting in eutrophication.

Q3.13.6. These bacteria are important in the conversion of ammonium to nitrite and nitrite to nitrate. If their actions were inhibited less nitrate would be available to plants and growth would be adversely affected.

$$\text{Ammonium} \xrightarrow{\text{Nitrosomonas}} \text{Nitrite} \xrightarrow{\text{Nitrobacter}} \text{Nitrate}$$

3.14. *Nitrogen balance for Lake Mendota, Wisconsin*

Q3.14.1.

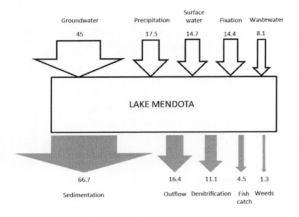

Q3.14.2. Groundwater.

Q3.14.3. Sedimentation.

Q3.14.4.

a. Organic matter from faeces.
b. Nitrogen fertiliser used in agriculture.

Q3.14.5. Nitrates are converted to nitrogen by denitrifying bacteria.

Q3.14.6. Excess nitrogen inputs into the lake—especially nitrates—are likely to cause eutrophication.

3.15. *The economics of fertiliser application*

Q3.15.1. Table 12.1

Table 12.1 The effect of nitrogen application on yield

Nitrogen applied (kg ha^{-1})	Yield (t ha^{-1})	Benefit–cost ratio	Net yield return (£)
0	2.9	–	–
50	3.4	2.0	50
100	4.3	2.8	180
150	4.7	2.4	210
200	4.8	1.9	180

Q3.15.2.

(a)

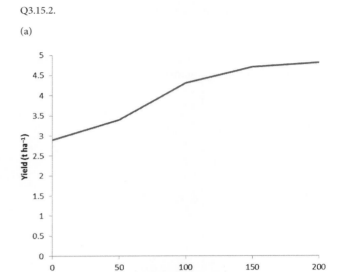

(b)

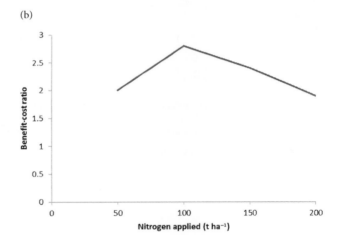

(c)

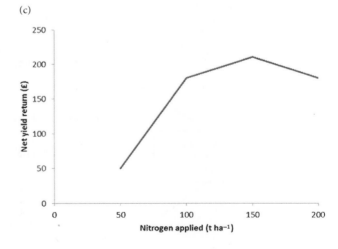

Q3.15.3. It would be costing more to add the fertiliser than the value of the increase in output achieved. Application of fertiliser would therefore give no economic benefit.

Q3.15.4.

Nitrogen applied (kg ha^{-1})	Yield (t ha^{-1})	Benefit–cost ratio	Net yield return (£)
0	2.9	–	–
50	3.4	1	0
100	4.3	1.4	60
150	4.7	1.2	45
200	4.8	0.95	−15

Q3.15.5. The cost of fertilising the crop is higher than the value of the fertilised crop so the farmer is making a loss. This is because there is very little increase in yield when the rate of fertilizer application is increased from 150 to 200 kg ha^{-1} (only 0.1 t ha^{-1}) but the cost of fertilizer increases by 33.3%.

3.16. *Sewage sludge cake as a fertiliser*

Q3.16.1.

Year	Year after first application					
	1	2	3	4	5	6
1	38	19	10	5	2	1
2	–	38	19	10	5	2
3	–	–	38	19	10	5
4	–	–	–	38	19	10
5	–	–	–	–	38	19
6	–	–	–	–	–	38
Total N kg ha^{-1}	38	57	67	72	74	75

Q3.16.2.

	Year after first application									
Year	1	2	3	4	5	6	7	8	9	10
1	38	19	10	5	2	1	0	0	0	0
2	–	38	19	10	5	2	1	0	0	0
3	–	–	38	19	10	5	2	1	0	0
4	–	–	–	38	19	10	5	2	1	0
5	–	–	–	–	38	19	10	5	2	1
6	–	–	–	–	–	38	19	10	5	2
7	–	–	–	–	–	–	38	19	10	5
8	–	–	–	–	–	–	–	38	19	10
9	–	–	–	–	–	–	–	–	38	19
10	–	–	–	–	–	–	–	–	–	38
Total N kg ha^{-1}	38	57	67	72	74	75	75	75	75	75

Q3.16.3.

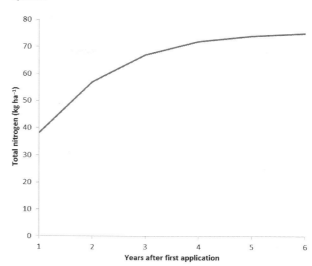

Q3.16.4. The use of sludge cake allows a build up of nitrogen in the soil which may be maintained with repeated applications. Inorganic nitrogen fertiliser tends to be leached downwards by water when it rains.

Q3.16.5. There is a risk that heavy metals and other toxins could be added to the soil by sludge cake and enter the food chain. There is also a risk that parasites could be transmitted. Wastewater treatment companies that produce sludge cake take care not to use sludge that contains high metal levels and the treatment of the sludge should remove most of the parasites.

3.17. *The effect of pH on crop growth*

Q3.17.1.

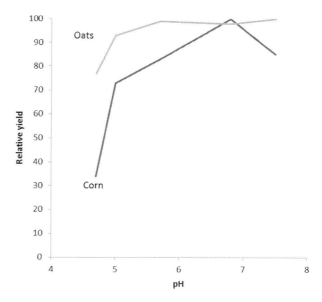

Q3.17.2. The range pH 6.8–7.5 produces the highest yields
(from 85% to 100% depending on the crop)

Q3.17.3.

a. Barley.
b. Alfalfa.
c. Corn.

Q3.17.4. pH 6.8.

Q3.17.5. Liming increases the pH, making it less acidic. Most nutrients
are optimally available at pH 6.5–7.5. If nutrient availability increases,
yield should increase.

Q3.17.6. Oats and barley would benefit. All of the others would not as
they already produce their optimum yield at pH 6.8

3.18. *Timanfaya National Park, Lanzarote: an opportunity
to study succession*

Q3.18.1. The park supports a very large number of lichen species; more
than the number of vascular plants. This is characteristic of a system
in the early stages of a primary succession before a mature soil has had
time to develop.

Q3.18.2. As the ecosystem matures we would expect to see soil
deepening and vascular plants increasing and becoming the dominant
feature of the vegetation.

Q3.18.3. Lichens and mosses would be very important in breaking
down the larva to form a more mature soil. As organisms die on the
surface the humus content of the soil should increase and the soil
should become deeper.

Q3.18.4. On Lanzarote, rainfall is low, temperatures are high and it
is frequently windy. Few trees can survive these conditions and most
of the vascular plants are succulents. There are xerophytes which
store water in their tissues and have a fleshy appearance, e.g., cacti. In
order to conserve water, leaf size may be reduced and leaves may be
protected by hairs. Stomata may be sunken and protected by enlarged
guard cells and possibly contain a waxy substance. Leaves and stems
may be covered by a thick cuticle. All of these features help to reduce
transpiration.

Q3.18.5. Timanfaya National Park preserves an area of great scientific
interest because the ecosystem is in the early stages of succession. It is
covered by volcanic soil and supports a delicate fauna and flora, the
vegetation consisting of many species of drought-resistant succulent plants,
well suited to the hot, dry climate. There are about 500 species of terrestrial
plants and nearly 300 marine animals in the coastal zone. Visitors are not
allowed to walk on the larva because it is brittle and easily damaged. The
visitor centre is located underground to preserve the unique character of
the landscape which contains few man-made structures. Access to most of
the Park is by coach only. This controls the numbers of visitors and restricts
them to a very small area thus minimising damage to the larva while still
allowing tourists to see the volcanoes at close quarters.

3.19. *Succession in bird species*

Q3.19.1.

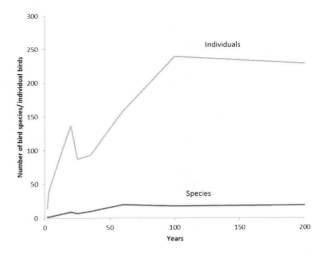

Q3.19.2. As the succession proceeds plants die and deepen the soil,
a greater variety of plants will invade the area providing more food
sources and creating new niches for birds and other species. Some of
these other species will provide food for birds. The system will become
more three-dimensional as larger plants create forests.

Q3.19.3. A primary succession occurs on land where soil has not
previously developed e.g., rocks of volcanic origin that have recently
been exuded onto the surface of the Earth. A secondary succession is
one that takes place on existing soil after the original ecosystem has
been disturbed in some way, e.g., after deforestation.

Q3.19.4. An oak–hickory climax forest is a forest that establishes at the end
of certain successions, where the dominant plants are oak and hickory. Such
a system is relatively stable compared with other stages in the succession.

Chapter 4 Determining Abundance and Distribution

4.1. *Recording the distribution of organisms*

Q4.1.1. This species may not have been recorded on islands J, K, and L
because (1) the species has never been present on these islands; (2) it
has been present in the past but is now locally extinct; (3) it is present
in such small numbers that it has never been recorded; (4) no recording
has ever been undertaken on the islands.

Q4.1.2.

a. This is likely to be a marine mammal that spends some of its time on
   the seashore, e.g., a seal species.
b. This type of map only shows the distribution on land. For a marine
   species such as a seal this is misleading because it will only record
   sightings made where animals haul out and at breeding places and
   not the distribution in the sea.

Q4.1.3. (1) It is definitely not present in the square; (2) the square has never been searched; (3) the human population in the square is so low that it has never been recorded; (4) the species is present in the square but is so rare that it has never been recorded.

Q4.1.4. Museums usually have a record of the locality from which each specimen was collected with the date of collection. Such information is very important in determining historical distributions.

Q4.1.5. Dot-maps only record presence in a 10-km square and tell us nothing about the distribution of recordings within each square. A cluster of recordings which cover less than $10\,km \times 10\,km$ could be placed in relation to the grid lines dividing up the map in such a way that they appear in four adjacent 10-km squares, giving the appearance of a much wider distribution on a dot-map (A). If all the recordings occur in the outside corners of a 2×2 grid of 10-km squares the central area would contain no individuals at all and yet each 10-km square would contain a dot, suggesting a contiguous distribution (B). The system only works well if all individuals are dispersed in a similar manner within each 10-km square (C).

Q4.1.6. Recording presence in 10-km squares tells us nothing about (1) the precise date when it was present; (2) its abundance within the square; (3) the habitat type occupied within the square; (4) the season when the species was recorded (migratory species will not be present all year).

Q4.1.7. Assuming sufficient data are available, dot-maps may be used to investigate (1) the progress of alien invasions over time; (2) the expansion or contraction of the range of a native species with time; (3) the history of local extinctions.

4.2. Problems in determining the historical distribution of organisms

Q4.2.1. Errors may occur due to (1) misidentification or (2) the presence of escapes from gardens into the wild.

Q4.2.2. Nothing. Just because one person did not find a particular species in the wild, this does not mean it was not present. By their very nature, rare or inconspicuous plants are difficult to find.

Q4.2.3. Many rich Victorians travelled widely and brought exotic plants and animals back to Britain from their trips. Once established in their gardens, the propagules of these exotic species were free to colonise the local area and in some cases have become widespread invasive species.

4.3. Community science: Moors for the Future

Q4.3.1. Climate change. *Sphagnum* grows in cool wet areas. If the climate warms areas covered by *Sphagnum* may decline.

Q4.3.2. It has no roots and absorbs water over the surface of its body.

Q4.3.3. There are 34 species of *Sphagnum* found in bogs in the UK and they are difficult to tell apart. It is better to have reliable information about the distribution and abundance of the genus *Sphagnum* than unreliable information about individual species which may have been misidentified.

Q4.3.4. Four: A, B and C together, F and G together, H; D and E are too far from the path to be counted. B and C are counted as a single patch, as are F and G because they are less than two metres apart.

Q4.3.5. Map references, GPS coordinates and photographs of the location.

Q4.3.6. *Sphagnum* helps to maintain healthy, resilient wet bogs that provide an important wildlife habitat.

4.4. Bird ringing

Q4.4.1. Bird ringing may be used to study many aspects of bird biology including mortality rates, dispersal from natal area, migration routes, and changes in body condition and growth.

Q4.4.2. Greylag geese (*Anser anser*) spend a lot of time swimming during which time their legs are underwater for most of the time. A collar is a more useful means of identification than a leg ring because it is visible when the bird is swimming and also when it is feeding in a large flock or in tall vegetation when a leg ring would be difficult to see.

Q4.4.3. Bird rings need to contain information about whom to contact when the ring is found so that data is not lost.

Q4.4.4. Bird ringing could be used to study the effects of climate change on bird populations by helping to monitor bird movements, changes in distribution, and changes in growth, body condition and survival as the climate changes.

4.5. Zonation on a rocky shore

Q4.5.1.

Code	Alga	1	2	3	4	5	6	7	8	9	10
A	Diatoma sp.										
B	Pelvetia canaliculata										
C	Porphyra umbilicalis										
D	Chaetomorpha										
E	Fucus spiralis										
F	Enteromorpha intestinalis										
G	Plocamium coccineum										
H	Ascophyllum nodosum										
I	Ulva lactuca										
J	Polysiphonia sp.										
K	Corallina officinalis										
L	Ectocarpus sp.										
M	Ceramium sp.										
N	Gracilaria sp.										
O	Polyides sp.										
P	Fucus serratus										
Q	Laurencia sp.										
R	Gigartina sp.										
S	Lomentaria articulate										
T	Chondrus crispus										
U	Lithothamnion sp.										
V	Rhodochorton sp.										
W	Scytosiphon sp.										
X	Laminaria digitata										
Y	Rhodymenia palmate										

Q4.5.2. Station 10 because it is at the low water mark. This is only exposed to the air when the tide reaches its lowest point on its way out immediately before the tide comes back in again.

Q4.5.3. Species X (*Laminaria digitata*) and Y (*Rhodymenia palmate*) are likely to be the least tolerant to drying out as they occur only at Station 10 and are therefore only exposed when the tide is turning.

Q4.5.4.

a. 6/25 = 24% of the algal taxa occur at the high water mark (Station 1).
b. 10/25 = 40% of the taxa occur at the low water mark (Station 10).

Q4.5.5. Draw a bar chart of the total number of taxa present at each station along the shore.

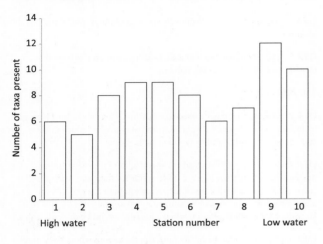

Q4.5.6. Stations 9 and 10 were occupied by the largest number of algal taxa. This section of beach supported taxa able to tolerate dessication and others that could not.

Q4.5.7. Plant succession is a temporal sequence of events from colonisation to a climax vegetation. Zonation is a permanent spatial pattern related to the position on the seashore in relation to the sea.

4.6. *The effect of sample size on population estimates obtained using quadrat sampling*

Q4.6.1.

Sample	Number of plants in quadrat	Mean number of plants per quadrat	Population estimate
A	3	3.00	3 × 25 = 75
B	7	(3 + 7)/2 = 5.00	5 × 25 = 125
C	5	(3 + 7 + 5)/3 = 5.00	5 × 25 = 125
D	3	4.50	113
E	7	5.00	125
F	2	4.50	113
G	4	4.43	111
H	3	4.25	106
I	6	4.44	111

Q4.6.2.

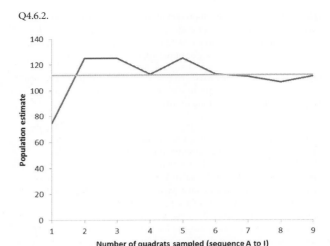

Q4.6.3. See graph above.

Q4.6.4.

Sample	Number of plants in quadrat	Mean number of plants per quadrat	Population estimate
I	6	3.00	75
H	3	4.50	113
G	4	4.33	108
F	2	3.75	94
E	7	4.40	110
D	3	4.17	104
C	5	4.29	107
B	7	4.63	116
A	3	4.44	111

Q4.6.5.

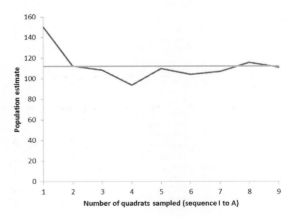

Q4.6.6. The last point on each graph represents the estimate of the population based on nine samples. This final estimate is not affected by the first eight estimates and is based on all nine samples regardless of the sequence in which they are collected.

Q4.6.7. The accuracy of any population estimate made from samples depends upon the number of samples taken and the extent to which these samples are representative of the population as a whole. Up to a point, the more samples used to calculate the estimate the more accurate it is likely to be. If only one sample is taken the estimate will be high if the density of organisms in the single quadrat is higher than average for the population as a whole. Conversely, if the first sample has a lower than average density of organisms, the estimate will be low. As the sample size increases the population estimate will approach the actual population size. By producing graphs like those shown here it is possible to determine how many samples should be taken in order to obtain an accurate estimate of population size.

4.7. Estimating the size of a population of mobile animals

Q4.7.1. The population estimate is $(21 \times 15)/9 = 35$.

Q4.7.2. If 3, 11 and 14 had lost their marks but still been recaptured the total recaptures now appears to be only 6 so the population estimate is:

$$(21 \times 15) / 6 = 52.5$$

Q4.7.3. If animals 1, 4 and 13 had died before the second trapping occasion they could not be in the total caught on the second trapping occasion so instead of 15 animals n_2 is now 12. If animals 3, 11 and 14 had lost their marks there would be just three marked animals left (7, 8 and 9) so r would be 3. The population estimate would now be:

$$(21 \times 12) / 3 = 84$$

Q4.7.4. If five unmarked animals had migrated into the population between the two trapping occasions and all of them had been caught on the second trapping occasion in addition to the 15 animals shown (and assuming none of these animal had lost its mark), n_1 and r would remain the same but n_2 would increase from 15 to 30. The population estimate would be:

$$(21 \times 20)/9 = 46.7$$

4.8. Estimating population size by removal trapping

Q4.8.1.

Sample number (day) (A)	Number captured in this sample (B)	Cumulative total previously captured (C)
1	38	0
2	40	38
3	33	78
4	31	111
5	27	142
6	30	169
7	24	199
8	22	223

Q4.8.2.

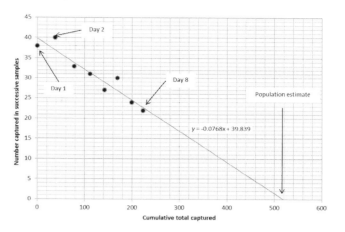

Q4.8.3.

The population estimate may be calculated using the equation of the line =

$$y = -0.0768x + 39.839$$
$$x = (y - 39.839) / -0.0768$$

When $y = 0$, $x = 518.7$ or approximately 519 animals.

Q4.8.4. The trapping effort each day needs to be kept constant because this affects the probability that an individual will be caught. For example, if the population size is 100 we might catch 10 individuals with trapping effort k. If trapping effort is doubled ($2k$) we might expect to catch twice as many animals (i.e., 20). The method requires that trapping effort is constant because it is based on the principle that as more and more animals are removed from the population it gets harder to catch the remaining individuals because the population is decreasing.

4.9. Studying animal populations using a calendar of catches

Q4.9.1.

Individual	Week									
	1	2	3	4	5	6	7	8	9	10
K ♂	●		●			●				●
L ♂						●	●	●		
M ♂	●			●			●			
R ♀	●	●	●				●	●	●	
S ♀	●	●			●		●			
T ♀	●	●	●			●	●		●	●
U ♀		●		●	●	●		●		●
New (+)	5	1	0	0	0	1	0	0	0	0
Disappeared (−)	0	0	0	0	0	0	1	1	0	0
Males	2	2	2	2	2	3	3	2	1	1
Females	3	4	4	4	4	4	4	4	4	4
Population size	5	6	6	6	6	7	7	6	5	5

Q4.9.2.

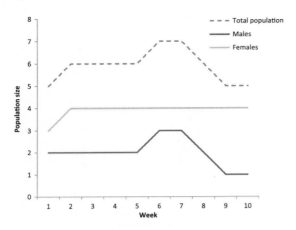

Q4.9.3. New animals may appear in the population due to births or immigration into the population from elsewhere.

Q4.9.4. Animals may disappear from the population due to deaths or emigration out of the population.

Q4.9.5. The pattern of capture of males may be different from that of females because the males have large home ranges and are moving in and out of the population whereas the females are resident and have small home ranges.

4.10. *Population dynamics of a zoo population of chimpanzees*

Q4.10.1.

Q4.10.2. Highest number of chimpanzees (7) was present between 2001 and 2009.

Q4.10.3. In 2010 there were two males in the population.

Q4.10.4. In 2012 there were three females in the population.

Q4.10.5. The table could be made more accurate by having a column for each month of the year. Although this would yield more accurate information, it would produce an unmanageable table; the current table would have 19 × 12 = 228 columns.

4.11. *Estimating population size indirectly: badgers and setts*

Q4.11.1. There are (1) 110 setts and (2) 365 badgers.

Q4.11.2. The ratio of badgers to setts in this area is calculated as follows:

The mean number of badgers/sett = 365/110 = 3.32, so there are 3.32 badgers to a sett.

Q4.11.3. If 76 setts were recorded in an area with similar ecological characteristics, the approximate size of the badger population would be 76 × 3.32 = 252.

Q4.11.4. Apart from direct observation of badgers entering and leaving a sett you could establish that it is in current use by (1) placing camera traps near the entrances; (2) spoil heaps or bedding outside entrances; (3) badger footprints and paths; (4) latrines; (5) badger hairs on fences or bushes; (6) scratching posts; (7) signs of digging for food. It should be borne in mind that some of these signs could persist for several weeks after the last use of a sett.

Name	Year																		
	96	97	98	99	00	01	02	03	04	05	06	07	08	09	10	11	12	13	14
Mitch ♂				A															
Jerry ♂	A													L					
Simon♂						B													
Mandy♀	A							D											
Clara♀	A																		
Jenny♀		A														D			
Eve♀									A										
Tessa♀			A																
Males arrived	1	0	0	1	0	0	0	0	0	0	0	0	0	0	0	0	0	0	0
Males left	0	0	0	0	0	0	0	0	0	0	0	0	0	1	0	0	0	0	0
Males born	0	0	0	0	0	1	0	0	0	0	0	0	0	0	0	0	0	0	0
Males died	0	0	0	0	0	0	0	0	0	0	0	0	0	0	0	0	0	0	0
Females arrived	2	1	1	0	0	0	0	0	1	0	0	0	0	0	0	0	0	0	0
Females left	0	0	0	0	0	0	0	0	0	0	0	0	0	0	0	0	0	0	0
Females born	0	0	0	0	0	0	0	0	0	0	0	0	0	0	0	0	0	0	0
Females died	0	0	0	0	0	0	0	1	0	0	0	0	0	0	1	0	0	0	0
Total males	1	1	1	2	2	3	3	3	3	3	3	3	3	3	2	2	2	2	2
Total females	2	3	4	4	4	4	4	4	4	4	4	4	4	4	4	4	3	3	3
Total	3	4	5	6	6	7	7	7	7	7	7	7	7	7	6	6	5	5	5

Q4.11.5. Badgers are generally absent from areas liable to flooding as water would fill their setts during storms.

Q4.11.6. Delayed implantation allows the mother to avoid unfavourable metabolic or environmental conditions so that she may give birth at a time that favours the survival of her young.

Q4.11.7. Badgers are largely nocturnal and move around at night. They may occupy large home ranges through which they travel searching for food and mates. They use traditional paths during their movements many of which may now be crossed by roads and country roads are generally unlit.

4.12. *Analysis of spatial distributions: clumped, uniform or random?*

Q4.12.1.

a. Mean
   $A = 16.40$
   $B = 9.70$
   $C = 9.10$.
b. Variance
   $A = 1.16$
   $B = 76.01$
   $C = 8.99$

Q4.12.2. Index of dispersion

$A = 0.07$ (Uniform or regular)
$B = 7.84$ (Clumped or contiguous)
$C = 0.99$ (Random)

Q4.12.3. Quadrat size may have an effect on the detection of distribution patterns. If a relatively large quadrat is used to sample a clumped distribution it may not be possible for it to fall in areas where there are few or no organisms. This would keep the variance of the samples taken relatively low and consequently produce a relatively low index of dispersion. For clumps to be detected the variance needs to be greater than the mean. This is more likely to occur when the quadrat is small enough to fall between the clumps, producing an index of dispersion greater than 1.

4.13. *Estimating the size of a large mammal population by transect sampling*

Q4.13.1.

No. of transects	Animals/ transect	Mean animals/ transect	Population estimate
1	5	$(A) = 5.00$	76.2
2	8	$(A + B)/2 = 6.50$	99.0
3	4	$(A + B + C)/3 = 5.67$	86.4

Q4.13.2. Actual population = 83. Estimate 1 is slightly low; estimate 2 is high (because transect B contained 8 animals); estimate 3 was reduced by the addition of only 4 animals in transect C. The estimate based on the mean density calculated from three transects is the most accurate (86.4). However, it is interesting to note that the estimate based on just one transect (A) was closer to the actual population size than the estimate calculated from two transects.

The average density per transect area ($105,000 \, \text{m}^2$) based on a population of 83 animals was approximately 5.4 individuals. In other words, for each $105,000 \, \text{m}^2$ of the total area there were 5.4 animals. Transect A had fewer than this so produced a low estimate; transects $A + B$ produced a mean of 6.5 animals and produced a high estimate.

Q4.13.3. Individual animals may be overlooked because they are (1) laying down in long grass or (2) hidden by trees or other animals.

Q4.13.4. Some of the advantages of using a vehicle when conducting transect counts are (1) the ability to cover distances faster than on foot; (2) the ability to see further from the transect centre line allowing a wider transect to be examined; (3) the potential to see animals that would be hidden from an observer located at ground level. Some of the disadvantages are that (1) if vehicles are confined to roads – as they are in many protected areas – transects cannot be located at random (some species may be attracted to, or avoid, roads); (2) vehicle noise may drive animals away from the road before the observers see the animals (and other vehicles may have already disturbed the animals); (3) the presence of a vehicle may change the behaviour of animals (e.g., elephants may group together to protect young).

4.14. *Sampling zooplankton populations*

Q4.14.1. The concentration of copepods is $750/5 = 150 \, \text{m}^{-3}$.

Q4.14.2. A mesh size of $70 \, \mu\text{m}$ (the smallest size) gives the most accurate information about the population densities of copepods and chaetognaths. All of the larger mesh sizes allow these small organisms to pass through and therefore give an underestimate of the numbers present.

Q4.14.3. The density of copepods at largest mesh size ($1000 \, \mu\text{m}$) is $7 \, \text{m}^{-2}$. At the smallest mesh size of $7 \, \mu\text{m}$ the density is $2500 \, \text{m}^{-2}$. The density of copepods obtained with the largest mesh size as a percentage of that obtained with the smallest mesh size is calculated as $(7/2500) \times 100 = 0.28\%$.

Q4.14.4. Mesh size does not appear to affect greatly the estimates of siphonophores because they are relatively large and therefore captured by all of the mesh sizes used in the study.

Q4.14.5.

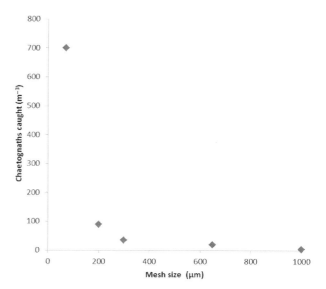

Q4.14.6. This causes a problem because their orientation when they contact the net will determine whether or not they are captured. This will affect the calculation of densities.

Q4.14.7. If a plankton net with a very fine mesh is used when zooplankton are very abundant it will quickly become clogged with organisms so that water will no longer pass through.

Q4.14.8. If the sampling net is clogged with zooplankton water will not flow through the net and this will reduce the number of organisms caught, resulting in an underestimate of the density. If the net is moved too quickly through the water it may eventually split.

4.15. *Estimating the size of whale populations*

Q4.15.1. The presence of a whale is detected by (1) a direct observation of the body at the surface of the sea (usually just a glimpse of the dorsal surface as it comes up for air but sometimes the whole body if it breaches); (2) a 'blow' or 'spout' as the whale exhales at the surface. The form and size of the spout may assist in the identification of the species.

Q4.15.2. If the location of the ship is known, the distance and angle (bearing) to the ship allows the calculation of the position of the whale using trigonometry.

Q4.15.3. Calculation of UTM position of the whale.

The easting position of the whale $=$ Easting position of boat $+ x_1$

where,

$$x_1 = d \times \sin \alpha$$

The northing position of the whale $=$ Northing position of boat $+ y_1$

where,

$$y_1 = d \times \sin \theta.$$

The distance x_1 is calculated as follows:

$$\sin \alpha = \frac{\text{opposite}}{\text{hypotenuse}}$$

so,

$$\sin 40° = \frac{x_1}{d}$$

Rearranged, this becomes:

$$x_1 = d \times \sin 40°$$

So,

$$x_1 = 500 \times 0.64278761$$
$$= 321\,\text{m}$$

Similarly,

$$y_1 = d \times \sin 50°$$

So,

$$y_1 = 500 \times 0.76604444$$
$$= 383\,\text{m}$$

Location of	Zone	Grid position	
		Easting (m)	Northing (m)
Boat	26V	507,693	7,168,942
Correction	26V	+321	+383
Whale	26V	508,014	7,169,325

Q4.15.4. In rough weather the surface of the sea is disturbed by the wind producing large waves. Surfacing whales are relatively easy to sea when the sea is flat because their bodies stand out above the flat surface. In high winds they disappear between the waves. In rough seas observers find it difficult to stand still and use binoculars. Icy conditions also make it difficult to count whales.

Q4.15.5. The 95% confidence interval is the range of values between which we are 95% confident that the true value lies. So, if the 95% confidence limit is 200–250 we are 95% certain that the true value is greater than or equal to 200 but less than or equal to 250.

Q4.15.6. There is no clear evidence that the grey whale population of the eastern North Pacific Ocean has changed in size between 1997/98 and 2006/07 because an actual population of 18,000 would fall within the 95% confidence limits for all of the surveys undertaken.

4.16. *Estimating population size using the Lincoln Index: a simulation*

Q4.16.1. When the proportion of marked animals in the population is very low it may be difficult to recapture marked animals on the second trapping occasion. This will lead to overestimation of the population size if r is much smaller than it should be. The estimates should become more accurate as the proportion of marked individuals is increased. However, the opposite effect may occur if it becomes difficult to 'catch' unmarked animals in the second sample because there are too many marks in the population.

Q4.16.2. This should be determined from your experiment and should be based on the minimum number you need to mark to obtain an accurate estimate.

Q4.16.3. Births between trapping occasions will reduce the chance that a marked animal is recaptured. This will reduce r and increase the population estimate. The estimate will reflect the new population size, not the size when the animals were marked.

Q4.16.4. If marks are lost the value of r is likely to be less than it would have been if the marks had remained in the population. If r is reduced the population estimate goes up.

4.17. *Population size and habitat selection in two species of gulls*

Q4.17.1. The completed table has been intentionally omitted as the reader is merely required to transcribe data from Fig. 4.23.

Q4.17.2. Herring gulls:

a. mean density in the dune slacks $= 10.45$, SD $= 8.73$.
b. mean density on the dunes $= 6.45$, SD $= 4.96$.

Lesser black-backed gulls:

a. mean density in the dune slacks $= 5.73$, SD $= 6.43$.
b. mean density on the dunes $= 2.55$, SD $= 3.34$.

Q4.17.3. Mean density of herring gulls   = 8.45 (n = 22).

Median density of herring gulls   = 6.5 (n = 22).
Mean density of lesser black-backed gulls   = 4.14 (n = 22).
Median density of lesser black-backed gulls   = 2.5 (n = 22).

Q4.17.4. Population estimates:

Herring gulls	(mean)	= 23,159.
	(median)	= 17,815.
Lesser black-backed gulls	(mean)	= 11,346.
	(median)	= 6852.
Total (both species)	(mean)	= 34,505.
	(median)	= 24,667.

Q4.17.5. The highest densities were recorded among the herring gulls in dune slacks, i.e., 10.45 per 20 m².

Q4.17.6. The lowest densities were recorded among lesser black-backed gulls on the sand dunes, i.e., 2.55 per 20 m².

Q4.17.7. The greatest variation in density was recorded among herring gulls in dune slacks, i.e., SD = 8.73.

Q4.17.8. Using the median density removes the effects of extreme values (very high or very low values) recorded in single quadrats. If the mean density is used, a small number of very high values could increase the population estimate; a small number of very low values would have the opposite effect.

Q4.17.9. Probably quadrat 13 because it contains an unusually high number of both species and could result in an overestimate of both species.

Q4.17.10.

a. a = 115; b = 71; c = 63; d = 28.
b. The null hypothesis is that there is no association between the two gull species and the two habitats.
c. $\chi^2 = 1.458$ (df = 1). This is not statistically significant: P > 0.05. To be significant at the 5% level the calculated value must exceed 3.841 (df = 1). The null hypothesis is accepted. This result suggests that the two gull species are distributed at random between dune and slack habitats.

4.18. *The factors affecting plant population estimates obtained by quadrat sampling: a computer simulation.*

Q4.18.1. There are very few population estimates calculated when the quadrat had sides of 5 units because no plants were counted at all until quadrat 14. This was because all other quadrats landed in spaces between the plants because the quadrat was so small. All of the estimates made from quadrat 15 onwards were based on a mean density calculated from one plant per quadrat. As the number of quadrats increased the mean density did not and the population estimate was calculated by using a progressively larger number of quadrats so each estimate was smaller than the last.

Q4.18.2. The number of quadrats sampled had little effect on the population estimate when the quadrat had sides of 50 units because the quadrat size was large relative to the area being sampled. Each sample included one ninth of the total area so even a single sample 'captured' a group of plants whose distribution and density was representative of the population as a whole.

Q4.18.3.

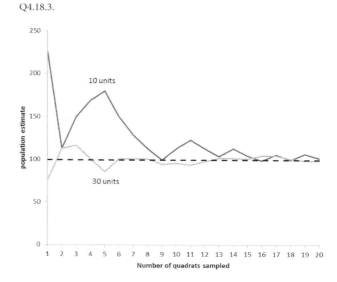

Q4.18.4. The minimum number of quadrats necessary in order to obtain an accurate estimate of the population size (i.e., within plus or minus 1% of the actual size) when the quadrat has sides of (1) 10 units was 9 (N = 100); (2) 30 units was 4 (N = 101).

Q4.18.5. As the number of samples is increased, the population estimate becomes more accurate. The accuracy improves more quickly (i.e., takes fewer samples) as quadrat size increases. This rule does not apply when the quadrat has sides of 50 units. This quadrat is so large in relation to the total area being sampled that, by chance, the first sample produces an estimate of 99. Subsequent estimates rise and fall between 89 and 99.

Chapter 5 Population Growth

5.1. *Exponential population growth*

Q5.1.1.

Time t (minutes)	Population size n (number of bacteria)	Log₁₀ number of bacteria
0	1	0
20	2	0.301
40	4	0.602
60	8	0.903
80	16	1.204
100	32	1.505
120	64	1.806
140	128	2.107

Q5.1.2.

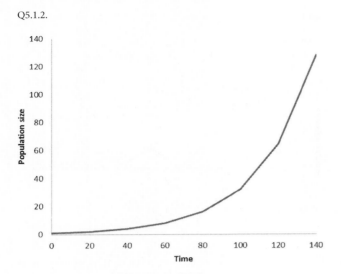

Q5.1.3.

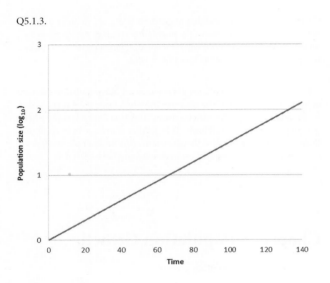

Q5.1.4. The curve becomes a straight line.

Q5.1.5. This model assumes that no deaths occur so the death rate is zero. Each bacterium divides into two and both of the new bacteria survive and in turn divide into two themselves.

Q5.1.6. If the environment became limiting we would expect the growth rate to reduce and possibly become zero and even negative if individuals began to die off.

Q5.1.7. Real environments are not unlimited so growth is limited by a lack of infinite resources.

5.2. *Boom and bust population growth*

Q5.2.1.

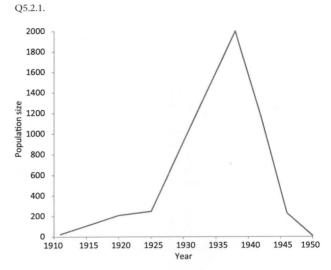

Q5.2.2. The annual rate of increase in reindeer between 1926 and 1936 is:

$$(1720 - 360)/10 = 136 \text{ animals/year}.$$

Q5.2.3. As the reindeer were introduced into the area and there were no natural predators the population was able to growth without restriction.

Q5.2.4. After 1938 the reindeer had overgrazed the area so there was insufficient food to support the large population and it went into a rapid decline as a result.

Q5.2.5. Most animal populations have evolved alongside other competing species and natural predators. As a consequence they do not have the opportunity to expand their populations in an uncontrolled manner.

5.3. *Logistic population growth*

Q5.3.1.

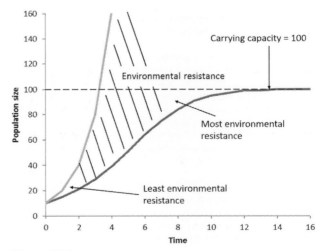

Figure 12.8

Q5.3.2. The growth rate (individuals/unit time) between time = 4 and time = 7 is

$$(75 - 39) / (7 - 4) = 36 / 3 = 12 \text{ individuals / unit time.}$$

Q5.3.3. Zero. The population has stopped growing and has stabilised.

Q5.3.4. 100.

Q5.3.5. See Fig. 12.8

Q5.3.6. See Fig. 12.8

Q5.3.7. See Fig. 12.8

5.4. *Density-dependent and density-independent factors and population control*

Q5.4.1.

Population density (m^{-2})	Species A		Species B	
	Population size	Mortality %	Population size	Mortality %
10	1	10.00	1	10
50	6	12.00	5	10
100	13	13.00	17	17.00
200	25	12.50	45	22.50
400	46	11.50	107	26.75
800	83	10.38	263	32.88

Q5.4.2.

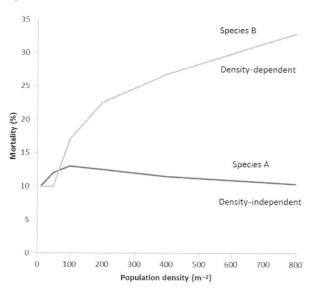

Q5.4.3.

a. Species B is exhibiting density-dependent mortality.
b. Species A is exhibiting density independent mortality.

Q5.4.4. Environmental factors that affect mortality in a density-dependent (DD) fashion are more likely to act as mechanisms of population regulation than are those that act in a density-independent (DI) fashion because as the population increases a DD factor causes the death of a higher proportion of the population (compared with when the population was lower in number). DI factors kill off more or less the same proportion of the population regardless of its density so have no greater effect on high populations than on low populations.

Q5.4.5.

Environmental variable	DD or DI?
Frost	DI
Predation	DD
Intraspecific competition	DD
Fire	DI
Parasitism	DD
Disease	DD
Road traffic	DI

5.5. *Using Leslie matrices to model population growth*

Q5.5.1.

Generation		f_0	f_1	f_2		Vector			N
1		0	6	10		10.0	0.0		
	p_0	0.27				0.0	2.7		
	p_1		0.53			0.0	0.0		
									2.7
2		0	6	10		0.0	16.2		
		0.27	0	0		2.7	0.0		
		0	0.53	0		0.0	1.4		
									17.6
3		0	6	10		16.2	14.3		
		0.27	0	0		0.0	4.4		
		0	0.53	0		1.4	0.0		
									18.7
4		0	6	10		14.3	26.2		
		0.27	0	0		4.4	3.9		
		0	0.53	0		0.0	2.3		
									32.4
5		0	6	10		26.2	46.4		
		0.27	0	0		3.9	7.1		
		0	0.53	0		2.3	2.0		
									55.5
6		0	6	10		46.4	63.0		
		0.27	0	0		7.1	12.5		
		0	0.53	0		2.0	3.8		
									79.3
7		0	6	10		63.0	112.7		
		0.27	0	0		12.5	17.0		
		0	0.53	0		3.8	6.6		
									136.3
8		0	6	10		112.7	168.4		
		0.27	0	0		17.0	30.4		
		0	0.53	0		6.6	9.0		
									207.8

Generation	Age class			Population size
	1	2	3	
0	10	0	0	10
1	0.0	2.7	0.0	2.7
2	16.2	0.0	1.4	17.6
3	14.3	4.4	0.0	18.7
4	26.2	3.9	2.3	32.4
5	46.4	7.1	2.0	55.5
6	63.0	12.5	3.8	79.3
7	112.7	17.0	6.6	136.3
8	168.4	30.4	9.0	207.8

Q5.5.2.

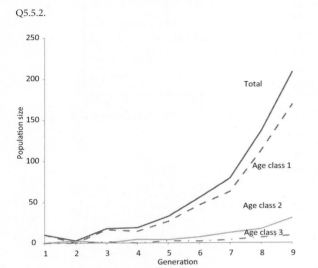

Q5.5.3. Age classes 2 and 3 remained low but age class 1 increased approximately exponentially thus causing the population as a whole to increase in a similar manner. This was the consequence of the high reproduction rate of age classes 2 and 3.

5.6. *Life table for the honey bee*

Q5.6.1.

Age (days) (x)	Survivors at start of age class x (l_x)	Deaths between age x and x + 1 (d_x)	Age specific death rate (q_x)
0–5	1000	1000 − 983 = 17	17/1000 = 0.017
5–10	983	983 − 960 = 23	23/983 = 0.023
10–15	960	960 − 943 = 17	17/960 = 0.018
15–20	943	33	0.035
20–25	910	183	0.201
25–30	727	272	0.374
30–35	455	316	0.695
35–40	139	93	0.669
40–45	46	40	0.870
45–50	6	6	1.00

Q5.6.2.

a. The lowest death rate is in age class 0–5 days.
b. The highest death rate is in age class 45–50 days.

Q5.6.3.

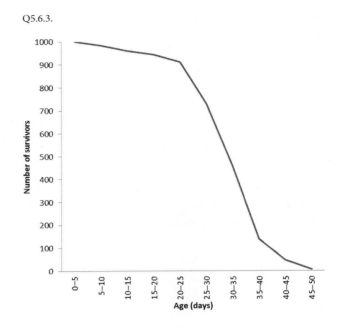

5.7. *Survivorship of Dall sheep*

Q5.7.1.

Age (years) x	Survivors at beginning of age class l_x	Deaths d_x	Mortality rate % q_x
0–0.5	1000	1000 − 946 = 54	54/1000 × 100 = 5.4%
0.5–1	946	145	15.33
1–2	801	12	1.50
2–3	789	13	1.65
3–4	776	12	1.55
4–5	764	30	3.93
5–6	734	46	6.27
6–7	688	48	6.98
7–8	640	69	10.78
8–9	571	132	23.12
9–10	439	187	42.60
10–11	252	156	61.90
11–12	96	90	93.75
12–13	6	3	50.00
13–14	3	3	100.00

Q5.7.2.

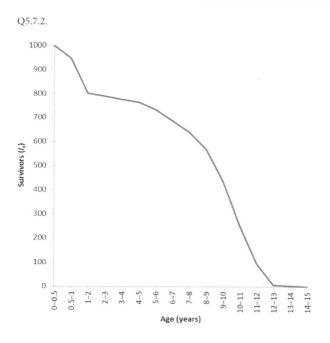

Q5.7.3.

a. The 13–14 year age class has 100% mortality; the 11–12 year age class has 93.75% mortality.
b. The 1–2 year age class has the lowest mortality at 1.5%.

Q5.7.4. This is a static life table because it was constructed from estimates of the age of dead sheep within a population rather than by following a cohort born at the same time and recording the age at death. These animals were not part of the same cohort.

5.8. *Types of survivorship curve*

Q5.8.1.

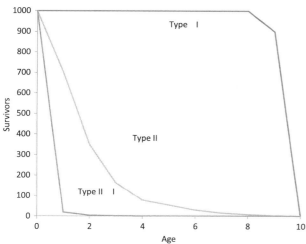

Q5.8.2. Type I shows 100% survival until age 8 then a very rapid decline. Type II shows a steady decline until age 7 then a small number survive with just one individual remaining at age 10. Type III exhibits a very rapid decline between zero and 1-year-old, with just a very small number surviving thereafter.

Q5.8.3.

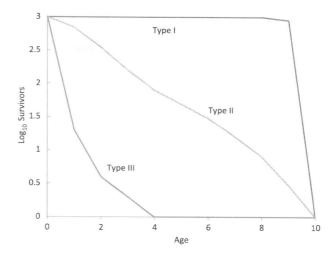

Q5.8.4. The Type I curve appears essentially the same shape. The most dramatic effect is on the Type II curve which becomes a more or less straight line. The Type III becomes less steep.

Q5.8.5.

a. Humans exhibit Type I survivorship.
b. Parasites exhibit Type III survivorship.
c. Songbirds exhibit Type II survivorship.

Q5.8.6. Type I, because almost all individuals die at the same age.

5.9. *The age structure of captive female African elephants*

Q5.9.1.

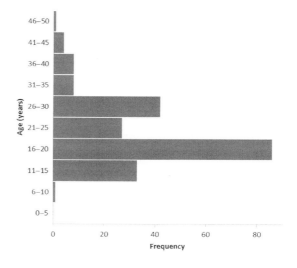

Q5.9.2. Ages range from 8 to 48 years, i.e., a range of 40 years.

Q5.9.3. Only three animals were less than 12 years old, so 207/210 × 100 = 98.6% of the elephants were theoretically capable of reproduction. In reality the actual percentage would have been much lower than this due to reproductive problems, behavioural problems, etc.

Q5.9.4. Birth rates have been historically low in populations of elephants living in zoos. Breeding has been difficult because it has often required animals to be moved between zoos as individual zoos have kept small numbers. Bulls have been relatively few in number as zoos have historically preferred to keep cows as they are considered less dangerous. Where births have occurred neonatal mortality rates have been high.

Q5.9.5. The population was in decline and not self-sustaining in the long run without the addition of elephants from elsewhere. It should be noted that births have increased in captive elephant populations in recent years because of the development of artificial insemination and increased calf survival rates.

5.10. *Opportunity or equilibrium: r-strategist or K-strategist?*

Q5.10.1.

a. The closer N gets to K the slower the growth rate.

b. Zero.

Q5.10.2.

Organism	Strategy (*r* or *K*)
African elephant	*K*
Parasitic insects	*r*
Dinosaurs	*K*
Grasses	*r*
Andean condor	*K*
Albatross	*K*
Bacteria	*r*

Q5.10.3.

Characteristic	*r*- or *K*-strategist
Long generation time	*K*
Inhabits stable environments	*K*
Exhibits high levels of migration	*r*
Low mortality rate	*K*
Colonises temporary habitats	*r*
Vulnerable to extinction	*K*

Q5.10.4. This organism is growing in number exponentially at a fast rate so is an *r*-strategist, e.g., a bacterium.

Chapter 6 Species Interactions

6.1. *Competition in two tree species*

Q6.1.1.

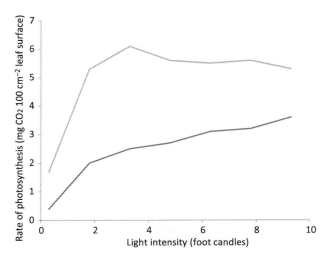

Q6.1.2. Eastern red oak reached its maximum photosynthetic rate at 3.3×10^3 foot candles. This is $3.3/9.3 \times 100 = 35.5\%$ of full sunlight. Loblolly pine reached its maximum photosynthetic rate at 9.3×10^3 foot candles. This is 100% of full sunlight.

Q6.1.3. The environment at the base of a forest is dark due to the density of the leaves. At all light intensities the oak seedlings photosynthesised at a higher rate than did the pine seedlings. Light conditions in the pine forest are too low for effective photosynthesis in pine seedlings, restricting root growth. During drought conditions this would be fatal to pine seedlings.

Q6.1.4. It would make the oak seedlings better able to absorb nutrients and water in soil filled with competing tree roots.

6.2. *Intraspecific competition in barley*

Q6.2.1.

Number of seeds sown per pot	Number of plants that germinated	Germination rate (%)	Total wet biomass of plants (g)	Mean wet biomass of plants (g)	Mean stem length (cm)
50	43	86	17.0	0.395	25.6
100	88	88	21.4	0.243	19.2
200	168	84	27.7	0.165	14.3
400	332	83	40.9	0.123	13.6

Q6.2.2.

(A)

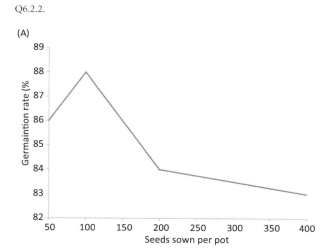

(B)

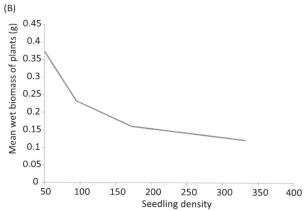

(C)

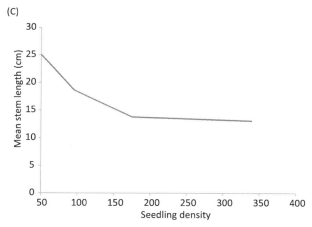

Q6.2.3. Germination rate fell with seed density. Mean wet biomass and mean stem length both decreased with seeding density. The most important factor here was probably intraspecific competition. In the seedlings this was probably for both light and nutrients.

Q6.2.4. Farmers need to be aware of the optimum seed density for their crops if they are to avoid wasting money sowing seeds that either will not germinate or will produce seedlings at densities at which they will not grow well due to intraspecific competition.

6.3. Competitive exclusion

Q6.3.1.

(A)

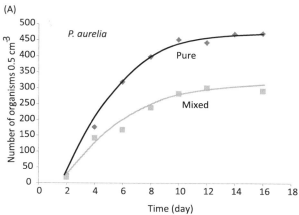

(B)

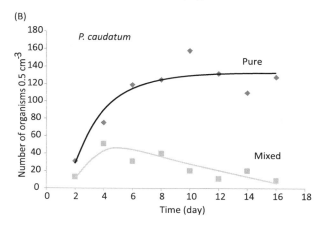

Q6.3.2. *Paramecium aurelia* grew at a faster rate and reached an asymptote at a higher population density than *P. caudatum* when each was grown in a pure culture.

Q6.3.3. In a mixed culture the growth rate of both species was reduced. *P. aurelia* levelled off at a much lower asymptote while *P. caudatum* survived at only a very low density after 16 days and eventually became extinct.

Q6.3.4. These differences were the result of interspecific competition for food (bacteria). *P. aurelia* grew more vigorously and out-competed *P. caudatum*.

Q6.3.5. The competitive exclusion principle states that complete competitors cannot coexist. When one has a slight advantage it will survive while the other must either become extinct or evolve to occupy a different niche.

6.4. *Competition in flour beetles*

Q6.4.1. The two climates in which the results of competition were completely consistent, i.e., the same species always won, were hot/moist and cold/dry.

Q6.4.2. The outcome of competition could not be predicted from the results obtained from growing single species cultures in the hot/moist and cold/moist conditions. In hot/moist conditions both species are equal in pure cultures but *casteneum* wins 100% in mixed cultures. In cold/moist conditions, *confusum* < *casteneum* in pure colonies, but *confusum* wins 71% in mixed cultures.

Q6.4.3. In temperate/dry conditions *confusum* wins 87% of the time. However, if we were to conduct an experiment with just a single culture we could not predict the outcome with certainty because the information we already have provides us with information about the probability of the outcome.

6.5. *Do cormorants and shags compete for food?*

Q6.5.1.

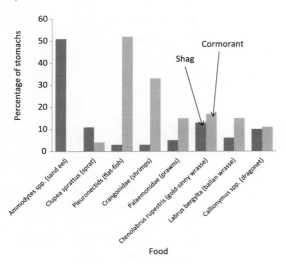

Q6.5.2. Yes. It is clear from the data that both birds eat the same food items.

Q6.5.3. Although both birds eat the same food types they differ in the extent to which they concentrate on particular foods. Shags ate a lot of sand eels and sprats but these were largely ignored by cormorants. Cormorants ate a lot of flat fish, shrimps and prawns; foods which were not very significant in the diets of shags.

Q6.5.4. The two species feed at different depths in the sea.

6.6. *Competition between ants and rodents*

Q6.6.1. The rodent population increased due to increased food availability.

Q6.6.2. The number of ant colonies increased due to increased availability.

Q6.6.3.

a. The same number of seeds was taken when either the ants were removed or the rodents were removed because the remaining guild increased in number in the absence of the other and consequently consumed the available food resource.

b. Once the ants and the rodents had been removed seeds continued to be produced by the plants but were not being removed by these animals.

Q6.6.4. Ants tend to take smaller seeds and rodents tend to take larger seeds. There must, however, be considerable overlap in the sizes of seeds taken by the two groups for competition to occur.

6.7. *Defining the niche*

Q6.7.1.

a.

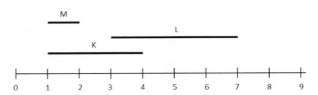

b. M overlaps with K, and K overlaps with L.

Q6.7.2.

a.

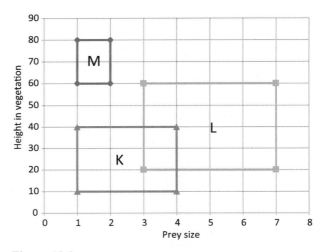

Figure 12.9

b. L and K overlap.

Q6.7.3.

a. A third axis (z) should be added to Fig. 12.9 creating cubes from each area
b. Adding a third axis shows that none of the species overlap.

Q6.7.4. For overlap to occur between any two species there must be an overlap in all three dimensions. As additional dimensions are added the apparent overlaps disappear. With one axis there appear to be two overlapping sections. With two axes there is a single overlapping area. When a third axis is added these overlapping areas are pulled apart.

Q6.7.5. Diagram A represents a tropical rainforest with a large number of narrow niches. Diagram B represents tundra with a small number of broad niches.

6.8. *Niche separation in warblers*

Q6.8.1.

| Height zone | Warbler species | | | | | | | |
| | Alpine | | Blue | | Carmine | | Diamond | |
	Number of recordings	Percentage of recordings	Number of recordings	Percentage of recordings	Number of recordings	Percentage of recordings	Number of recordings	Percentage of recordings
1	7	50.0	0	0	0	0.0	9	37.5
2	6	42.9	0	0	2	10.5	11	45.8
3	0	0.0	0	0	14	73.7	1	4.2
4	1	7.1	7	33.3	2	10.5	2	8.3
5	0	0.0	14	66.7	1	5.3	1	4.2
Total	14	100	21	100	19	100	24	100

Q6.8.2.

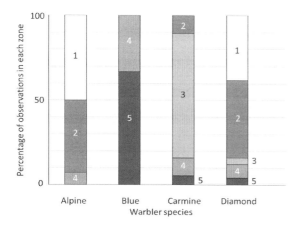

Q6.8.3. The four species utilise different height zones in the forest. Although there is some overlap between species they coexist by each having one zone they use more than any other.

Q6.8.4. It may be that species avoid each other by using the same places but at different times of the day as well as by using different height zones at the same time. This could only be determined if the times that recordings were made were noted.

6.9. *Niche separation in tropical monkeys*

Q6.9.1.

| Forest layer | Species | | | | | | | |
	A	B	C	D	E	F	G	H
Tall trees		X						X
Medium sized trees		X	X	X				X
Small trees			X		X			X
Shrubs	X		X			X	X	
Low herbage			X				X	

Q6.9.2. Medium-sized trees and shrubs; both are used by four species.

Q6.9.3. It reduces competition for food, sheltering places, etc.

Q6.9.4. The great variety of flowering plants provides a wide range of leaves, fruits and seeds for the monkeys, thereby reducing competition for food.

Q6.9.5. They communicate by vocalisation.

6.10. *Predator–prey simulation*

Q6.10.1.

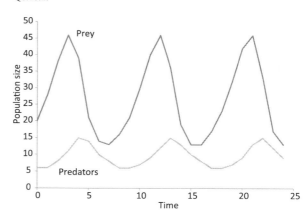

Q6.10.2. Neither the prey nor the predator population dies out; they coexist.

Q6.10.3. The peaks in:

a. the prey population occur 9 time units apart.
b. the predator population also occur 9 time units apart.

Q6.10.4. The increase in the predator population is a response to the prior increase in the prey numbers. When prey numbers are high, predator reproduction and survival is likely to be increased due to the abundance of food. Then prey numbers fall, predators are short of food and numbers decline

Q6.10.5. Most predators feed on a number of different prey species and most prey species are likely to have a number of different predators. When one prey species becomes rare, a predator is likely to switch to a commoner species. The interactions between predators and their prey are complex and situations where the relationship between the numbers of a predator and those of a single prey species is the sole determinant of their numbers are probably very rare in nature.

6.11. *Predation of foxes on rabbits*

Q6.11.1.

Soil	Number of warrens		Percentage of total attacked	Number of nests destroyed by foxes
	Total	Attacked by foxes		
Stony banks	829	0	0	0
Loam	871	66	7.58	78
Stony hills	806	11	1.36	12
Stony pediments	371	46	12.40	58
Desert loam	324	148	45.68	255
Sand dunes	3309	2339	70.69	2550

Q6.11.2.

Soil	% total warrens attacked
Stony banks	0.00
Stony hills	1.36
Loam	7.58
Stony pediments	12.40
Desert loam	45.68
Sand dunes	70.69

Q6.11.3. Approximately 9 times (70.69/7.58 = 9.33).

Q6.11.4. Some warrens contained more than one nest.

Q6.11.5. It appears from the data that stony banks and stony hills are the safest places for rabbits to construct their warrens. This may be because foxes find it more difficult to dig in these soils than in those with a lower stone content.

6.12. *The effects of disease, climate and vegetation change on large mammal populations*

Q6.12.1.

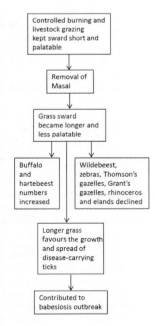

Figure 12.10

Q6.12.2. An increase in longer less palatable plants in the grass sward favoured some other grazers but caused a decline in rhinoceros. (See Fig. 12.10)

Q6.12.3. Lion numbers were held down by canine distemper virus (CDV), babesiosis, inbreeding, increased human population and a plague of biting flies.

Q6.12.4. Invasive weeds have reduced the available food for herbivores. Attempts have been made to control them by burning and mowing.

Q6.12.5. It suppressed the numbers because the species is highly susceptible.

Q6.12.6. They are major carriers of CDV.

Q6.12.7. Flooding of Lake Magadi has caused an increase in soil salinity; fungal infection and elephants have damaged trees.

6.13. *Biological control*

Q6.13.1.

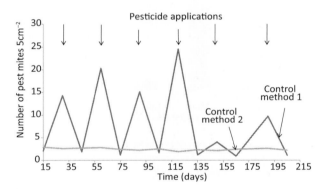

Figure 12.11

Q6.13.2. Control method 2 is biological control; the pest population is being held more or less constant at a relatively low level by the presence of predators. Control method 1 uses pesticides; these are applied when pest numbers become high causing a dramatic fall in pest numbers which later recover, requiring another application of pesticide.

Q6.13.3. See Fig. 12.11.

Q6.13.4. See Fig. 12.11.

Q6.13.5. The predatory mites' numbers would have been stable but at a lower population level than that of the prey.

Q6.13.6. Unlike many chemical pesticides, biological control methods do not cause pollution of the environment. If used properly they can maintain the target organisms at very low levels. Biological control methods do not suffer from the problem of resistance seen in pesticides. They can be economical and long-lasting in their effect.

Q6.13.7. Biological control methods depend upon the availability of a suitable predator, parasite or disease that can be used to control the target pest organism. In many cases such organisms are not available. Pesticides are easy to apply and readily available. Sometimes biological control agents have attacked nontarget native species. Great care needs to be used in determining the possible unintended effects on nontarget species and the general ecology of areas where biological control is used.

6.14. Seals and phocine distemper virus

Q6.14.1.

a. The null hypothesis is that there is no association between the PDV antibody and either of the seal species.
b. $\chi^2 = 0.459$ ($df = 1$).
c. The results are not statistically significant, so we must accept the null hypothesis. ($P > 0.10$).

Q6.14.2.

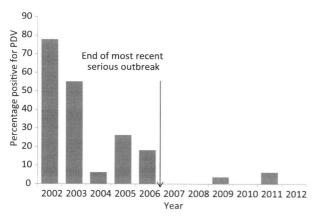

Figure 12.12

Q6.14.3. See Fig. 12.12.

Q6.14.4. The last outbreak of PDV occurred in 2002. At the end of this year 18 out of 35 juveniles tested positive for antibodies. 18/35 × 100 = 51.4%

Q6.14.5. Antibodies were detected in many seals in 2002 and 2003 but after 2003 antibodies were detected only in pups and adult seals that probably had survived the 2002 outbreak. As the majority of seals do not carry antibodies and are therefore not immune a future outbreak is likely.

Q6.14.6. The seals examined in this study were not a random sample so may not be representative of the populations as a whole.

Chapter 7 Behavioural Ecology and Ecological Genetics

7.1. Identifying individual animals

Q7.1.1. Size, shape, tears and holes, size of overturned area at top.

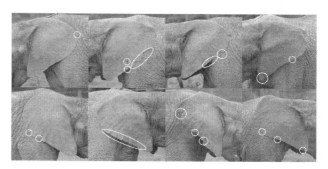

Q7.1.2. Ears change over time because as the elephant ages the amount of the top that is turned over will increase, holes and tears may change size and shape, new holes and tears may appear.

Q7.1.3. Recording both ears should allow the individual to be identified from either side. If an elephant was walking from left to right only the right ear would be visible unless it turned to face the other direction.

Q7.1.4. Tusk length, thickness, shape/tail length, shape (especially if kinked), number of hairs, length of hairs/ body shape, height, build.

Q7.1.5. Usually females Asian elephants do not have tusks but some may have a tush (a very short stump of ivory).

Q7.1.6. Each family begins with a different letter of the alphabet. All of the elephants with names beginning with the letter 'A' are part of one family; all of the elephants with names beginning with the letter 'T' are part of a different family.

Q7.1.7. The family whose names begin with 'A' has two members whose names could potentially begin with ALI. Moss used a three-letter identification code so by spelling Alyce with a 'y' she could code her name as ALY and use the code ALI for Alison.

Q7.1.8. Wart Ear was so named because she had a large wart on her ear. She was part of the A family but had been named before the A family was assigned names so her original name stuck.

7.2. Radio-tracking animals

Q7.2.1.

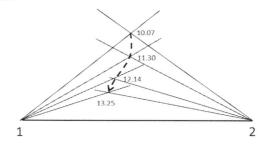

Figure 12.13

Q7.2.2. See Fig. 12.13.

Q7.2.3. South.

Q7.2.4. This method assumes that the animal does not move while the researcher moves from location 1 to location 2. Any movement would almost certainly invalidate the bearing measured from location 1.

Q7.2.5. Battery life. Although battery size has been reduced and battery life has been extended due to developments in battery technology, eventually a battery will fail and as a result radio transmissions will cease.

Q7.2.6. When there is a chance that the animal could be able to remove a collar from around its neck

Q7.2.7. In order to fit a radio collar a suitable animal has to be located, trapped, restrained and then released. This process may cause considerable stress in some species and the use of an anaesthetic dart fired from a gun may result in the death of a large mammal if does not receive the antidote after an appropriate amount of time. Care must be taken to ensure that the radio collar does not adversely affect the animal's survival by making it more likely to be detected by its potential predators or prey. Radio collars should not be used where they may interfere with locomotion or burrowing.

7.3. *Activity budgets in Asian elephants*

Q7.3.1.

(A)

Symbol	Behaviour	Day 1		Day 2		Day 3	
		No.	%	No.	%	No.	%
AGG	Aggression	0	0	0	0	0	0
DG	Digging	0	0	0	0	0	0
DT	Dusting	0	0	1	2.1	2	4.2
F	Feeding	7	14.9	3	6.3	6	12.5
L	Walking (locomotion)	2	4.3	3	6.3	3	6.3
S	Standing	9	19.1	10	20.8	7	14.6
SEX	Sexual behaviour	0	0.0	0	0.0	0	0.0
ST	Stereotyping	29	61.7	31	64.6	30	62.5
0	No recording made	1		0		0	
Total recordings made		47	–	48	–	48	

(B)

Symbol	Behaviour	Day 1		Day 2		Day 3	
		No.	%	No.	%	No.	%
AGG	Aggression	0	0	0	0	1	2.1
DG	Digging	0	0	0	0	1	2.1
DT	Dusting	0	0	1	2.1	2	4.2
F	Feeding	14	30.4	2	4.2	6	12.5
L	Walking (locomotion)	12	26.1	4	8.3	7	14.6
S	Standing	16	34.8	11	22.9	26	54.2
SEX	Sexual behaviour	0	0.0	0	0.0	0	0
ST	Stereotyping	4	8.7	30	62.5	5	10.4
0	No recording made	2		0		0	
Total recordings made		46	–	48	–	48	–

Q7.3.2.

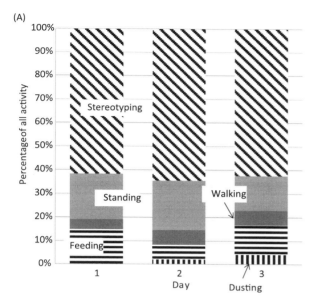

(A)

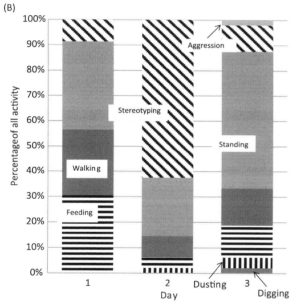

(B)

Q7.3.3. The behaviour of elephant A was more consistent between days than that of elephant B. A spent more than 60% of her time stereotyping each day, 6–15% feeding and 14–21% standing still. Elephant B's behaviour was more variable. She spent 62.5% of her time stereotyping on day 2 but on other days this only took up approximately 9–10% of her time. Elephant B spent more time standing and walking than elephant A.

Q7.3.4. Elephant A.

7.4. *Feeding strategies in lizards*

Q7.4.1.

Characteristic	Sit-and-wait	Widely-foraging
Prey type	Active	Sedentary
Morphology	Stocky	Streamlined
Brain size	Small	Large
Volume of prey captured/day	Low	High
Endurance capacity	Limited	High
Learning ability	Limited	Enhanced learning and memory
Daily metabolic cost	Low	High

Q7.4.2. The sit-and-wait lizard moves very little with time, whereas the widely-foraging lizard is constantly searching for prey.

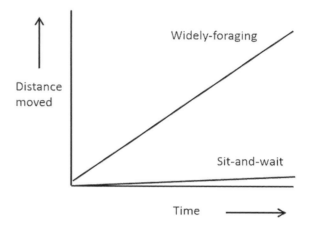

Q7.4.3.

a. The sit-and-wait lizard would encounter predators least often because it is not moving around much so chance encounters with predators would be rare.
b. Widely-foraging lizards are most vulnerable to those that sit-and-wait because they are constantly moving and likely to encounter them by chance.

Q7.4.4.

a. The wriggling tail distracts the predator while the lizard escapes.
b. By calculating the frequency of lizards that have shed their tails it would be possible to estimate the extent to which the population has been subjected to attacks by predators.

Q7.4.5. If a lizard sheds its tail when the tail is bitten by a venomous snake this would prevent the venom from reaching its body and allow it to escape unharmed.

7.5. *Roaring contests in red deer*

Q7.5.1. Stag A – 160 roars; stag B – 71 roars.

Q7.5.2. Stag A – 160/40 = 4 min⁻¹; stag B – 71/40 = 1.78 min⁻¹.

Q7.5.3.

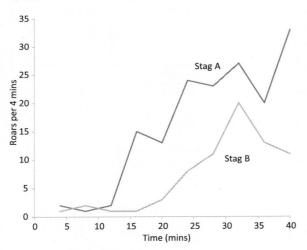

Q7.5.4. Stag A because it continued roaring at a high rate while Stag B achieved a maximum roaring rate of 20 per 4 mins before its rate declined to 11 when stage A was peaking at 33.

Q7.5.5.

a. No.
b. Yes.

Q7.5.6. Fighting has the potential to be very expensive in terms of energy use and injury, and it may possibly result in death. Conversely, roaring allows each animal to assess the strength of his opponent with minimum risk.

7.6. *Territoriality in feral cats*

Q7.6.1. Many male cats were never seen together and others were rarely seen together. Values for the association index over 0.5 were only recorded after neutering when males began behaving more like females, i.e., instead of wandering between groups of females they tended to remain within a particular group.

Q7.6.2.

Cat	No. of associates	
	Before neutering	**After neutering**
27	7	7
28	7	6
34	6	7
44	8	6
50	2	2
52	3	1
53	7	3
49	7	6
68	1	2
23	7	6
29	7	6
65	2	2
Mean	5.33	4.50

Q7.6.3. After neutering the nomadic males moved between the social groups less than was the case before neutering. This had the effect of reducing the number of male associates for many males. This may have been the result of removing their testes: the source of the testosterone which influences much of the behaviour of males.

Q7.6.4. Male cats 50 and 68 were not seen associating when entire (before being neutered) but were closely associated when neutered.

Q7.6.5. Male cat 68 had the smallest number of male associates before he was neutered: one.

7.7. *Safety in numbers: flocking as an antipredator device*

Q7.7.1.

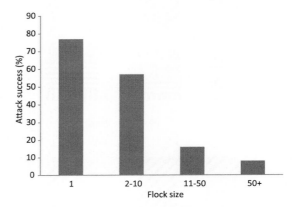

Q7.7.2. As flock size increases, hunting success falls.

Q7.7.3. Large flocks provide far more choice of prey than do small flocks and therefore more distraction for the predator. This reduces predation success.

Q7.7.4. Hunting pressure would act as an agent of natural selection. Assuming a genetic component to flocking behaviour, those individuals less likely to exhibit flocking behaviour would be more likely to be removed from the population than those that flocked, so the proportion of individuals engaging in flocking behaviour would increase with successive generations.

Q7.7.5. Assuming an inherited component to hunting ability, we would expect good hunters to have a higher survival rate and leave more young (with similar genes to their own) than would poor hunters thereby causing the proportion of good hunters in the population to increase with time.

7.8. *The effect of prey density on territory size in an avian predator*

Q7.8.1.

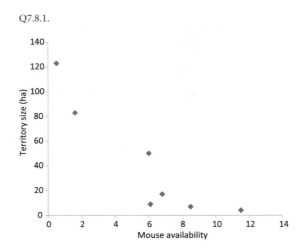

Q7.8.2. Only active prey could be caught. Those hiding away were not available for capture so, effectively, were not present.

Q7.8.3. When prey density is low the harrier increases the size of its territory.

Q7.8.4. If mouse availability throughout the area was 1.6 individuals per 0.25 ha territory size would be 83 ha. The number of territories that could fit into this area is 1000/83 = 12.

Q7.8.5. If mouse availability rose to 8.5 individuals per 0.25 ha the territory size would be just 7 ha. The number of territories that could be supported would be 1000/7 = 143.

7.9. *Flexibility in the social behaviour of pied wagtails*

Q7.9.1.

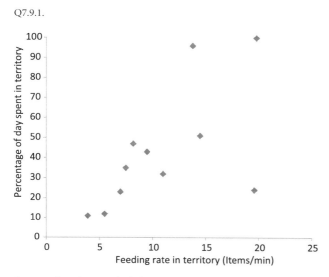

Q7.9.2. When there was little food available on the territory the wagtail still spent some of its time there, although it was more cost-effective in terms of time to look for food elsewhere with the flock, because it still needed to defend its territory.

Q7.9.3. Davies used patterns in the birds' plumage to tell them apart.

Q7.9.4. $500 \times (12 \times 0.9 \times 60) \times 15 = 4,860,000$.

7.10. *Optimal foraging theory: using game theory to study ecological strategies.*

Q7.10.1.

(a) All fast attacks		
Predator strategy	**Prey type**	**Energy gain**
Fast	Fast	$10.0 - 5.0 = 5.0$
Fast	Fast	$10.0 - 5.0 = 5.0$
Fast	Slow	$10.0 - 5.0 = 5.0$
Fast	Slow	$10.0 - 5.0 = 5.0$
Total		20.0

(b) All slow attacks		
Predator strategy	**Prey type**	**Energy gain**
Slow	Fast	$0 - 2.5 = -2.5$
Slow	Fast	$0 - 2.5 = -2.5$
Slow	Slow	$10 - 2.5 = 7.5$
Slow	Slow	$10 - 2.5 = 7.5$
Total		10.0

(c) Fifty per cent fast attacks and fifty per cent slow attacks		
Predator strategy	**Prey type**	**Energy gain**
Fast	Fast	$10.0 - 5.0 = 5.0$
Slow	Fast	$0 - 2.5 = -2.5$
Fast	Slow	$10.0 - 5.0 = 5.0$
Slow	Slow	$10 - 2.5 = 7.5$
Total		15.0

Q7.10.2. If half of the prey are of the slow type and half are of the fast type, the predator should always attack fast in order to maximise its energy gain. When all fast attacks are used, although the predator wastes energy running fast after the slow prey, it always catches its prey so the net gain is higher than in the other strategies where prey sometimes escape.

Q7.10.3. Yes. If the predator could predict the speed of its prey it could adjust the speed of its attack accordingly and thereby reduce the amount of energy lost by using a fast attack on a slow prey. All attacks on slow prey would be slow and all attacks on fast prey would be fast.

7.11. *Evolutionarily stable strategies: reproduction in dung flies*

Q7.11.1.

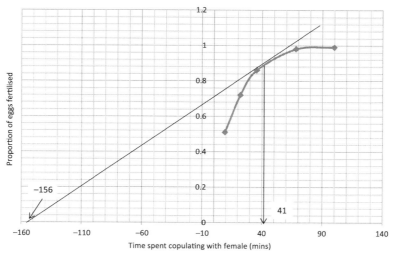

Figure 12.14

Q7.11.2. See Fig. 12.14.

Q7.11.3. See Fig. 12.14.

Q7.11.4. The optimum copulation time recorded by Parker and Stuart in the field was 35.5 minutes. The predicted value should be around 41 minutes.

7.12. *Industrial melanism: hiding in plain sight*

Q7.12.1.

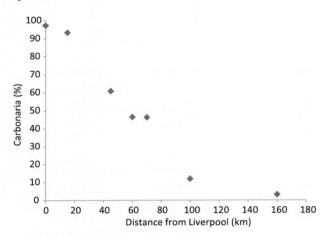

Q7.12.2. The proportion of *carbonaria* in the populations decreased with distance from the industrial areas of Liverpool.

Q7.12.3. Prevailing winds in this area are westerly so pollution from Liverpool blows towards the east. The air in North Wales was comparatively clean as a result of this and because of the absence of concentrations of heavy industry.

Q7.12.4. You should expect to find that bird predators had taken more of the *carbonaria* morphs on trees in rural areas because they were easy to distinguish from the pale background and that more of the *typica* morphs had been taken in industrial areas because they were easy to see against dark backgrounds.

7.13. *Zinc tolerance in Agrostis capillaris.*

Q7.13.1.

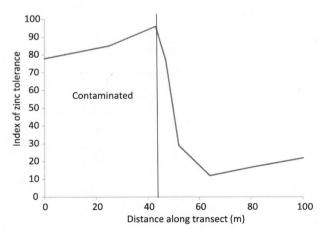

Q7.13.2. The boundary occurs between 43 m and 52 m along the transect (i.e., in the middle) with the mine on the left and the pasture on the right of the graph.

Q7.13.3. Abrupt.

Q7.13.4. It prevents cross-fertilisation between the tolerant and nontolerant populations, encouraging inbreeding within the tolerant population and reducing gene flow. Inbreeding increases the chance that tolerant individuals will have tolerant offspring.

Q7.13.5. Increase in self-fertility as this also encourages inbreeding.

Q7.13.6. There is much less potential for asexually reproducing species to produce variation. Sexual reproduction causes much mixing of genes and is therefore much more likely to result in the development of tolerance.

Q7.13.7. Metal tolerant grasses can be used to reclaim derelict land contaminated with the metals to which they have evolved tolerance.

7.14. *Inbreeding and homozygosity*

Q7.14.1.

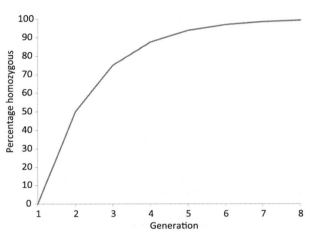

Q7.14.2. Generation 1 consists of a single heterozygote, so 100% of the population is heterozygous. The second generation consists of four individuals with the genotypes, AA, Aa, Aa and aa, so now 50% of the population is heterozygous. The AA individual now produces four AA offspring in generation 3. Likewise, the aa individual produces four aa offspring. The two Aa individuals produce four offspring with the genotypes AA, Aa, Aa, and aa. If we count the genotypes present in generation 3 we find 6 × AA, 4 × Aa and 6 × aa. So of the 16 individuals, 4 (25%) are heterozygotes (Aa) and 12 (75%) are homozygotes (i.e., either AA or aa). If we continue with this process, the number of heterozygotes continues to halve with each generation so that by generation 8 the population is 99.2% homozygous. Inbreeding of this type produces two homozygous lines and practically eliminates the heterozygotes.

Q7.14.3. $100 - 99.2 = 0.8\%$.

Q7.14.4. In generation 7, 98.4% of the population is homozygous. Half of these are homozygous recessive so $98.4/2 = 49.2\%$ have the disease.

Q7.14.5. Inbreeding is a potential problem for zoos because, for some species (especially rare species) the populations they hold are small so the pool of potential mates is also small. Captive breeding programmes are carefully managed by studbook keepers who determine which animals will mate and keep detailed records of individual matings and the relationships between the individuals within the population. This avoids breeding between close relatives, although the captive populations of some rare species have been produced from a small founder population (e.g., California condor (*Gymnogyps californianus*) and Przewalski's horse (*Equus ferus przewalskii*)).

7.15. *The problem of genetically isolated populations: inbreeding in lion and black rhinoceros*

Q7.15.1. The dominant male had sired the majority of rhinoceroses in the crater so inbreeding could cause problems in the future. This could result in abnormalities, reduced fertility, increased susceptibility of the population to disease, etc.

Q7.15.2. These animals were brought in to increase the genetic diversity of the population and to reduce the future threat of inbreeding.

Q7.15.3. Two rhinoceros died of the disease but it was unclear if they were residents or part of the group that had been translocated as authorities disagree.

Q7.15.4. None appears to have been reported.

Q7.15.5. The population is isolated with few immigrants. The strong coalitions formed by males may have prevented immigrants from entering the crater thereby increasing inbreeding.

Q7.15.6. There is evidence from computer simulations of the genetics of the population.

7.16. *Calculating effective population size*

Q7.16.1.

Breeding males	Breeding females	Actual population size	Effective population size
0	100	100	0
10	90	100	36
20	80	100	64
30	70	100	84
40	60	100	96
50	50	100	100
60	40	100	96
70	30	100	84
80	20	100	64
90	10	100	36
100	0	100	0

Q7.16.2.

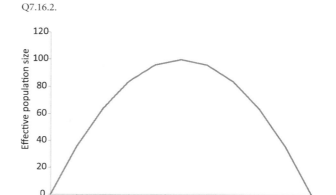

Q7.16.3. The highest effective population size is achieved with a ratio of males to females of 50:50.

Q7.16.4.

a. Artificial insemination may be used to increase the effective population size of a captive population of animals held in zoos where individuals of the species are difficult to move and using only captive specimens by collecting semen from captive males and transporting this to females. This avoids the risks associated with transporting animals; it allows a wide range of possible matings; and it removes the problems associated with mate selection whereby some pairings of individuals do not result in successful matings because the individuals are incompatible.
b. Zoo breeding programmes can increase genetic diversity in zoo populations by collecting sperm from wild individuals without taking them into captivity permanently.

7.17. *Charles Darwin careful scientist or clumsy naturalist?*

Q7.17.1. Darwin ate some of his tortoise specimens while on board HMS Beagle and threw their carapaces overboard believing them to be of no importance, whereas, in fact, they were the most important features distinguishing the various species.

Q7.17.2. Darwin relied upon experts at the Zoological Society of London to identify most of his specimens and he openly admitted to having no idea what species his plant specimens represented.

Q7.17.3. Darwin was not only poor at species identification and classification but he also failed to label many of his specimens. He disposed of important specimens because he failed to recognise their significance.

Q7.17.4. Possibly. Although Darwin collected these finches, it was David Lack who described them in detail and explained their true evolutionary significance.

Q7.17.5. College textbooks often oversimplify scientific concepts to explain them to inexperienced students. Unfortunately, textbooks tend to copy each other's mistakes so incorrect information may pass from one textbook to another, perpetuating misconceptions and errors for many years and many generations of students. However, some of the errors discussed here are historical errors rather than scientific oversimplifications. Some textbooks acknowledge that Darwin was inexperienced and that his handling of his specimens and understanding of their significance was less than perfect, but most ignore this and give an idealise account of his work. Whether or not the theory of evolution by natural selection would eventually have been proposed by someone else is impossible to know for certain but it is clear that others (notably Alfred Russell Wallace) were collecting similar evidence to that collected by Darwin and their thoughts were heading in a similar direction. It is certainly true that Darwin eventually made an enormous contribution to our understanding of evolution. Acknowledging his early errors would help to put this achievement in its historical context.

7.18. *Weight distribution and natural selection in the horse chestnut tree*

Q7.18.1.

a. Mean = 6.08 g.
b. Standard deviation = 2.47 g.
c. Range = 14-66–1.12 = 13.54 g.

Q7.18.2.

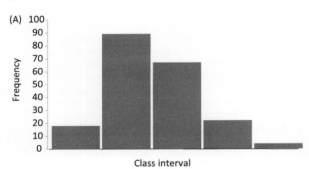

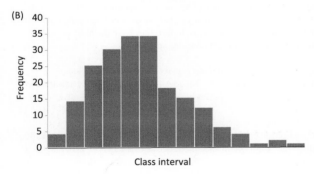

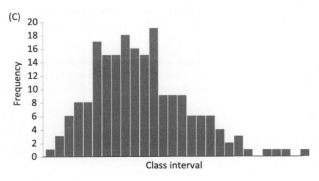

Q7.18.3. Large class intervals produce a small number of columns which may not be sufficient to display a discernable distribution. Very small class intervals often result in gaps where no individuals were recorded in certain classes. This makes the distribution look fragmented and incomplete. Choosing an appropriate class interval is largely a matter of experience and trial-and-error. In this case, graph (b) is probably the most suitable.

Q7.18.4. The distribution looks more-or-less normal but appears slightly skewed, i.e., the apex of the distribution is not in the centre.

Q7.18.5. Children almost certainly take the larger conkers because they should be heavier and therefore better for playing 'conkers' than smaller, lighter seeds. If this is true, and if children had been 'predating' the populations from which this sample was taken, we would expect the distribution of weights to be skewed to the right, or positively skewed (i.e., the long 'tail' is to the right of the peak).

Chapter 8 Environmental Pollution and Perturbations

8.1. *The biological concentration of DDT residues in food chains*

Q8.1.1.

	DDT residues ppm
Water	0.00005
Plankton	0.04
Silverside minnow	0.23
Sheephead minnow	0.94
Pickerel (a predatory fish)	1.33
Needlefish (a predatory fish)	2.07
Heron (feeds on small animals)	3.57
Tern (feeds on small animals)	3.91
Herring gull (a scavenger)	6.00
Fish hawk (osprey) egg	13.80
Merganser (a fish-eating duck)	22.80
Cormorant	26.40

Q8.1.2.

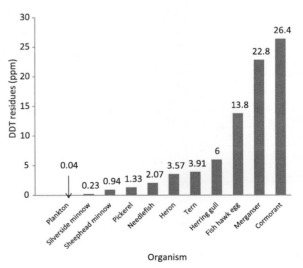

Q8.1.3.

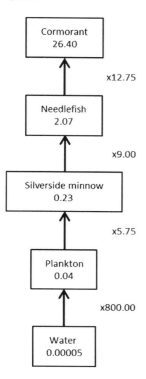

Q8.1.4. 26.40/0.00005 = 528,000.

Q8.1.5. From their food and from the water.

Q8.1.6. Although DDT residues may exist in the physical environment at very low concentrations, they are concentrated as they pass up the food chain so that by the time they reach the top carnivores they may be present in concentrations many hundreds of thousands of times greater than those in the environment.

Q8.1.7. Calcium is an important constituent of eggshells in birds. Exposure to DDE results in thinner, weaker eggshells. When birds sit on their eggs to incubate them they crack before the chick is properly developed and the chick dies. As a consequence of low egg survival the populations of birds of prey declined when DDE was widespread in the environment.

8.2. *Pesticides and eggshells*

Q8.2.1.

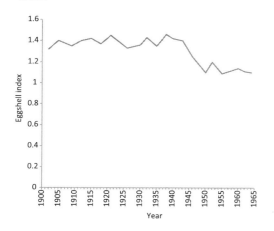

Q8.2.2. Eggshell thickness began to decline from around 1938 when the index was 1.46. By 1950 it was only 1.1.

Q8.2.3. This decline in eggshell thickness corresponded with the Second World War when DDT was widely used to kill disease-bearing insects.

Q8.2.4. The ban on the use of DDT prevented further additions of the chemical into food chains. This led to a gradual reduction in the amount of DDT passing from other organisms to the top predators, including birds of prey. This led to an increase in eggshell thickness and a consequent increase in the survival of eggs and chicks. The decline in birds of prey numbers caused by DDT was therefore arrested.

8.3. *Toxicity of pesticides and herbicides*

Q8.3.1.

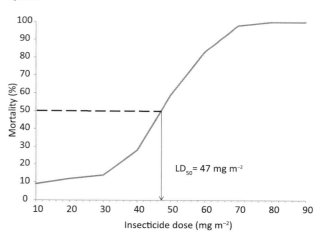

Q8.3.2. 47 mg m^{-2}.

Q8.3.3.

a. Least toxic to mammals is Picloram.
b. Most toxic to mammals is sodium arsenite.

Q8.3.4. All of the animals in this table are birds and therefore belong to the class Aves. In some species the insecticide is lethal while in others there is no discernable effect. It is therefore not possible to make any generalisation about the effect of this particular chemical on birds.

8.4. *Cultural eutrophication*

Q8.4.1. Sewage contains organic material. The nutrients in this waste fertilise the water and cause an increase in plant material. Plants contain chlorophyll so the amount of chlorophyll present acts as a bioindicator of the level of organic contamination.

Q8.4.2.

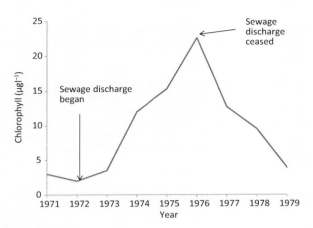

Figure 12.15

Q8.4.3. See Fig. 12.15.

Q8.4.4.

Variable	Normal condition (unpolluted)	Eutrophic condition
Biological oxygen demand	Low	High
Dissolved oxygen	High	Low
Fish numbers	High	Low
Water transparency	High	Low
Submerged plant numbers	Low	High
Phosphates	Low	High
Nitrates	Low	High
Sedimentation rate	Low	High
Primary productivity	Low	High
Bottom fauna species diversity	High	Low

Q8.4.5.

Detergent	Yes	Dairy waste	Yes	Lead	No
Paper waste	Yes	Zinc	No	Farm slurry	Yes
Fertiliser	Yes	Herbicide	No	Mercury	No

Q8.4.6. Nutrients washed into a stream are likely to be carried downstream and away from the pollution source. Nutrients washed into a lake will tend to accumulate because the turnover of water takes much longer here than in a section of flowing water in a stream or river.

8.5. *Organic pollution of freshwater ecosystems*

Q8.5.1. Site B is contaminated with organic pollution. This site has reduced biodiversity compared with site A and taxa that thrive in polluted conditions are present: *Asellus, Chironomus, Tubifex*.

Q8.5.2. *Tubifex* contains haemoglobin which allows it to obtain oxygen when it is in short supply as it is when organic pollution is present.

Q8.5.3. *Baetis* is a genus of mayflies. Their nymphs live in streams and are highly susceptible to pollution. They only survive in unpolluted water.

Q8.5.4. *Gammarus* numbers at site B are only 10% of the numbers at site A, so there has been a 90% decrease.

Q8.5.5.

Variable	Site A	Site B
BOD	Low	High
Dissolved oxygen	High	Low
Diversity index	High	Low
Fish numbers	High	Low

Q8.5.6. The purpose of the measurement is to determine how much oxygen is being used up by bacteria. BOD is measured in the dark to prevent photosynthetic organisms in the water sample from producing oxygen as a result of photosynthesis and distorting the results.

8.6. *The effects of acid rain on invertebrate diversity in Norwegian lakes*

Q8.6.1.

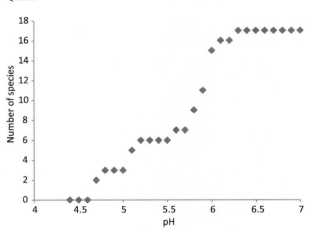

Q8.6.2. pH 4.6.

Q8.6.3. pH 6.3.

Q8.6.4. Between pH 5.9 and 6.0.

Q8.6.5. Acid rain caused by pollution travelling from the UK on westerly winds.

Q8.6.6. It interferes with the ability of molluscs to obtain the calcium they need to produce their shells.

Q8.6.7. Limestone (calcium carbonate).

8.7. *Dippers and acid rain*

Q8.7.1.

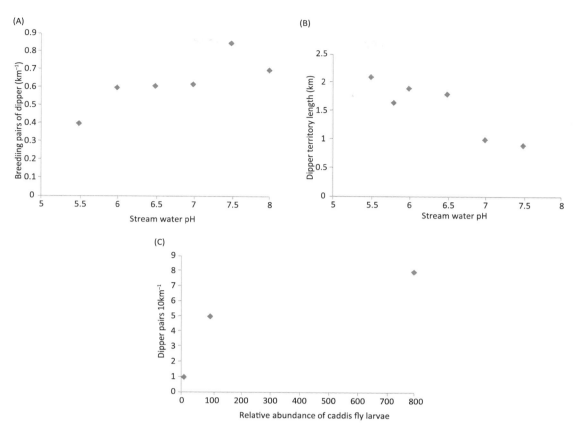

Q8.7.2.

a. It allows a simple standardisation whereby the value 10.0 represents the abundance of caddis flies when there is one dipper pair per 10 km. The value 100 represents 10 times this number without the need to specify the actual number. Actual numbers may have been difficult to assess.

b. This measure is dimensionless. For example, if we say A is twice as long as B there is no unit of length specified but we still know the relative lengths of A and B: A is twice as long as B, or B is half the length of A.

Q8.7.3. Acid rain appears to be causing a decrease in the availability of caddis flies. As pH falls the number of breeding pairs decreases per km and territory size increases (because they need more space to find enough food). As caddis fly numbers increase, so to do dipper numbers.

8.8. *Monitoring river pollution using diversity indices*

Q8.8.1.

Organism	Site A	p_i	p_i^2	Site B	p_i	p_i^2
Ancylus	8	0.043716	0.001911	–	0	0
Asellus	17	0.092896	0.00863	9	0.333333	0.111111
Baetis	52	0.284153	0.080743	3	0.111111	0.012346
Chironomus	2	0.010929	0.000119	5	0.185185	0.034294
Ecdyonurus	19	0.103825	0.01078	2	0.074074	0.005487
Elmis	11	0.060109	0.003613	2	0.074074	0.005487
Gammarus	31	0.169399	0.028696	1	0.037037	0.001372
Hydropsyche	7	0.038251	0.001463	4	0.148148	0.021948
Limnaea	11	0.060109	0.003613	–	0	0
Perla	12	0.065574	0.0043	1	0.037037	0.001372
Polycelis	13	0.071038	0.005046	–	0	0
Tubifex	–	0	0	–	0	0
Total	183	1	0.148915	27	1	0.193416

a. Simpson's index of diversity (D)
  Site A   1-0.148915 = 0.851085
  Site B   1-0.193416 = 0.806584.

b. Menhinick's index of diversity (d).
  Site A $11/\sqrt{183}$ = 0.813143
  Site B $8/\sqrt{27}$ = 1.539601.

Q8.8.2. Site B is polluted. This pollution has the effect of reducing the value of Simpson's index but increasing the value of Menhinick's index.

Q8.8.3. Nothing. The index does not consider the relative abundance of particular species. The same values remain in the table but in different positions. This does not affect the calculations at all.

Q8.8.4. The indices use different assumptions so cannot be compared. It is only possible to compare values obtained with the same index.

8.9. *Thermal pollution and water fleas*

Q8.9.1.

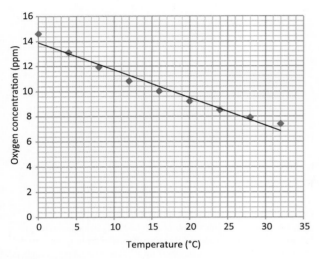

Figure 12.16

Q8.9.2.

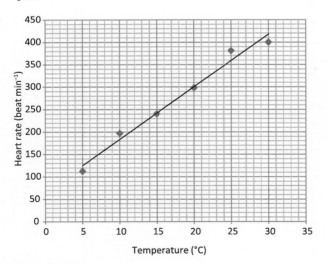

Figure 12.17

Q8.9.3. Fig. 12.16 shows a negative correlation; Fig. 12.17 shows a positive correlation

Q8.9.4. Poikilotherm.

Q8.9.5. Oxygen concentration in the water would fall from 10.8 to 8.2 ppm. The heart rate of the *Daphina* would increase from 230 to 370 beats per minute.

Q8.9.6. This temperature increase is likely to be damaging because the *Daphnia*'s demand for oxygen (due to increased activity) will increase at a time when the oxygen concentration has decreased due to the warming of the water.

8.10. *Temperature inversions and air pollution*

Q8.10.1.

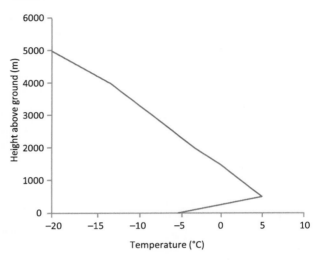

Q8.10.2. This is appropriate, because it makes more sense to show height running from the bottom to the top of the *y*-axis than running from left to right on the *x*-axis. The graph then looks like a vertical section through the atmosphere.

Q8.10.3.

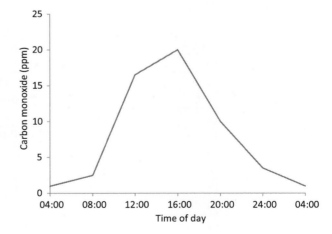

Q8.10.4. Vehicle exhausts, because carbon monoxide is produced as a result of the incomplete combustion of fuels.

Q8.10.5. A temperature inversion would prevent the vertical movement of gases and hold pollutants near the ground.

Q8.10.6. Carbon monoxide levels are high from 12.00 noon and 20.00 h with a peak at 16.00 h.

8.11. The effect of sulphur dioxide pollution on growth in ryegrass

Q8.11.1. The percentage decrease in productivity in the plants exposed to the higher level of sulphur dioxide compared with that achieved at the lower level of exposure were:

Number of tillers	41.1%
Number of living leaves	44.7%
Dry weight of leaves	50.9%
Leaf area	51.2%

Q8.11.2. All the indicators of yield that were measured showed a decrease of between approximately 40 and 50% when plants were exposed to 191 μg m^{-3} sulphur dioxide.

Q8.11.3. Sulphur dioxide will enter through the stomata in the leaves, i.e., the route through which carbon dioxide enters.

Q8.11.4. Stomata tend to close when plants are buffeted by the wind to reduce water loss. This response would reduce the uptake of gaseous pollutants. Wind would also tend to replace the air around the plants quickly and blow the pollutants away.

Q8.11.5. The sort of changes described here are subtle and 'invisible'. They would be difficult to detect in the field without conducting a controlled experiment.

Q8.11.6. The tight packing of the plants may have made it more difficult for the sulphur dioxide to penetrate the spaces between the individual plants. Leaves exposed directly to the SO_2 would have experienced surface damage.

Q8.11.7. It would be difficult to create a controlled atmosphere in a field situation. There are other pollutants in the air in addition to sulphur dioxide. Other environmental variables, such as light intensity, water stress, temperature, relative humidity, and soil sulphur content, may also affect plant growth.

8.12. Trends in greenhouse gas emissions in the UK

Q8.12.1. The percentage drop in carbon emissions between 1900 and 2013 for each sector were:

Energy supply	32.0%
Transport	4.0%
Business	21.2%
Residential	3.7%
Agriculture	18.6%
Waste management	67.4%
Industrial processes	78.7%
Public	29.6%
LULUCF	232.5%
Total	29.8%

Q8.12.2.

a. LULUCF.

b. Residential.

Q8.12.3. Removal of carbon from the atmosphere.

Q8.12.4. Absorption of carbon by trees (increases as planted trees mature); less intensive arable agriculture.

Q8.12.5. In cities electric vehicles are useful because they do not emit harmful exhaust gases and so may contribute to improvements in air quality. However, their batteries must be charged and as long as electricity is produced as a result of the combustion of fossil fuels greenhouse gases will continue to be emitted into the atmosphere.

Q8.12.6. The construction of a sufficient number of wind, solar and tidal power facilities to replace existing coal- and gas-powered electricity generation stations would be extremely expensive and would take a considerable amount of time. The generation of electricity from solar and wind power facilities is unreliable in many parts of the world and such facilities use up large amount of space. Tidal power schemes face opposition from ecologists because of their effect on coastal fauna and flora.

8.13. Decay and concentration of radioisotopes

Q8.13.1.

Number of half-lives	Percentage of original sample remaining
0	100.000
1	50.000
2	25.000
3	12.500
4	6.250
5	3.125
6	1.563

Q8.13.2.

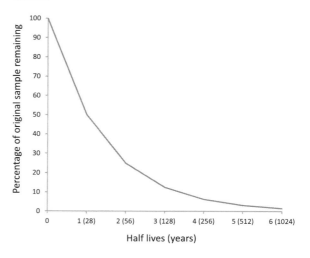

Q8.13.3.

a. Uranium-238.
b. Iodine-131.

Q8.13.4. A period of 73,000 years represents three half-lives for plutonium-339 so 1/8 would still be present, i.e., 12.5%.

Q8.13.5. This requires six half-lives i.e., 8.6 days × 6 = 51.6 days.

Q8.13.6.

a. Concentration between the soil and the grass = 2.5/0.12 = × 22.32.
b. Concentration between the grass and the sheep bone = 80/2.5 = × 32.

Q8.13.7. The strontium has been concentrated along the food chain and would be concentrated further in humans who ate meat from sheep.

8.14. The Hubbard Brook study: deforestation and nutrient loss

Q8.14.1. It would slowly have dripped from the trees over a period of time instead of falling directly onto the soil. This would have allowed it to be absorbed slowly by the soil instead of running quickly over the surface. The water would therefore have been held in the forest for a much longer period. Some of the water would also have evaporated from the surfaces of the trees and other plants in the forest.

Q8.14.2. Dead animals and plants that live in or fall into the water, animal faeces, soil, rock and rainwater.

Q8.14.3. When trees are cut down the amount of material decaying on the forest floor increases. These break down and the nutrients from the breakdown of organic matter are washed into the streams and rivers. The lack of trees to intercept the rain means that rainwater quickly runs off the soil surface into water courses carrying with it nutrients leached from the soil.

Q8.14.4. Negative values indicate a net gain in nutrients.

Q8.14.5. Nitrogen is lost as nitrate and ammonium.

Q8.14.6. Nitrate.

Q8.14.7. Modern forestry operations do not usually involve the clear felling of large areas of forest. Smaller areas are cleared so that the wildlife and general ecology do not suffer too much disturbance.

8.15. Interactions between urbanisation, forestry and hydrology

Q8.15.1.

Watts Branch: 1910/3.7 = 516.2 tons $mi^{-2} y^{-1}$.
Little Falls Branch: 9530/4.1 = 2324.4 tons $mi^{-2} y^{-1}$.

Q8.15.2. Little Falls is an urban area. Much of the surface is impermeable because it is covered in roads and buildings. Consequently there is more runoff and this will carry more sediment.

Q8.15.3. From 50 to 400 tons mi^{-2} $year^{-1}$ is an increase of 800% in average annual sediment discharge when forest cover was reduced from 80% to 20%.

Q8.15.4. The soil is now being washed from the forest because there are no trees to retain it. This increases the sediment load.

Q8.15.5. The nutrient status would become much poorer as nutrients are washed out.

Q8.15.6. Lead deposits from leaded fuel spilled on the roads.

Q8.15.7. Urban areas retain heat in buildings and roads. During a storm some of this heat is transferred to rainwater and carried to watercourses.

8.16. Lichens as monitors of air pollution

Q8.16.1.

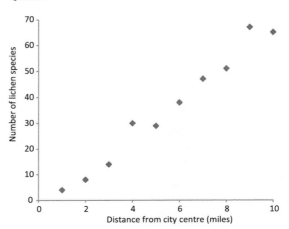

Q8.16.2. Some lichen species are highly susceptible to air pollutants, especially sulphur dioxide. Levels of pollution decrease with distance from urban areas. Consequently lichen biodiversity increases with distance from city centres.

Q8.16.3. Lichens are long-lived organisms and are permanently present in the environment so will respond to pollution levels present over a period of time. However, chemical methods allow us to determine a precise level of particular pollutants at a specific point in time.

Q8.16.4. The effect of the pollution from the city would extend further to the east of the city centre than to the west because the pollution is being blown to the east by the wind. For any given distance from the city centre we would expect to find more lichen species to the west than to the east.

8.17. Feral pigeons as bioindicators of lead levels in the environment

Q8.17.1.

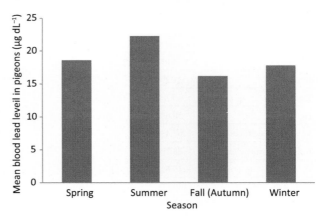

Q8.17.2.

a. Pigeons are more active in summer.
b. Lead levels are higher in summer.
c. Pigeons are able to incorporate lead more quickly into their soft tissue or bone in summer.
d. Some combination of a, b and c.

Q8.17.3.

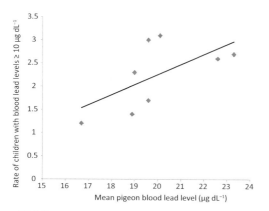

Figure 12.18

Q8.17.4. See Fig. 12.18.

Q8.17.5. Correlation coefficient = 0.626. This value is statistically significant ($P < 0.05$).

Q8.17.6. There is a positive correlation between the blood lead levels in pigeons and those in children so lead levels in pigeons is a good predictor of lead levels in children.

Q8.17.7. Yes.

Q8.17.8. Lead is known to affect the nervous system. It may cause lethargy, blindness, seizures, locomotion and balance problems.

Q8.17.9. Blood Pb $\geq 10\,\mu g\,dL^{-1}$ was considered to represent the nationally acceptable exposure to lead at the time.

Q8.17.10. Historical contamination of roads with lead.

Q8.17.11. Previous use of lead paint in buildings.

8.18. *Where has all the ice gone?: polar bears and climate change*

Q8.18.1.

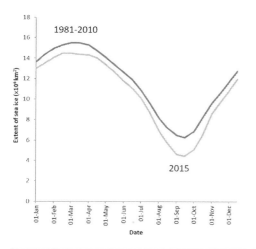

Q8.18.2. $4.447/6.277 \times 100 = 70.8\%$.

Q8.18.3. $(14.469 - 11.733)/92 = 0.03$ millions $km^2\,day^{-1}$.

Q8.18.4. $(14.469 - 4.447)/14.469 \times 100 = 69.3\%$.

Q8.18.5. $14.469 - 4.447 = 10.022$ millions km^2.

Q8.18.6. The data for 1981–2010 has been averaged so there is no way of determining what the minimum value was in any of the years from which the average was calculated. The means may be hiding some very high and some very low values but there is no way of knowing without inspecting the original data for each year.

Q8.18.7. Polar bears must move about on ice to find suitable prey. Although they can swim very well they need to search large areas to find sufficient food, especially before they hibernate. They can only do this efficiently from the ice.

Chapter 9 Conservation Biology

9.1. *Counting threatened species: the IUCN Red List*

Q9.1.1.

Taxon	Described species	Number of threatened species (IUCN categories CR, EN and VU)	
		1996–98	**2016**
Mammals	5536	1096	1208
Birds	10,424	1107	1375
Reptiles	10,450	253	989
Amphibians	7538	124	2063
Fishes	33,300	734	2343
Subtotal	67,248	3314	7978
Invertebrates	1,305,250	1891	4383
Total	1,372,498	5205	12,361

Q9.1.2.

1996–98: $5205/1372498 \times 100 = 0.38\%$.
2016: $12361/1372498 \times 100 = 0.90\%$.

Q9.1.3. Invertebrates: 2492.

Q9.1.4. We would need to know the total number of species described in 1996–98 and also which of the species classified as threatened in 2016 had been described in 1996–98. If a large number of threatened species had been described for the first time since 1998 (and immediately classified as threatened) this would distort the data and make it look like there had been more of a deterioration in the status of described species than was, in fact, the case.

9.2. *The biology of the giant panda*

Q9.2.1. The greatest threat to wild giant pandas is habitat loss. There have been recent improvements in the situation in China as a result of greater habitat protection and afforestation projects. However, climate change is likely to be a significant threat to panda habit in the future.

Q9.2.2. Giant pandas are specialist feeders and their diet consists almost entirely of bamboo.

Q9.2.3. Dependence on a particular type of plant restricts the range of habitats they are able to occupy. Pandas in the wild may be limited by the availability of high quality bamboo.

Q9.2.4. Giant pandas are solitary except during the mating season. When densities are low in the wild they will find it difficult to find mates. Females are only fertile for four days a year and usually only have one cub at a time. This makes it difficult for populations to recover quickly.

Q9.2.5. They occur in low densities, are shy and difficult to find.

Q9.2.6. It is very rare to see a giant panda outside China. A small number of zoos 'rent' pandas from the Chinese government but if they have any offspring while on loan to these zoos they also become the property of the Chinese government. This means that there is no opportunity for zoos outside China to acquire pandas for a breeding programme of their own. When zoos rent pandas they generally see an increase in visitors, and some zoos, e.g., Edinburgh Zoo, have had to ask visitors to prebook their visit to the panda enclosure because the demand was so high. The increased revenue received by zoos when they have pandas may be used to support the conservation activities of a zoo.

9.3. *Conserving tropical plants: why bother?*

Q9.3.1. The data show that a wide range of species of plants are useful to humans. There would undoubtedly be further discoveries of useful plants if the forests were to be conserved and the properties of the plant species investigated.

Q9.3.2. Medicines, as each would need to be rigorously tested to establish its efficacy.

Q9.3.3. Some species may have more than one use.

Q9.3.4. A large number of species occurs per unit area. If unknown species are present they could easily become extinct (at least locally) as a result of deforestation.

Q9.3.5. $(375 \times 7.4)/23 = 121$ trees ha^{-1}.

Q9.3.6. In order to be preserved in a seed bank seeds need to be capable of surviving once they have been dried out. Drying will kill some types of seeds.

Q9.3.7. Seeds kept in a seed bank will not live forever, although drying and freezing both slow down the ageing process. Seeds need to be germinated periodically so that new plants may be produced which will eventually reproduce and create new seeds.

9.4. *Gullypots: a threat to small vertebrates*

Q9.4.1.

Taxon	Alive	Dead	Total	Mortality (%)
Toads	716	92	808	11.39
Frogs	136	21	157	13.38
Newts	48	4	52	7.69
Mammals	0	101	101	100.00
Total	900	218	1118	
%	80.50	19.50		

Q9.4.2. Mammals.

Q9.4.3. Newts.

Q9.4.4. They probably drowned. The taxa with higher survival rates were all amphibians.

Q9.4.5. No. Although more than five times as many toads were trapped than frogs this may reflect the relative abundance of the two taxa rather than indicating that toads have a greater propensity to fall into drains than do frogs. We would need to establish the relative numbers of toads and frogs in the area to draw any firm conclusions.

Q9.4.6. Specially designed 'wildlife kerbs' allow small vertebrates such as newts to pass along a recess and avoid falling into the drain. Sloping ladders can be added to the design of the drain so that animals can climb out of the gullypot.

9.5. *Black rhinoceros poaching in Tanzania*

Q9.5.1. The severe drought of 1959–60 caused increased mortality in the Masai herds. They turned to poaching rhinoceros to supplement their incomes.

Q9.5.2. Between 1963 and 1965, 82% of crater rhinoceros were adults but this had declined to just 28% by 1978. This may be evidence for heavy poaching, as commercial poachers select adult rhinoceros with large horns.

Q9.5.3. Increased ranger patrols and cash rewards for information leading to the conviction of poachers.

Q9.5.4. In the 19th century, wildlife was seen by European hunters and explorers as an unlimited resource. Trophy hunting was a popular sport and animal specimens were widely taken for private collections and museums. Before Africa was opened up by European colonial powers the exploitation of wildlife by indigenous peoples would have had little impact on animal numbers. The actions of Baumann's party in killing what is now an endangered species cannot be judged by reference to modern ethical standards relating to wildlife and environmental protection.

9.6. *Conserving fish stocks by mesh regulation*

Q9.6.1.

a. 2.2×10^7 fish
   1.16×10^4 metric tons.
b. 2×10^7 fish
   1.665×10^4 metric tons.

Q9.6.2. 100 mm.

Q9.6.3. They escape through the mesh at 100 mm but were caught when it was only 80 mm.

Q9.6.4. $((12.6-5)/5) \times 100 = 152\%$.

Q9.6.5.

(A)

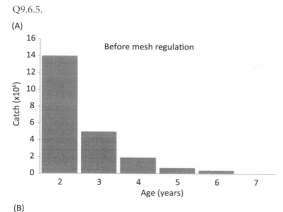

(B)

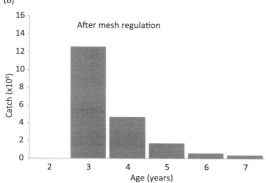

Q9.6.6. The increase mesh size has allowed the 2-year-old fish to escape and become 3-year-old fish. There are more older fish being caught because more of the young fish are escaping capture and continuing to grow.

9.7. *Game ranching in Zimbabwe*

Q9.7.1.

Q9.7.4.

(A)

(B)

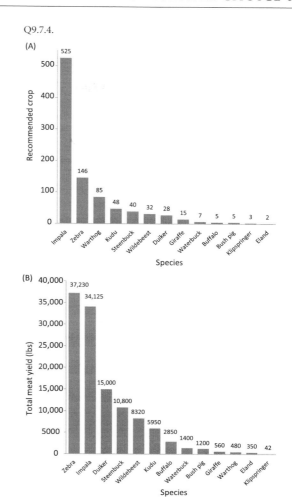

Species	Estimated number	Recommended crop	Carcass weight (lbs)	Total meat yield (lbs)	Gross value (£)	Mean value/ lb
Impala	2100	525	65	34,125	2133	0.063
Zebra	730	146	255	37,230	1168	0.031
Steenbuck	200	40	12	480	36	0.075
Warthog	170	85	70	5950	297	0.050
Kudu	160	48	225	10,800	540	0.050
Wildebeest	160	32	260	8320	416	0.050
Giraffe	90	15	1000	15,000	600	0.040
Duiker	80	28	20	560	42	0.075
Waterbuck	35	7	200	1400	70	0.050
Buffalo	30	5	570	2850	119	0.042
Eland	10	2	600	1200	60	0.050
Klipspringer	10	3	14	42	3	0.071
Bush pig	10	5	70	350	17	0.049

Q9.7.2. The most valuable species per pound are steenbuck and duiker.

Q9.7.3.

Total meat yield for all species = 118, 307 lbs
Gross value for all species = £5,501
Total meat yield for all species per square mile = 118, 307 / 50 = £2366.14
Gross value for all species per square mile = £5,501 / 50 = £110.02.

Q9.7.5. Those species with the highest crop do not necessarily produce the largest amount of meat because of differences in the sizes of the animals. The total yields from impala and zebra are very similar but the total recommended crop of impala is much higher than that of zebra. Zebra yield more meat than impala because of their much greater size.

Q9.7.6.

Type of ranching	Gross profit per year (£)	Expenses per year (£)	Net profit per year (£)
Beef	5198	4692	506
Game	5500	2300	3200

9.8. *Use it or lose it: sport hunting as conservation*

Q9.8.1.

	Zimbabwe dollars (Z$)	Percentage of total income
Income		
Sport hunting	10,307,342	89.98
Tourism	163,677	1.43
PAC* hides and ivory	243,614	2.13
Other	739,905	6.46
Total income =	11,454,538	
Allocation		Percentage of total allocated
District councils	1,339,302	13.91
Wildlife management	2,532,843	26.30
Ward/village/house	5,459,554	56.70
Other	297,588	3.09
Total allocated =	9,629,287	

Q9.8.2. Sport hunting.

Q9.8.3. Individual wards, villages and houses.

Q9.8.4.

Species	Total shot	Total income (Z$)	Percentage of total income
Elephant	56	1,996,400	62.11
Elephant PAC	5	74,065	2.30
Buffalo (male)	230	540,270	16.81
Leopard	39	145,353	4.52
Sable antelope	27	94,014	2.93
Lion (male)	15	56,745	1.77
Other	-	307,199	9.56
Total		3,214,046	100

Q9.8.5. Elephant.

Q9.8.6. 1,996,400/56 = Z$ 35,650.

Q9.8.7. A ban on hunting would probably have done more harm than good because the rural communities would have been deprived of an income and they would have had no economic interest in preventing elephant poaching so elephant populations would probably have suffered.

9.9. *How does shape affect a protected area?*

Q9.9.1.

The square reserve has a perimeter of $10 \times 4\,km = 40\,km$
The radius of the circle = $\sqrt{(100/\pi)} = 5.642\,km$
The perimeter of this circle = $2\pi \times 5.642 = 35.45\,km$

The circular reserve has the shortest perimeter so should be better than the square reserve at retaining animals.

Q9.9.2.

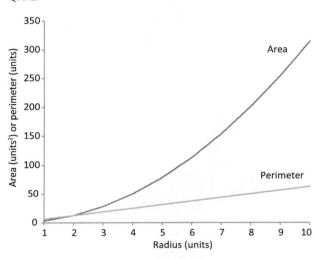

Q9.9.3.

a. Area increases geometrically with increasing radius.
b. Perimeter increases in a linear fashion with increasing radius.

Q9.9.4. Large because large reserves have proportionately shorter perimeter boundaries than do smaller reserves.

Q9.9.5.

a. A single circular reserve of $100\,km^2$ has a boundary of $35.45\,km$ (see answer to Q9.9.1).
b. A single circular reserve of $25\,km^2$ has a boundary of:
   i. radius = $\sqrt{(25/\pi)} = 2.821\,km$
   ii. perimeter = $2\pi \times 2.821 = 17.725\,km$
   iii. The total perimeter of four areas this size = $17.725 \times 4 = 70.9\,km$.

The total combined perimeter of the four areas is approximately twice that of the single large area. The total perimeter length per unit area is $35.45\,km$ for the single area of $100\,km^2$ ($0.355\,km.km^{-2}$) and $70.9\,km$ for the four separate areas (total = $100\,km^2$) ($0.709\,km.km^{-2}$). The single large area therefore has a much smaller perimeter over which to lose organisms than the total perimeter length of the four smaller areas combined.

Q9.9.6. The best is C followed by A then B. All reserves are the same size but those in C are linked by corridors. The reserves in A are arranged so that each is near the other two, so individuals lost from one reserve may move to another nearby. B is the worst arrangement because they are not linked and the two at each end are only near one other reserve.

Q9.9.7. Size, shape and proximity to (or links with) other protected areas.

Q9.9.8. Fig. A is better at retaining organisms than is design B because it has less perimeter per unit volume over which to lose animals.

Q9.9.9. D is better at retaining organisms than C because there are corridors linking the individual areas.

Q9.9.10. A fire in forest in C2 is likely to be more serious than one in D2 because it can travel along the corridors to C1 and C3.

9.10. *Animal reintroductions*

Q9.10.1. Studies need to be conducted to determine if suitable habitat is available including appropriate nesting sites, suitable and sufficient prey organisms, absence of predators or competitors which would hamper the reintroduction etc.

Q9.10.2. Landowners, anglers, farmers, naturalists and conservation groups, water companies, foresters, local government, owners of land used for game bird shoots.

Q9.10.3. This species could attract tourists, particularly bird watchers and other naturalists. It would also increase biodiversity. Eagle owls may take mice and rats but also conservation-sensitive species including other birds of prey. Local farmers may be concerned about possible predation on lambs and other livestock and consequent economic losses. They may also prey on pets such as cats and rabbits. Eagle owls may take fish and game birds of economic interest.

Q9.10.4. You will almost certainly need to apply to the local authorities for a licence to take these birds from the wild. Before a licence is issued you may have to demonstrate that taking these birds will not jeopardise the long-term survival of the source population, possibly by using a computer simulation of the effects on this population. This might involve, for example, demonstrating that chicks taken would have died anyway or only taking chicks from large clutches.

Q9.10.5. The genetics of the source population should be as similar as possible to the genetics of the original inhabits of the reintroduction area.

Q9.10.6. The keepers should provide the food from behind a wooden screen so that the birds do not associate the provision of food with the presence of people. Such an association could cause problems after release if the birds sought out humans in the hope of obtaining food.

Q9.10.7. Release of nonnative species generally requires a licence from the relevant statutory conservation agency. A licensing system prevents the uncontrolled release of animals (and plants) that could cause ecological damage and helps to ensure the welfare of the animals being released.

Q9.10.8. After release the dispersion, health, reproduction, causes of mortality, movements and other aspects of the biology of the species should be monitored in order to judge the success of the project. Such information should be used to inform any further reintroductions.

9.11. *Causes of manatee mortalities in Florida*

Q9.11.1.

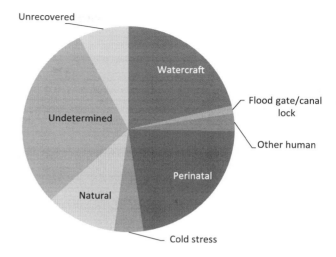

Q9.11.2. $29/371 \times 100 = 7.82\%$

Q9.11.3.

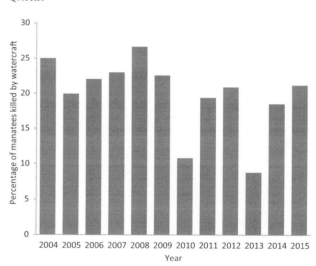

Q9.11.4. Certain areas could be set aside where the use of watercraft is limited or completely prohibited. Water speed zones could be introduced. Boat users could be educated about manatees. Marked channels for boats could be introduced. The use of polarised sunglasses while operating a boat could be encouraged to assist the operator in seeing below the water's surface. Boat users could be encouraged to speak quietly and to lower anchors carefully.

9.12. *The conservation role of zoos*

Q9.12.1. The Directive does not apply to this zoo because it is a privately owned collection of animals to which the public do not have access on at least 7 days a year.

Q9.12.2. All zoos covered by the Directive must have an education function. It appears that Zoo B does not comply with this requirement as it does not have an education programme or even accurate signs on its enclosures.

Q9.12.3. No. Undertaking research from which conservation benefits accrue is just one of a list of alternative means of complying with the Directive. Note the use of the term 'and/or' in Article 3.

Q9.12.4. Zoos need to know which species (and possibly subspecies) they keep, the sex of each individual, its location in the zoo, its age, and its veterinary and breeding history. This is essential if the animals are to be of conservation value and to keep the populations healthy.

Q9.12.5. Small animals and those that are difficult to count, e.g., insects and other small invertebrates, small fishes.

Q9.12.6. The reintroduction of animals to the wild is one of several activities zoos may undertake in order to comply with the directive but it is optional. It would not make sense to require all zoos to reintroduce species because they may not have the expertise or access to suitable animals for such projects.

9.13. *Zoos and evolutionarily significant units (ESUs)*

Q9.13.1. The definition of an ESU appears to be being applied by zoos inconsistently between species because some breeding programmes are for subspecies and others are for entire species.

Q9.13.2.

Vernacular name	Scientific name	European Endangered species Programme (EEP)	Species Survival Plan Programme (SSP)
Asian lion	*Panthera leo persica*	●	
Siberian Tiger	*Panthera tigris altaica*	●	●

(Continued)

(Continued)

Vernacular name	Scientific name	European Endangered species Programme (EEP)	Species Survival Plan Programme (SSP)
Indochinese Tiger	*Panthera tigris corbetti*		●
Sumatran tiger	*Panthera tigris sumatrae*	●	●
Sri Lanka leopard	*Panthera pardus kotiya*	●	
Amur leopard	*Panthera pardus orientalis*	●	
Persian leopard	*Panthera pardus saxicolor*	●	
North Chinese leopard	*Panthera pardus japonensis*	●	

Q9.13.3. The subspecies for which captive breeding programmes have been established are all considered to be geographically and genetically isolated from other subspecies within the same species. There must be a sufficiently large number of representatives of these subspecies in captivity to form a viable captive population. This approach maximises the amount of genetic variation that can be preserved within the species.

Q9.13.4.

a. There are too few elephants in captivity to sustain viable populations of subspecies.
b. There is disagreement among experts regarding the status of the suggested subspecies.
c. There has already been considerable hybridisation within zoos and these hybrids would be useless for breeding if programmes were established for subspecies.

Q9.13.5.

Region	Species					
	African elephant (*Loxodonata africana*)			Asian elephant (*Elephas maximus*)		
	Bulls	Cows	Sex ratio	Bulls	Cows	Sex ratio
Europe	40	126	1:3.15	51	188	1:3.69
North America	22	125	1:5.68	23	101	1:4.39

Q9.13.6. Bulls are underrepresented in these data. Zoos and circuses have traditionally preferred to keep cows because they are thought to be less dangerous than bull. Bulls experience periods of musth during which they become very aggressive and difficult to manage. Many former circus elephants have previously been sold to zoos. Breeding and calf survival in zoos have been low.

Q9.13.7. In recent years breeding and calf survival in zoos have both improved so more bulls have been entering the zoo population. The development of artificial insemination for elephants has increased the birth rate and also helped to redress the imbalance in the sex ratio.

9.14. *Returning grey wolves to their former habitats in the USA*

Q9.14.1. In the early parts of the 20th century large predators were seen as pests and little value was given to wildlife. Predators such as wolves were killed to protect livestock.

Q9.14.2. Grey wolves were once widespread across the USA and conservationists believe we have an obligation to return them to the wild wherever possible. They were the top predators in many ecosystems and some herbivore populations have grown out of control in the absence of wolves. Wolves can be important in attracting tourists to areas where there is little employment.

Q9.14.3. The designation of reintroduced wolves as an 'experimental population' by the USFWS was used by the ranchers in the Yellowstone area in an attempt to have them removed. They argued that the presence of these wolves threatened the wild wolves that had migrated into the area from Canada because it was not possible to tell them apart. This meant that protected wolves could be killed unintentionally.

Q9.14.4. Ranchers could legally kill the reintroduced wolves if they attacked their livestock.

Q9.14.5. The original decision of the court to have the wolves removed was stayed because it was clear that the decision would be appealed.

Q9.14.6. The reintroduced wolves attract large numbers of tourist to the Yellowstone area.

9.15. *Do fences make good neighbours? Are wildlife fences part of the problem or part of the solution?*

Q9.15.1. Two benefits of fencing a protected area: keeps animals in, keeps nonnative predators, livestock and poachers out.

Q9.15.2. Two disadvantages of erecting wildlife fences: prevents natural dispersal of wildlife, may interfere with migration routes, animals may become entangled and trapped in the fence.

Q9.15.3. Young dingoes are able to pass through these small holes rendering it ineffective.

Q9.15.4. Dingoes were originally a genetically isolated population and some scientists believe that they should be classified as a distinct species. Recent hybridisation with domestic dogs makes their taxonomic status unclear. Hybrids between species are generally considered to be of very little conservation value.

9.16. *Ethics in ecological research and ecosystem management*

Q9.16.1. Although cetaceans are protected by many national and international laws the International Whaling Commission still permits 'scientific whaling'. This has been conducted by Japanese whaling ships for many decades. While the study of the ecology and behaviour of whales is important to their conservation most – and probably all – of what we need to know about these animals can be obtained by nonlethal means. Most scientists and conservationists believe that so-called scientific whaling is really commercial whaling disguised as a scientific endeavour.

Q9.16.2. Whether or not this study can be justified depends partly on the context in which it occurs. If it was part of normal forest operations, then the damage to the forest would be occurring anyway and it would make sense for scientists to take the opportunity to study the effects. If the felling of trees was undertaken purely for the sake of the study it could be justified if the area concerned was not of high conservation value and if the study was likely to generate information that could inform future forestry operations and lead to better forest management.

Q9.16.3. Early radio collars were large and cumbersome due to the size of the electronic components, especially the size of the batteries. Improvements in battery performance and reductions in their size have allowed the development of much smaller units in recent times. Biologists would not obtain useful information from radio collars if they altered the behaviour of collared animals in any significant way. There is therefore a disincentive for biologists to use any equipment that would adversely affect a study animal. However, some animal welfare advocates believe that we should not attach equipment of this type to wild animals.

Q9.16.4. The capture, marking and release of small mammals in order to calculate their population size is a well-established method of population estimation. Capturing animals is often stressful for them. Small mammals are usually captured in small traps which contain some bedding and food. Small mammals require a great deal of food for their size in order to maintain their body temperature. It is, therefore, imperative that traps are frequently inspected so the animals are not held in the traps for an unnecessarily long period, run out of food and die. Provided that the marks used are humane and do not adversely affect the animals, and they are handled as little as possible, this type of study should be able to be completed without harming the animals.

Q9.16.5. Depending upon the nature of the vegetation, this activity could produce a small long-lasting scar or be covered and repaired naturally within a very short time. Whether or not it is justified depends upon the purpose and importance of the study and the sensitivity and conservation status of the organisms removed and destroyed.

Q9.16.6. Capturing wild animals can be very stressful for them and may even result in death. Individual animals do not care if their species becomes extinct, so when we take animals from the wild to prevent extinction we are doing this for the benefit of our species and not theirs. Some animal welfare advocates argue that the welfare of the individual animals is paramount and they would prefer to see a species become extinct rather than have animals taken into captivity. Conservationists cannot care solely about welfare if they are to establish captive breeding programmes. Captive animals are likely to suffer some reduction in their welfare if held in cages or enclosures. However, we must not assume that the welfare status of a wild individual is always going to be higher that of a captive individual of the same species as, clearly, wild animals suffer from disease, attacks by competitors and predators etc., and generally receive no veterinary treatment.

Q9.16.7. The release of captive bred animals into the wild as part of a reintroduction programme is an important component of many conservation programmes. In some early reintroduction programmes survival rates of released animals were low and individual animals may have suffered unnecessarily due to starvation or by being unable to avoid predators because they had been inadequately prepared before release. Modern reintroduction programmes are managed on a scientific basis, often using soft release methods in which animals are given supplementary food while they adapt to wild conditions. After release they are monitored for disease, survival rates, etc.

Q9.16.8. The clipping of the toes of small mammals as a means of marking them in ecological studies used to be widely practiced. It has the advantage that the 'marks' cannot be lost and by clipping different toes on different days the mark can contain information about when the animal was captured. This marking method clearly causes unnecessary pain to the animals and is unjustifiable.

Q9.16.9. African elephant populations have been in decline across the continent as a whole for many decades. However, they have become locally abundant and wildlife managers have decided to control them using culling or contraception in a number of countries, especially where they have done excessive damage to forests. Their intelligence, social structure and other aspects of their biology have meant that elephants have received special treatment by animal welfare groups for many years. Some campaigners will never accept that elephants should be culled because they take an animal welfare approach to dealing with these animals. Conservationists are more likely to consider the health and survival of entire ecosystems rather than focus on a single species or particular individual animals within a species.

Q9.16.10. Feral cats have been responsible for the demise of many island species of animals, especially ground-nesting birds. Cat problems have been tackled by such methods as shooting and introducing disease and dogs to islands. Some of these methods have been inhumane and this has attracted adverse public attention partly, no doubt, because most people only encounter cats as companion animals and are not aware of the devastating effect that they can have on wildlife. From a purely conservation point of view cats, especially introduced feral cats, are undesirable predators and need to be eradicated to protect native fauna.

Chapter 10 Statistics

10.1. *Sewage sludge parasites*

Q10.1.1. The probability of an embryo developing from:

a. an *Ascaris* ovum is $21/33 = 0.636$
b. a *Toxocara* ovum is $34/64 = 0.531$.

Q10.1.2.

a.

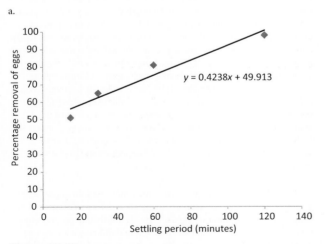

Figure 12.19

b. See Fig. 12.19. The formula of the line is $y = 0.4238x + 49.913$.
c. After 45 minutes there would be 69% removal.

10.2. *Ozone levels in a city*

Q10.2.1. For the ozone levels at each location:

Statistic	A	B
Mean	401.727	387.667
Standard deviation	21.583	13.379
Median	405.000	388.000

Q10.2.2.

a. A two-tailed test.
b. The null hypothesis is that there is no difference between the mean ozone level at the two sites.
c. $t = 2.163$ ($df = 18$).
d. Significant, $P < 0.05$.
e. Reject the null hypothesis.

10.3. *Lead pollution in gulls*

Q10.3.1. Draw a graph of the distribution of lead levels in the seawater samples listed in Table 10.4.

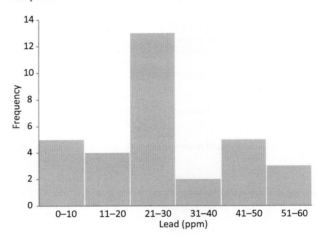

Q10.3.2. For the data on lead levels in the sea, calculate:

Statistics	Sample 1	Sample 2
Mean	88.933	69.857
Mode	N/A	56.000
Median	88.000	72.000
Standard deviation	15.406	16.792

Q10.3.3.

a. A one-tailed test.
b. The null hypothesis is that there is no difference between the means of the two samples.
c. $t = 3.181$ ($df = 27$).
d. Significant, $P < 0.005$.
e. Reject the null hypothesis.

Q10.3.4. Samples should be taken at random.

10.4. *Body length and mass in humpback whales*

Q10.4.1.

a. A one-tailed test.
b. The null hypothesis is that there is no correlation between body length and body mass in humpback whales.
c. $r = 0.818$.
d. Significant, $P < 0.005$.
e. Reject the null hypothesis.

Q10.4.2.

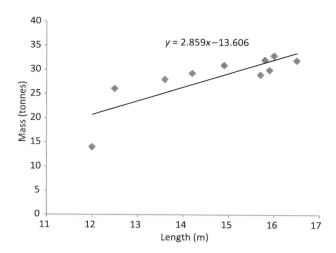

Q10.4.3. The equation of the line is: $y = 2.859x - 13.606$.

10.5. *Seals and disease*

Q10.5.1.

a. The null hypothesis is that the sex ratio does not differ from 1:1.
b. $\chi^2 = 1.28$ $(df = 1)$
c. Not significant, $P > 0.10$.
d. Accept the null hypothesis.

Q10.5.2.

a. The null hypothesis is that the virus does not preferentially attack either of the seal species.
b. $\chi^2 = 6.738$ $(df = 1)$
c. Significant, $P < 0.01$.
d. Reject the null hypothesis.

10.6. *BOD and Tubifex*

Q10.6.1.

a. A one-tailed test.
b. The null hypothesis is that there is no correlation between BOD and the number of *Tubifex* worms present.
c. $r = 0.996$.
d. Significant, $P < 0.005$.
e. Reject the null hypothesis.

Q10.6.2.

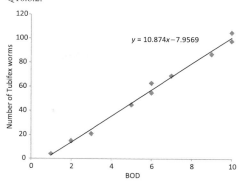

Q10.6.3. The equation of the line is: $y = 10.874x - 7.9569$.

10.7. *Weather measurements*

Q10.7.1.

Statistic	A	B
Mean	31.273	28.667
Standard deviation	3.608	3.808
Median	30.000	28.000

Q10.7.2.

a. A two-tailed test.
b. The null hypothesis is that there is no difference in the mean temperatures at the two sites.
c. $t = 1.559$ $(df = 18)$.
d. Not significant, $P > 0.05$.
e. Accept the null hypothesis.

10.8. *Lichen diversity and air pollution*

Q10.8.1.

a. A one-tailed test.
b. The null hypothesis is that here is no correlation between the number of lichen species recorded and the distance from the city centre.
c. $r = 0.988$.
d. Significant, $P < 0.005$.
e. Reject the null hypothesis.

Q10.8.2.

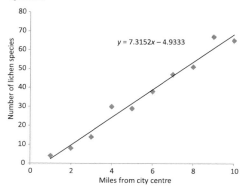

Q10.8.3. The equation of the line is: $y = 7.3152x - 4.9333$.

Answers to multiple choice questions

Q	Ch1	Ch2	Ch3(1)	Ch3(2)	Ch4	Ch5	Ch6	Ch7	Ch8	Ch9	Ch10
1	a	a	c	b	d	a	d	d	a	b	a
2	b	d	b	b	c	d	a	b	d	b	b
3	c	d	a	c	b	c	a	a	b	b	b
4	a	d	c	d	a	b	d	a	a	d	d
5	c	b	d	a	b	c	a	b	b	a	a
6	d	a	a	c	d	a	b	c	d	a	b
7	d	c	a	a	a	d	b	b	c	c	d
8	b	b	a	a	a	c	b	d	c	c	c
9	a	a	d	c	c	b	a	a	b	d	d
10	c	d	a	d	c	a	b	c	c	b	b
11	d	d	c	c	c	c	d	b	d	c	a
12	d	a	a	b	a	b	c	c	b	d	c
13	b	b	c	d	b	d	b	b	a	c	b
14	a	d	a	b	d	a	a	a	d	b	a
15	c	d	c	c	b	a	d	b	c	b	c
16	b	c	d	c	d	c	c	c	c	a	b
17	d	c	b	d	b	a	c	a	d	c	c
18	b	a	d	b	a	a	d	d	c	b	a
19	c	a	b	d	c	b	c	a	a	b	d
20	d	d	b	a	b	c	a	b	d	d	b

GLOSSARY

Abiotic factor A physical factor, e.g., temperature.

Acid rain Rain with a pH less than 5.6.

Activity budget A description of the amount of time an individual animal spends on various activities during the day (e.g., feeding, sleeping, resting, walking, etc.) as defined by an ethogram. Usually expressed as a percentage (or proportion) of the total amount of time the animal is observed. Data is often collected by instantaneous scan sampling.

Aerobic respiration The biochemical process by which cells extract the energy from sugars using oxygen. This begins with the splitting of glucose into two three-carbon sugars in the process of glycolysis, thereby producing ATP. Each of these sugars enters the Krebs' cycle and produces additional molecules of ATP. The NAD (an electron carrier) produced during glycolysis and the Krebs' cycle generates further ATP via an electron transport system within the mitochondria. The entire process produces 36–38 molecules of ATP from a single molecule of glucose. $C_6H_{12}O_6 + 6O_2 \rightarrow 6CO_2 + 6H_2O$ + energy

Anaerobic respiration The cellular process by which energy is released from food molecules (e.g., glucose) in the absence of oxygen. ATP is generated from an electron transport system.

Assimilation Within an organism, the process of absorbing small molecules and converting them into the more complex molecules that make up that organism

Autotomy The voluntary self-amputation of a body part (usually a limb or tail) often as a result of attack and to escape capture. Occurs in the claws of some crustaceans and the tails of some lizards.

Autotroph A primary producer. An organism capable of obtaining its energy from inorganic sources by photosynthesis or chemosynthesis.

Behavioural ecology The scientific study of the ecological and evolutionary basis of animal behaviour, and its role in adapting an organism to its environment, particularly the way in which behaviour contributes to reproduction and survival.

Binomial system The naming system devised by Carl Linnaeus which assigns scientific names to organisms consisting of the genus and the species.

Biodiversity A contraction of 'biological diversity'; the component organisms in a particular place e.g., a locality, habitat, ecosystem, or the whole of the globe.

Biological control A means of controlling pest species using natural predators.

Biomass The mass of biological material. May refer to the mass of a particular organism or the mass present in all or part of a particular ecosystem, or a specific area e.g., the biomass of trees in an area of tropical forest.

Biome A terrestrial ecosystem of a characteristic type: arctic tundra, northern coniferous forest, temperate forest, temperate grassland, chaparral, tropical rain forest, tropical savannah grassland, and desert.

Biotic factor A biological factor; a living thing that affects the ecosystem or a population of another organism.

Boom-and-bust growth Population growth that is initially exponential but then crashes to a low level once the environment is no longer able to supply the resources necessary to sustain it.

Botany The scientific study of plants.

Chemotroph An organism that obtains it energy from chemical reactions.

Climate The meteorological conditions pertaining in a particular place such as the rainfall, temperature, humidity, hours of sunshine, etc.

Climate change A significant change in the state of the climate that persists for an extended period of time. Usually used to refer to human-induced change, e.g., global warming.

Climax community The relatively stable group of organisms that exists in equilibrium with existing environmental conditions present at the final stage of an ecological succession.

Coadaptation The parallel evolution of two species that allows them to coexist, e.g., predators and their prey.

Community A group of species that occur together, e.g., sand dune community, grassland community.

Competition The process that occurs when two or more animals are using a resource which is in short supply, e.g., food, space, nesting places, etc. Competition may be between individuals of the same species (intraspecific) or between individuals of different species (interspecific).

Competitive exclusion principle In ecology, the theory that species that are complete competitors cannot coexist because they require exactly the same resources from the environment. This should result in one species surviving and the other becoming extinct. Evolution has produced species that have adapted so that they can coexist.

Decomposer A bacterium, fungus, or invertebrate involved in the decomposition of organic matter.

Deforestation The clearance of forest, usually by large companies for commercial gain.

Detritivore An organism that feeds on detritus.

Detritus Dead plants and animals and material derived from them and their decomposition e.g., leaf litter and other material on a forest floor.

Diapause A period of suspended development, especially in insects and some other invertebrates, but also applied to mammals.

Dichotomous key An identification key which uses a series of questions to which there can only be two possible answers, e.g., Does the insect have one pair of wings or two? Does the bird have a white breast (yes or no)?

Ecology The scientific study of the interactions between organisms and their environment, including other organisms.

Ecophysiology The branch of physiology that is concerned with the responses of the body to the environment.

Ecosystem A biological community and the physical environment in which it occurs.

Energy flow The movement of energy from one component to another along food chains.

Entomology The scientific study of insects.

Environment The totality of all of the factors that affect an organism (and its survival) including the physical (abiotic) factors (such as temperature and humidity) and the biological (biotic) factors (such as the presence of competitors and predators).

Ethogram A list and description of the behaviours exhibited by a species that is used for recording behaviours, especially in activity budgets.

Eutrophication The enrichment of a body of water with excessive nutrients causing an overgrowth of plant material and leading to a deficiency in oxygen.

Evapotranspiration The sum of the evaporation of water from the land and water making up the Earth plus the water lost in transpiration by plants.

Evolutionarily stable strategy An inherited strategy (usually behavioural) which, if practised by most members of a population, cannot be supplanted during evolution by a different strategy.

Exponential growth Growth of a population in which the number being added at any point in time is proportional to the number already present.

Extinction The process by which organisms die out, especially the disappearance of species.

Food chain A sequential linear representation of the organisms that feed upon each other in an ecosystem resulting in a simplistic representation of the flow of energy from the sun to a green plant, herbivore, carnivore, and then top carnivore. Species at the top of food chains tend to be generalists in relation to feeding, while those at the bottom tend to be specialists.

Fundamental niche The range of conditions and resources that an organism free of interference from other species could utilise.

Game ranching The exploitation of wild animals (usually ungulates) for food using a system similar to cattle ranching.

Game theory The branch of mathematics concerned with the analysis of strategies used by individuals in situations where the behaviour of one individual depends upon the behaviour of another (e.g., interactions between a predator and its prey).

Heavy metal A metal with a specific gravity greater than approximately 5.0. Often refers to one that is poisonous, e.g., lead, mercury, cadmium.

Herpetology The scientific study of retiles and amphibians.

Heterotroph An organism that requires organic compounds as a source of food.

Homiotherm An animal that maintains its body temperature physiologically, independent of fluctuations in the environment and within a relatively narrow range, i.e., mammals and birds.

Hybrid An individual that has been produced by the mating of organisms of two different species, varieties, races, or breeds.

Hydrology The scientific study of the distribution, movement, and quality of water.

Ichthyology The scientific study of fishes.

Insolation The quantity of solar radiation received by a given surface area of the Earth in a given period of time.

Interspecific competition Competition between individuals of different species for resources, e.g., food, sheltering places.

Intraspecific competition Competition between individuals of the same species for resources e.g., food, mates, sheltering places.

Keystone species A species which has a disproportionately large effect on the biological community in which it occurs, thereby helping to maintain local biodiversity.

Kinesis A movement which is proportional to the strength of a stimulus but the response is not directional. For example, woodlice are more active in dry atmospheres than in humid atmospheres. They are therefore more likely to remain in humid environments and more likely to move quickly through dry environments.

Lichen A composite organism consisting of an alga and a fungus living symbiotically.

Life table A table of data showing the mortality rates of different age classes within a population of organisms. Used to produce a survivorship curve.

Limnology The scientific study of the biological, physical, geological, and chemical aspects of freshwater systems (lakes, ponds, and rivers).

Lincoln Index A method of estimating the size of an animal population in the wild by marking a sample, releasing the marked animals, and then capturing a second sample. The proportion of animals in the second sample which is marked should be the same as the proportion marked in the entire population.

Logistic growth A type of population growth in which numbers initially grow exponentially and then stablise at the carrying capacity of the environment (the asymptote). Also known as S-shaped growth. May occur when a species invades or is introduced to a new habitat and there are no significant competitors or predators.

Mammalogy The scientific study of mammals.

Melanism An overdevelopment of dark pigmentation (melanin) in the skin. In some industrial areas where pollution has darkened the surface of tree bark, walls, and buildings some species have evolved melanic forms as a result of natural selection, whereby they have been favored over lighter forms which stand out against these dark surfaces. This phenomenon is known as industrial melanism and has occurred in the peppered moth *Biston betularia*.

Model In mathematics, a model is an equation or group of equations which allows events to be simulated, e.g., population viability analysis allows theoretical changes in a population to be simulated. Some models are deterministic and the outcome is the same each time the model is run. Others are stochastic and include random elements so the outcome is different each time.

Niche The role an organism plays in an ecosystem and the range of environmental conditions in which it is able to survive.

Niche separation The phenomenon whereby closely related species that utilise the same habitats (sympatric species) avoid interspecific competition by evolving so that they occupy different niches, e.g., feeding on different species (resource partitioning) or in different places.

Nitrogen cycle The movement and exchange of nitrogen between different components of the environment.

Nutrient cycle The movement and exchange of organic and inorganic materials between different components of the environment, living and nonliving.

Organic pollution Environmental contaminants that are derived from biological materials, e.g., sewage sludge, dairy waste.

Ornithology The scientific study of birds.

Parasitology The scientific study of parasites.

Photosynthesis A biochemical process by which green plants produce glucose and oxygen from carbon dioxide and water using the energy from sunlight.

$$6CO_2 + 6H_2O + \text{light energy} \rightarrow C_6H_{12}O_6 + 6O_2$$

Phytoplankton Very small plants that live in water.

Pioneer species A hardy species which colonises new habitats at the very beginning of an ecological succession, e.g., lichens.

Poikilotherm An animal that is not capable of regulating its body temperature physiologically. Internal temperature fluctuates considerably and is dependent upon temperature changes in the environment. However, it can be raised or lowered by moving into the sun or shade. All animal taxa apart from mammals and birds.

Pollution The introduction of contaminants to the environment that cause harm, usually by humans.

Population A group of organisms of the same species, living at the same time and in the same place. In statistics, the complete set of values of a particular variable in a given situation.

Predation The act of one animal killing and feeding upon another. The term usually, but not always, refers to the behaviour of carnivores.

Primary production The total amount of organic matter synthesised by autotrophs (usually green plants) in a particular place over a particular period of time.

Primary succession A succession which begins on a bare substrate before any soil has been created

Productivity The rate at which biomass is generated in an ecosystem.

Quadrat A square used for sampling in ecology. It may be made of wood or metal (e.g., a one-metre square) or a large square area marked out on the ground.

Realised niche That part of an organism's fundamental niche that is actually occupied due to the effect of the various limiting factors present in the environment.

Respiration The biochemical process by which organisms obtain energy by breaking down food. Also used to mean breathing in some contexts.

Savannah A dry tropical or subtropical grassland habitat characterised by drought-resistant vegetation dominated by grasses with scattered trees and inhabited by large grazing mammals.

Secondary production The total amount of organic matter accumulated by herbivores, carnivores, top carnivores, fungi, etc.; the formation of heterotrophic biomass with time. Sometimes used to refer only to production by herbivores.

Secondary succession A succession which occurs on a preexisting soil after the primary succession has been disrupted, e.g., after deforestation.

sp. Abbreviation for a species. After the name of a genus it indicates a single unspecified species, e.g., *Canis* sp.

Species A group of interbreeding natural populations that are reproductively isolated from other such groups. This is known as the biological species concept. A subdivision of a genus.

spp. After the name of a genus, this indicates several members of that genus, e.g., *Panthera* spp. The plural of sp.

Succession A gradual and orderly change in an ecosystem, whereby plants and animal communities replace each other over time.

Survivorship curve A graph showing the pattern of mortality within a population with age, usually corrected so that the starting population is 1000.

Taxon A group of related organisms, e.g., a species, genus, family, order, class, phylum, etc.

Taxonomy The scientific classification of organisms into groups related by evolution.

Tertiary production Sometimes used to refer to production by carnivores.

Thermocline A distinct layer in a body of water where temperature decreases rapidly with depth.

Transect A line or path across a habitat along which organisms are sampled. May be used to sample plant populations or count animals.

Transpiration The loss of water from plants, primarily through stomata (pores) in the leaves.

Trophic level The feeding level in an ecosystem (e.g., herbivores, carnivores).

Tundra A biome that consists of a treeless plain which experiences low temperatures and rainfall, and has low biological diversity and a simple vegetation structure, typically including lichens. It occurs in the Arctic and Antarctic.

Vernacular name The common name of a species, e.g., red fox, American robin, silver birch.

Water cycle The hydrological cycle: the movement of water on, above, and below the surface of the Earth which connects, rivers, the atmosphere, the oceans, etc.

Zoology The scientific study of animals.

Zooplankton Very small animals that live in water.

INDEX

Note: Page numbers followed by "*f*" and "*t*" refer to figures and tables, respectively.

Printed in the United States
By Bookmasters